Food quality management

Food quality management

Technological and managerial principles and practices

Pieternel A. Luning
and
Willem J. Marcelis

Wageningen Academic
Publishers

ISBN 987-90-8686-116-3

First published, 2009
Reprint, 2011

Table of contents

Preface		**13**
Authors		**17**
1	**Introduction food quality management**	**19**
1.1	Food quality management	19
1.1.1	Quality in the agri-food chain	19
1.1.2	Complexity food quality management	21
1.2	Techno-managerial approach in food quality management	23
1.2.1	Techno-managerial approach	24
1.2.2	Food quality relationship	26
1.2.3	Food quality management functions	26
1.3	Quality	29
1.3.1	Quality definitions	30
1.3.2	Quality concepts	31
1.4	Structure of the book	36
2	**Food quality**	**39**
2.1	Introduction	39
2.2	Quality perception	39
2.3	Quality attributes	42
2.3.1	Safety	44
2.3.2	Health	49
2.3.3	Sensory	50
2.3.4	Shelf life	53
2.3.5	Convenience	55
2.3.6	Extrinsic attributes	55
2.4	Factors influencing food behaviour	57
2.4.1	Variable food composition	57
2.4.2	Dynamic food processes	61
2.4.3	Packaging conditions	69
2.4.4	Food processing equipment conditions	70
2.4.5	Food plant and environmental conditions	72
2.5	Food quality from a chain perspective	72
2.5.1	Primary production	73
2.5.2	Animal production	73
2.5.3	Vegetable production	77
2.5.4	Processing and distribution	79
2.5.5	Retail and final food handling	81
2.6	Environmental factors	83
Appendix 2. Overview of common packaging materials for agri-products and food		90

3	**Quality management**	**93**
3.1	Introduction	93
3.2	Management	95
3.2.1	Management concerns	95
3.2.2	Decision-making	97
3.2.3	Factors influencing decision-making	99
3.2.4	Decision-making models	104
3.2.5	Administrative concepts	107
3.3	Quality management	108
3.3.1	Quality management history	108
3.3.2	The quality gurus	110
3.3.3	Quality management activities	115
3.4	Food quality management	116
3.4.1	Quality assurance	116
3.4.2	Food quality management functions	117
3.4.3	Food quality decisions	119
3.5	Factors influencing quality behaviour	120
3.5.1	Quality behaviour and leadership	121
3.5.2	Human dynamics	124
3.5.3	Administrative conditions	128
3.5.4	Organisational structure design	132
3.6	Quality management from a chain perspective	138
3.6.1	Supply chain management	139
3.6.2	Customer-supplier relationships	140
4	**Quality design**	**143**
4.1	Introduction	143
4.2	The quality design process	144
4.2.1	Quality design activities	145
4.2.2	Quality design decisions	149
4.3	Variables in product and process design	149
4.3.1	New food products	150
4.3.2	Product design variables	151
4.3.3	Process design variables	154
4.4	Customer-oriented design	160
4.4.1	Customer-oriented design processes	160
4.4.2	Customer information	163
4.4.3	Quality function deployment	165
4.5	Tools to support quality design	167
4.5.1	Methods and techniques to assess product characteristics	168
4.5.2	Modelling to support product and process design	172
4.5.3	Failure mode and effect analysis (FMEA) in process design	174
4.5.4	Hygienic design concept	176
4.5.5	Taguchi method for product and process design	180

4.6	Managing the design process	183
4.6.1	Product development strategy	183
4.6.2	Collaboration in design processes	185
4.6.3	Project management	190
4.7	Quality design in practice	193
5	**Quality control**	**197**
5.1	Introduction	197
5.2	The quality control process	198
5.2.1	Quality control activities	199
5.2.2	Quality control decisions	202
5.2.3	Variation	202
5.3	Quality control in food production	205
5.3.1	Supply control	205
5.3.2	Production control	210
5.3.3	Distribution control	213
5.4	Tools and methods for quality control	217
5.4.1	Acceptance sampling	218
5.4.2	Statistical process control	223
5.4.3	Quality analysis and measurement	229
5.4.4	Visual inspection	235
5.5	Managing the control process	237
5.5.1	Quality controlled logistics	237
5.5.2	Forms of organisational control	240
5.5.3	Costs and benefits of control	242
5.6	Quality control in practice	243
6	**Quality improvement**	**245**
6.1	Introduction	245
6.2	Quality improvement process	247
6.2.1	Quality improvement activities	247
6.2.2	Quality improvement decisions	250
6.2.3	Complexity of food production	250
6.2.4	Learning	253
6.3	Quality gurus on improvement	254
6.4	Tools and methods for quality improvement	258
6.4.1	Tools for improvement	259
6.4.2	Methods for improvement	261
6.5	Managing the quality improvement process	265
6.5.1	Basic conditions for quality improvement	265
6.5.2	Training	267
6.5.3	Teamwork	269
6.5.4	Project approach	272
6.5.5	Organisational change strategies	273
6.6	Quality improvement in practice	277

7 Quality assurance 279
7.1 Introduction 279
7.2 Quality assurance process 280
7.2.1 Quality assurance activities 282
7.2.2 Quality assurance decisions 283
7.3 Primary quality assurance guidelines and standards 283
7.3.1 Principles of basic codes and guidelines 285
7.3.2 Principles Hazard Analysis and Critical Control Point (HACCP) 290
7.3.3 Principles ISO 9001:2000 standard 302
7.4 Tools and methods supporting quality assurance 306
7.4.1 Quantitative risk assessment 306
7.4.2 Predictive food microbiology 313
7.4.3 Traceability systems 316
7.5 Managing quality assurance 320
7.5.1 Quality assurance in the organisation 321
7.5.2 Quality auditing and certification 322
7.6 Quality assurance from a chain perspective 325
7.6.1 Quality assurance standards with a chain focus (ISO 22000, SQF) 325
7.6.2 Quality assurance demands in the retail sector (GFSI, BRC, IFS) 328
7.6.3 Quality assurance guidelines in animal production 330
7.6.4 Quality assurance standards for vegetables and fruits production 333
7.6.5 Quality assurance standards/guidelines in flower/plant sector 335
7.7 Quality assurance in practice 336
Appendix 7.1. Hygiene issues that are covered in the 'Codex general principles of food hygiene' 339
Appendix 7.2. Overview of prerequisite programmes 348
Appendix 7.3. Overview of items for which requirements are described in global standard for food safety from the British Retail Consortium 351

Chapter 8

Quality policy and strategy 355
8.1 Introduction 355
8.2 The quality policy and strategy process 357
8.2.1 Quality policy and strategy activities 357
8.2.2 Quality policy and strategy decisions 359
8.2.3 SWOT analysis and strategic alternatives 359
8.3 Factors influencing quality policy 365
8.3.1 Quality philosophies 365
8.3.2 Organisational maturity 367
8.3.3 Technological complexity 367
8.3.4 Quality costs 370
8.4 Total quality management 373
8.4.1 Policy orientations 373
8.4.2 Total quality management 375
8.4.3 Total quality control and total quality learning 379
8.4.4 Implementing TQM 380

8.4.5 Quality policy evaluation 381
8.5 Quality policy from a chain perspective 384
8.5.1 The value system 384
8.5.2 Collaboration in the food supply chain 385
8.6 Quality policy in practice 388

Chapter 9

Food quality management in perspective **391**
9.1 Dynamics in the agri-food production chain 391
9.2 Core developments in food quality management 392
9.3 Food quality management research 395

References **397**

Index **417**

Preface

Both in literature and in practice, the management of food quality is often perceived as the development and application of quality systems. In the food production chain, much attention is commonly paid to the performance of implemented HACCP principles and basic requirements, like PRP's and GMP, in effective food safety management systems. However, a screening of the literature on problems experienced when implementing food safety management systems in practice revealed typical observations, such as:

- Problems with identifying and prioritising chemical/microbial hazards, and assessing critical points on a scientific and quantitative basis.
- Problems with monitoring systems, incorrect plant lay-out, and poorly designed equipment.
- Problems with operators, e.g. inappropriate corrective actions and insufficient record-keeping and documentation.
- Problems with understanding the system, ineffective training, misconceptions about safety risks, and lack of management commitment (Luning and Marcelis, 2006).

These problems refer to both the dynamics of food and food production, and the dynamic behaviour of the people involved. We concluded from these findings that dealing with food quality (safety) management as a straightforward system, risks ignoring all kinds of influencing factors that cannot be easily predicted. As a consequence, important factors like unpredictable behaviour of raw materials resulting in production problems, or conflicting individual interests of people resulting in unexpected behaviour, are not systematically taken into account when analysing the problem situation.

Food and food production systems are really dynamic! We have to deal with living materials with large variation in product properties, which also change over time. For example, levels of glucosinolates in processed *Brassica* vegetables upon consumption can vary up to 100 fold, due to variation in raw materials, in industrial processing and storage and in household preparation. The same goes for people. They show dynamic behaviour every day! One should therefore accept that people have their own way of dealing with problems, in a more or less subjective way. It depends on personal characteristics, motivation and ability, but also on organisational conditions, conflicting interests, availability of information and power. As a consequence human behaviour is rather variable as well.

Taking into consideration the above-mentioned aspects, our conclusion is that food quality management research should start with understanding the behaviour of food and human systems when trying to predict process outcomes in terms of quality. This is contrary to what most people are actually doing. When people are confronted with complex problems, they tend to reduce them to a simple system level, which enables a straightforward choice of solutions. They try to ignore complexity, or make sense of it in a way that reflects their own perception of the situation rather than the situation itself. These insights have lead to our basic idea of the techno-managerial approach as a way of analysing complex food quality management issues.

In 2002 we published the first edition of our book 'Food Quality Management, a techno-managerial approach'. We introduced the techno-managerial research approach to food quality management in acknowledgement of the complexity of food quality and quality management, and in order to analyse problems from the point of view that food quality is dependent on both food behaviour and human behaviour. Since 2002 we have carried out a wide range of MSc thesis and PhD research projects using this viewpoint, and we have published various scientific articles on this topic. Moreover, we have gathered a lot of experience from educational situations and critical discussions with our students and colleagues at Wageningen University and Research Centre. The research projects and experiences have provided a strong basis for the new edition of our book.

This new book has been fully redesigned according to a new structure, including new and redefined concepts, with the objective of making the approach more clear and understandable. The introductory Chapter 1 explains the main principles behind the techno-managerial approach to food quality management. Chapter 2 is an introduction to food quality providing a technological knowledge basis for understanding the major mechanisms in food systems and the technological processes that affect them. Similarly, Chapter 3 is an introduction to quality management providing a managerial knowledge basis for understanding human behaviour and the managerial processes that influence it.

The next five chapters are about the five food quality management functions, quality design, quality control, quality assurance, quality improvement, and quality policy and strategy. These chapters are similar in structure. They start with a description of the basic objectives and activities of the specific food quality management function. Then process steps and typical topics of the FQM function are mentioned and explained. Thereafter, tools and methods, which are useful in supporting the FQM function, are explained in detail. Finally, the function is described within its organisational context, with the focus on how these functions are executed by managers. The last chapter is about considerations on food quality management in the near future.

While describing the FQM functions more clearly we also included new insights and concepts, for example:

Chapter 1: A better defined description of the T-M approach and theoretical assumptions behind it.

Chapter 2: New defined concepts for food quality are included, and the basic technological principles of food systems and technological conditions are explained in more detail.

Chapter 3: A description of people's judgement processes and behaviour is included.

Chapter 4: Design processes and variables are more concisely described.

Chapter 5: New sections on variation and visual inspection are included.

Chapter 6: New sections on learning and training are included.

Chapter 7: A clear distinction between assurance and control, new sections on traceability and predictive models are included, and all quality assurance standards and guidelines are updated.

Chapter 8: New sections on total quality management and chain collaboration are included.

Chapter 9: A new section on research methodology is included.

The book as presented to you is the basis of the two-year Master's programme in Food Quality Management at Wageningen University but is also useful as a handbook for practising QA managers. The FQM programme includes food quality management courses that integrate the technological and managerial knowledge. Students can specialise in quality management, food and quality assurance, or food logistics management. Students with a background in different disciplines, such as food technology, agricultural engineering, bio-product management, animal production, plant production, veterinary sciences, management and consumer sciences, can follow this MSc programme.

We have further developed our food quality management approach by continuous improvement. We have tried to be customer focused and stay ahead in this area of scientific research. This book has also been written in a context of empowerment and teamwork. These four pillars of total quality management will hopefully ensure an enjoyable read!

Wageningen, May 2009

Pieternel. A. Luning
Willem. J. Marcelis

Authors

Dr. Ir. Pieternel A. Luning, Associate Professor Food Quality Management, Wageningen University and Research Centre, Chair Group Product Design and Quality Management.

Pieternel Luning studied Food Chemistry and Microbiology at Wageningen University in the Netherlands. After graduating in 1988, she was involved in various different projects in the area of food flavour research. She worked at the University for two years on bread flavour. From 1990-1995 she was employed as project manager at the Agrotechnological Research Institute ATO-DLO, where she finished her PhD thesis on the flavour of fresh and processed bell peppers. Immediately thereafter, she was involved in a flavour project on vegetable oils for Unilever. In May of 1996 she started work at TNO Research and Nutrition Institute as a product manager of 'Innovative packaging'. In July 1998, she was employed as an assistant professor in Food Quality Management Science at the Product Design and Quality Management group at Wageningen University. Together with Dr. ir. W.J. Marcelis and Prof. dr. W.M.F. Jongen she published the book 'Food Quality Management, a techno-managerial approach'. The principles behind food quality management research have been published by Luning and Marcelis in a range of three coherent papers. Within the field of food quality management she also edited the book 'Safety in agri-food chains' (2006) in collaboration with Prof. Devlieghere and Prof. Verhé from Ghent University in Belgium. She was sub-coordinator of two Socrates curriculum development programmes: 'Total food quality management' and 'Safety in agri-food chains'. The curricula were developed in collaboration with 15-20 partners from various European countries, and both books formed the basis of the respective European curricula.

At Wageningen University she was, together with Dr. Marcelis, responsible for the development of the MSc programme Food Quality Management, which started in 2003. Approximately 25 students from various different countries worldwide have now joined this programme each year.

After the development of the education programme and in collaboration with Dr. Marcelis, Dr. Luning helped design a coherent research programme on Food Quality Management which is now being executed by MSc thesis students, PhD students (approx. 10), post-docs, and guest researchers. Since 2005, she is involved in a large 5-year EU project called 'Pathogen combat', in which the techno-managerial approach is used to analyse and improve food safety management systems. In this project she closely collaborates with the University of Ghent (Belgium) and Burgos (Spain), which has resulted in a diagnostic instrument and microbiological assessment scheme to systematically assess the performance of food safety management systems independent of the quality assurance standards and guidelines that have been implemented. The instruments have been applied in various sectors (meat, dairy, fish, tofu, tea, food service establishments) in various countries (Belgium, Spain, Greece, the Netherlands, Indonesia, Philippines) in micro, small, medium and large food companies. The principles and applications are now being published in various international journals. In 2006 she has become associate professor of Food Quality Management.

Dr. Ir. Willem J. Marcelis, Associate Professor, Food Quality Management, Wageningen University and Research Centre, Chair Group Management Studies.

Willem Marcelis finished his degree in Mechanical Engineering, with a specialisation in management studies, at Wageningen University, the Netherlands, in 1972. Since then he has conducted research on maintenance management and published two books and various articles on that topic. He finished his PhD in 1984 and has been a management consultant in the field of maintenance management for many years, working for companies like Unilever, Heineken, AKZO-Nobel, Heinz, Friesland Foods and Philips. He collaborated with Professor A.A. Kampfraath in developing a management theory based on administrative processes. A book on this topic was published in 1981.

In 1974 he was employed as a teacher and researcher in management at Wageningen University and Research Centre, and became an associate professor in 1985.

In the mid-eighties he started working on the initial educational programme on Food Quality Management. Since then he has collaborated with the different technological departments on extending and improving the food quality management programme.

In 2002, together with Dr. Ir. P.A. Luning and Prof. Dr. W.M.F. Jongen, he published the book 'Food Quality Management, a techno-managerial approach'. Together with Dr. Luning he has written various articles and book chapters on food quality management.

In 2003 he and Dr. Luning developed the two-year Master's programme in Food Quality Management at Wageningen University. After the development of the education programme, and in collaboration with Dr. Luning, Dr. Marcelis designed a coherent research programme on Food Quality Management which is now being executed by MSc thesis students, PhD students (approx. 10), post-docs, and guest researchers.

Since 2004 he has been working within the food quality management research programme on topics like food safety management, critical quality points and quality controlled logistics.

Chapter 1

Introduction food quality management

In the last few decades quality has become extremely important within society. Agribusiness and food companies are nowadays increasingly judged on their overall quality performance not just their financial performance. To understand the role of quality in the agricultural sector one needs to have an idea about the importance and complexity of managing quality in the agri-food production chain (section 1.1). In order to deal with the complexity of food quality management a techno-managerial approach (section 1.2.1), a conceptual food quality relationship (section 1.2.2) and a food quality management functions model (section 1.2.3) have been developed, which are the pillars of the theoretical framework for analysing food quality management issues. Since quality is not a univocal concept but multidimensional, dynamic and user dependent, an overview is provided of typical quality definitions and concepts/models (section 1.3). Section 1.4 explains how the basic concepts of FQM are reflected in the structure of the book.

1.1 Food quality management

Quality attracts much attention in agribusiness and the food industry due to raised consumer concerns and the increased demands of stakeholders, like government, interested parties and retailers (Luning *et al.*, 2002; Van der Meulen and Van der Velde, 2004; Jongen and Meulenberg, 2005). This section discusses the importance of quality in the agri-food chain (section 1.1.1) and the complexity of managing food quality (section 1.1.2).

1.1.1 Quality in the agri-food chain

Characterisation of the agri-food production chain

The agri-food production chain refers to the chain wherein agricultural products are transformed into food products for use and consumption. Nowadays this chain is approached in a holistic way, e.g. the stable-to-table, or farm-to-fork, or seed-to-food approach (Knura *et al.*, 2006). This means that it covers the whole food supply chain starting with the production of animal feed for animal derived products, and seed production for vegetable-based and flower/plant products. *Agricultural products* refer to initial products in the chain (e.g. animal feed, seeds), raw (unprocessed) materials and ingredients, but also flowers and pot plants; the latter are not directly suitable for human consumption. *Food products* concern fresh produce, animal products, and manufactured food, which can be consumed immediately or after processing at home.

Various authors have emphasised that agriculture and food products and their production chains are different from others (Luning *et al.*, 2002; Trienekens and Van der Vorst, 2006). Agriculture and food products:

- are *living systems*;
- usually have a *restricted shelf life*;
- are harvested *seasonally*;
- are often *heterogeneous*;

- have a high *variability* (10% is common);
- and may contain *hidden risks*.

These characteristics have consequences for activities such as storage technology and conditioning, packaging, tracking and tracing, and quality control and assurance systems. Moreover, agri-food production is characterised by:

- *complex production chains*, e.g. variety of converging and diverging production processes;
- complex *networks with many suppliers* and *customers*;
- unintentional and unwanted *by-products*;
- *high turn-over* volumes;
- *low-value* products;
- substantial *environmental impact*, e.g. packaging material, surplus products, used product;
- and, last but not least, major public *interest* of organisations in all links of the chain, such as health and safety concerns.

Importance of quality

The importance of quality, and how to manage quality along the agri-food chain, has increased due to various concerns and changes in the last few decades. Several serious food crises have occurred in the agri-food chain, such as Bovine Spongiform Encephalopathy (BSE) and classical swine fever (CSF) in 1997, the Belgian dioxin affair in 1999, foot and mouth disease (FMD) in 2001, the nitrophen and medroxyprogesterone acetate (MPA) incidents in 2002, outbreaks of Avian Influenza in Asia since 2005, various large *Salmonella* outbreaks in USA in 2008, and the large melamine contamination scandal in China in 2008. These incidents have enhanced consumer and governmental concerns on food quality and safety (Lin *et al.*, 2005; Verbeke *et al.*, 2006; Bruhn *et al.*, 2007). As a consequence, the European food law has changed considerably, putting more demands on the assurance of quality and safety of agriculture and food products and their production system characteristics (Van der Meulen and Van der Velden, 2004, 2007). Other concerns of consumers and government, like the environmental burden, climate change, but also the increasing number of people with obesity, have consequences for managing quality in food production (e.g. Parry *et al.*, 2004; Lang and Rayner, 2005; Bhaskaran *et al.*, 2006; Tukker and Jansen, 2006; Weber and Matthews, 2008).

Moreover, changes in society have stimulated agribusiness and the food industry to pay more attention to quality. Modifications in demographic situations (e.g. more elderly people, different cultures), social situations (e.g. more double incomes, higher education levels), consumption behaviour and lifestyle (e.g. increased outdoor consumption, more travelling and exposure to unsafe foods, growing consumption of imported products, changes in food preparation habits) and globalisation of markets have resulted in higher demands on quality management (e.g. Grunert, 2005). Last but not least, changes in food and feed processing techniques offer new opportunities to comply with increased quality requirements (e.g. less intensively processed) (Boom *et al.*, 2005), but put new demands on the management of food production.

The typical food production characteristics and the increased concerns and changes in society have stimulated the interest in food quality management in agribusiness and the food industry worldwide.

1.1.2 Complexity food quality management

Food quality (FQ) and quality management (QM) are the two basic components of food quality management. It deals with the physical aspects of quality as well as managing the people who have to realise it. Figure 1.1 illustrates that many different aspects affect food quality management, including the characteristics of food production and their production chains, the demands and concerns of consumers and stakeholders, and people within their organisational and chain context.

Food quality management is thus rather complex. Luning and Marcelis (2006) used the systems hierarchy theory of Boulding (1956) to obtain a better understanding of why it can be considered as complex. The systems hierarchy of Boulding (1956) provided fundamental insights into systems thinking. He distinguished nine system levels, each higher level includes all lower levels but new functions are added and more information is needed (Figure 1.2). The levels are briefly described below and characterise the different system levels involved in food quality management.

1. *Frame-work system*: a static structure, e.g. a stone. These systems are also described as skeleton systems.
2. *Clock-work system*: a system with predetermined motion, e.g. a clock. The time factor has been added, resulting in simple dynamic systems.
3. *Cybernetic system*: closed-loop control system, e.g. a thermostat. At this level feedback information ensures that one or more variables are kept at a constant level.
4. *Open system*: structurally self-maintaining, self- production, e.g. biological cells. Here life starts with self-maintenance and self-reproduction. At this level total information is passed through the living system.
5. *Lower organism system*: differentiation in organs, predetermined growth, reproduction, e.g. plants. At this level the systems are able to group cells to organs with predetermined final forms and functions.

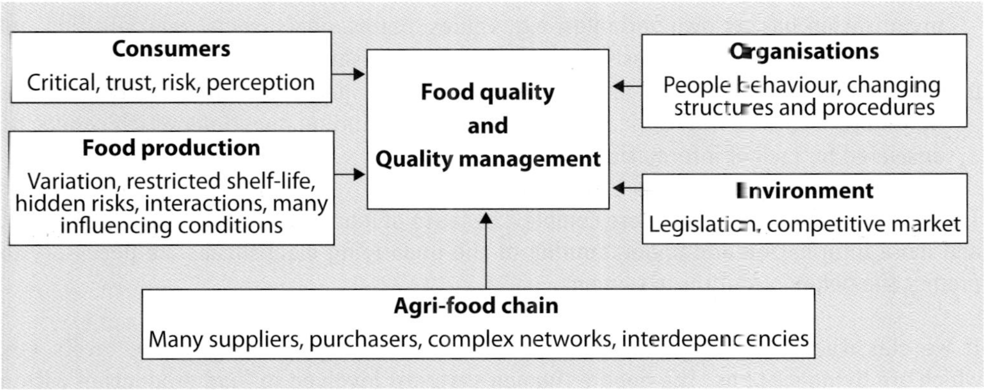

Figure 1.1. Factors influencing food quality management.

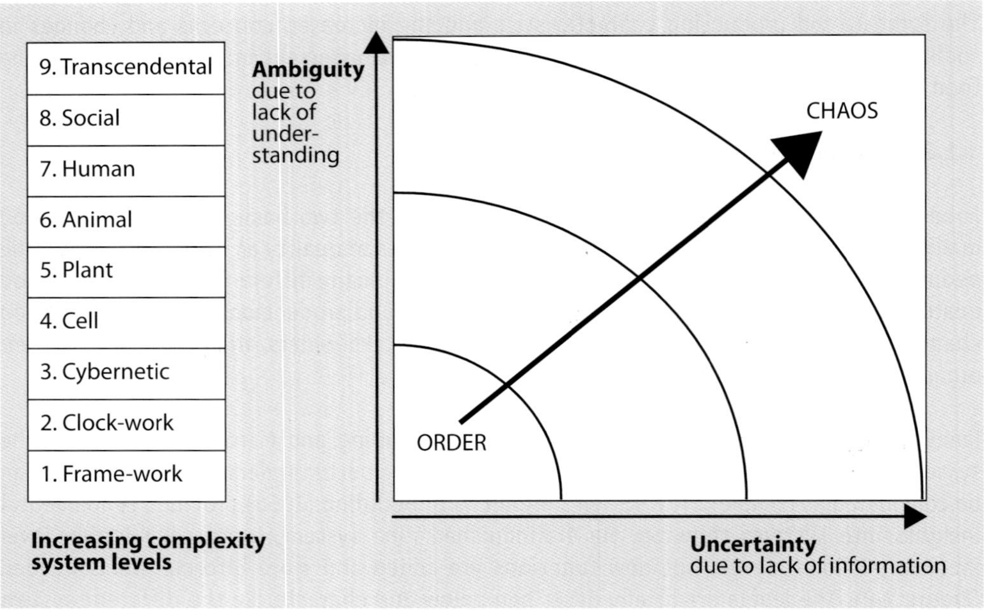

Figure 1.2. Food quality management complexity dependent on ambiguity and uncertainty (from Luning and Marcelis, 2006).

6. *Animal system*: consciousness, information collection, brain-guided behaviour, i.e. animals. The animal system has an extra ability to the plant system which is the capacity to process and store information. Special organs for information retrieval are formed (ear, eye). Consciousness arises in the form of images, knowledge structure and an overview of the environment as a whole.
7. *Human system*: self-reflection, knowledge, anticipation, and symbolic language, i.e. human beings. At this level self-reflection is added. Human beings not only know, but they know that they know. This ability is connected with the ability to form a language and to symbolise. Human beings are also aware of time and are thus able to anticipate.
8. *Social system*: relationships, interests, communication, culture, e.g. an organisation. Humans are only human in a social situation, where they play different roles. The social organisation has its own variables, e.g. values, norms, goals, social relationships and different interests. Culture arises within social organisations.
9. *Transcendental system*: non-material systems, e.g. the idea of divinity. Finally, there are non-material systems of logic, axioms and belief, including questions which cannot be answered for lack of information.

The higher systems levels are more complex, because of more interrelations and variability, and more information and understanding of the underlying mechanisms are necessary to predict the behaviour of these systems.

If we classify food quality management, then food systems belong to the levels 4-6, which are living systems. The people (human systems) involved in food production within a certain organisational context can be typified as high system levels 7-8. According to

Boulding's system hierarchy (1956) food quality management deals with high system levels. However, what often happens in practice when having to deal with complex systems, is that one tends to reduce it to a lower system level, which enables a straightforward choice of solutions. Likewise, despite its complexity, food quality management is often assumed to be a rather controllable process on the cybernetic system level. Consequently, straightforward mechanisms of control are introduced to manage the variation in outcome. This is illustrated by the huge focus on quality assurance guidelines and standards (like HACCP, BRC, SQF, ISO 22000) that have been introduced into company-specific quality (safety) management system in agribusiness and the food industry (e.g. Efstratiadis *et al.*, 2000; Ropkins and Beck, 2000; Botterbrodt, 2003; Fielding *et al.*, 2005; Luning *et al.*, 2008). These quality (safety) management systems are commonly based on procedures and control circles as a mechanism to control and assure quality. After implementation, however, it often appears in practice that the intended results are not obtained, and that there is a lack of understanding of the reasons why.

In fact, food quality management is full of ambiguity and uncertainty. Ambiguity is described as a situation where there is no clear interpretation of phenomena or events; available evidence supports more than one interpretation. Uncertainty is the situation in which there is lack of information about phenomena or events. Figure 1.2 shows that systems at higher levels cause more ambiguity due to a lack of understanding of underlying factors and mechanisms, because they are more complex and dynamic. Uncertainty is assumed to be only dependent on lack of availability of information and or data. In a situation of low ambiguity and low uncertainty, order exists because the systems are fully understood and the people implementing them are fully informed about them. A situation of order can be characterised by aspects such as linearity, rationality and stability. More ambiguity and uncertainty will result in more chaos. The ultimate situation of chaos is one of non-linearity, irrationality and instability (Luning and Marcelis, 2006).

So, food quality management research and analysis should assume a situation of relative chaos and not order, and should recognise ambiguity and uncertainty. However, at the same time it must realise predictive and guaranteed quality levels. Both food and human systems have their own dynamic behaviour and are therefore not completely predictable. However, conditioning will generate organised and patterned behaviour (Kampfraath and Marcelis, 1981; Teale *et al.*, 2003). Conditioning aims at reducing variation in dynamic behaviour and creates boundaries to constrain extremes in behaviour outcomes.

1.2 Techno-managerial approach in food quality management

Food quality management issues cannot just be approached from one discipline because of the dynamic and complex characteristics of both food production systems and the people involved in the production of agriculture and food products. This section provides the basic models and concepts that support an integrated and extended analysis of food quality management issues.

1.2.1 Techno-managerial approach

From a research point of view we need a more sophisticated approach to prevent the simplification of food quality management to cybernetic systems (Luning and Marcelis, 2006, 2009). Based on the characteristics of food and humans systems and the character of food quality management, we proposed a techno-managerial approach *to food quality management research* (Figure 1.3).

Within this approach we described a *food system* as the whole range of product properties (e.g. content of proteins, sugars, pH, a_w) and dynamic food processes (e.g. enzyme activity, microbial growth rate). *Food behaviour* is the result of the outcomes of the food system over time. A food system is subject to a *food production system*, which is the whole range of production activities and technological conditions that are necessary to realise the desired product properties. The complexity and variability of these food production systems are ascribed to:

- Heterogeneity of food products due to their complex structure, heterogenous distribution of constituents, and compartments.
- Variation in composition and level due to cultivar and breeding differences, seasonal influences, weather and harvesting conditions.
- Continuous decay of quality attributes due to a wide range of food processes, like microbial, physical, biochemical, chemical and physiological processes.

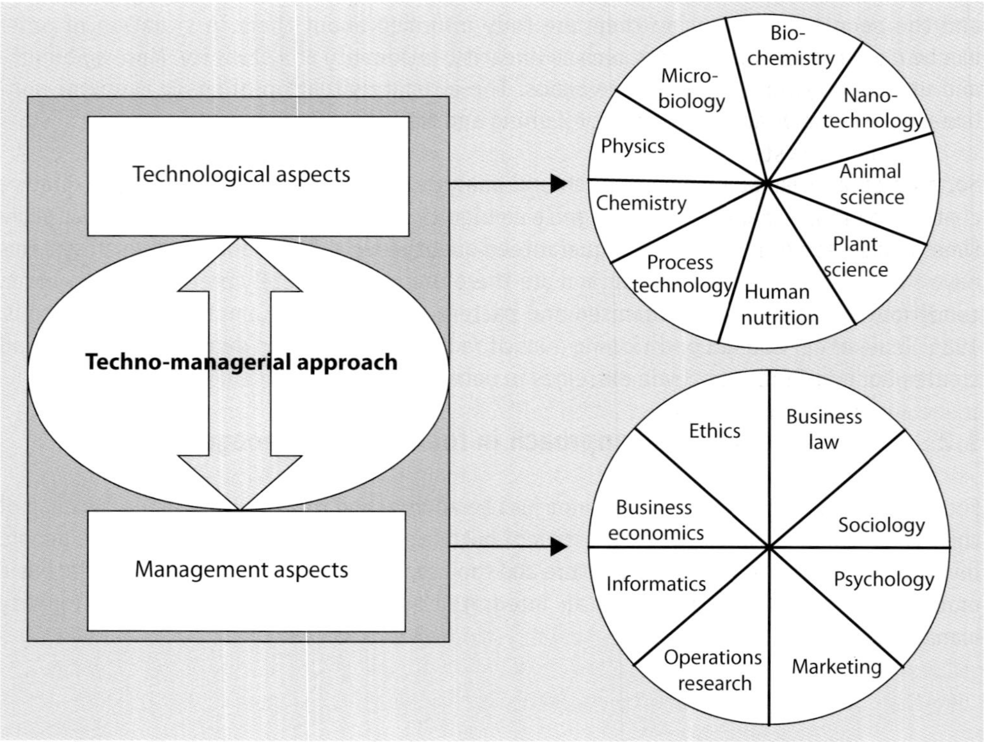

Figure 1.3. Techno-managerial approach.

- Interactions between food processes.
- Interaction of food materials with contact materials like equipment.
- Dependability of food properties on process design characteristics.

A food system is thus a complex system with a dynamic and variable behaviour.

Human systems relate to people with certain individual characteristics taking decisions in order to reach goals. *Human behaviour* is the result of the decision-making outcomes of people (human system) over time. Human systems are subject to a *management system*, which is described as the whole range of managerial activities and administrative conditions (e.g. organisational relationships, available information) influencing decision-making activities in order to obtain the desired decisions. The resulting decisions are meant to initiate actions towards people and towards the food production system in order to manage them so that food quality can be attained.

The characterisation of the decision-making reveals it to be a complex and dynamic process. Decision-making can be influenced by various aspects like type of information, personal characteristics, motivation and ability but also organisational conditions, conflicting interests, availability of information and power. Besides the complex and dynamic character of decision-making of people, organisational conditions also have their own dynamics and life cycle, due to a continuously changing environment that requires organisations to adapt. Therefore, under various organisational conditions, decision-making will result in different outcomes.

The complex and dynamic characteristics of both systems underpin the importance of analysing them systematically to understand and predict the quality outcome. The *techno-managerial approach* advocates the importance of understanding the behaviour of both food and human systems and their interdependency. It includes the integrative use of technological and managerial theories, wherein technological theories are needed to explain food behaviour and managerial theories to explain human behaviour, and integrative use refers to integrating these theories (Figure 1.3).

Technological disciplines, such as microbiology, chemistry, process technology, physics, human nutrition, plant science and animal science, are needed for understanding variability and complexity of food (and agricultural) systems, in order to control and predict them. Management studies incorporates psychology, sociology, economics, mathematics, and legal science to understand peoples' behaviour within their organisational and chain context. As a consequence, food quality management embraces the integrated use of technological disciplines as well as the integrated use of managerial sciences.

A techno-managerial approach differs from a managerial or a technological one. Typical of a managerial approach is that technological aspects are considered as given facts, whereas in a technological approach, management aspects are considered as non-changeable boundary restrictions.

1.2.2 Food quality relationship

The techno-managerial approach forms the basic approach for analysing food quality management (FQM) issues, but is not a very concrete way to explain what to analyse in detail. Therefore, a conceptual food quality (FQ) relationship has been developed indicating which underlying mechanisms of food and human behaviour may affect food quality (Luning and Marcelis, 2006, 2007, 2009), as shown in Figure 1.4.

In this relationship it is assumed that both food and human systems contribute to food quality. The outcomes of these systems over time reflect their behaviour, i.e. food and human behaviour respectively. The relationship shows that food behaviour (FB) is dependent on food dynamics (FD, i.e. variability in properties due to e.g. variable compositions, enzyme activities, or level of pathogens) and applied technological conditions (TC, like process conditions, equipment and building circumstances). Likewise, human behaviour (HB) is dependent on human dynamics (HD, i.e. variability in decisions due to e.g. variable perceptions, attitudes, or choice intentions) and administrative conditions (AC, like, procedures, responsibilities, and (information) systems).

Table 1.1 provides some typical examples to illustrate what is meant by the different factors of the FQ relationship model. This model is aimed at supporting the selection of relevant theories and models to understand the dynamics of the two systems, the impact of conditioning, and how these systems may interfere or interact.

1.2.3 Food quality management functions

The above-mentioned FQ relationship and TM approach indicate which underlying mechanisms may contribute to food quality and what type of theories should be considered, but it does not answer the question as to what kind of activities are needed to make products that satisfy quality requirements. Food quality management deals with how to realise food quality. It consists of goal-oriented decisions resulting in activities which contribute to meeting or even exceeding the quality requirements of customers and or consumers. A Food Quality Management (FQM) functions model has been developed showing which

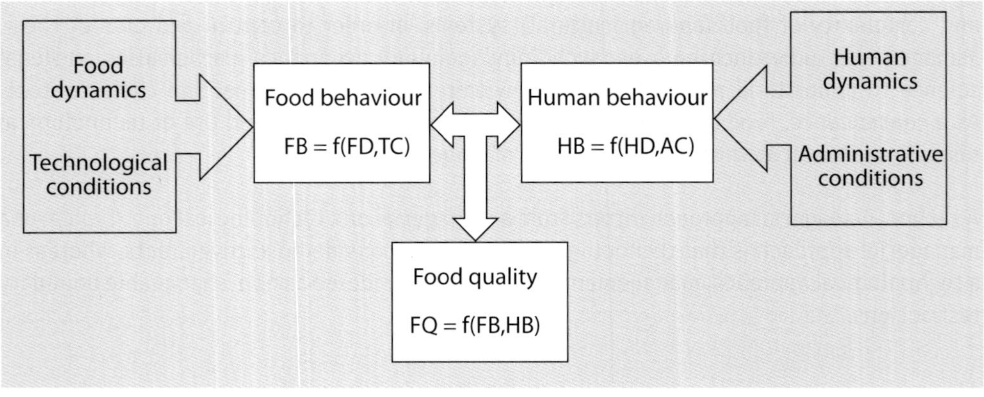

Figure 1.4. Food quality relationship.

Table 1.1. Examples to illustrate the factors influencing food quality based on the FQ relationship model.

Element FQ relationship model	Examples
Food dynamics	Inherent variability in product composition (i.e. concentrations of macro and micro ingredients, pH, a_w) Dynamic food processes in time, like enzyme activities, growth rates of micro-organisms, and respiration rates
Technological conditions	Process conditions (process parameters, type of process), equipment (design, lay-out, type), and buildings (infrastructure) as applied in production and distribution
Human dynamics	Inherent variability in people with respect to their quality perceptions, attitudes, interests, and other personal factors influencing decision-making on food quality
Administrative conditions	Procedures and rules, responsibilities and tasks, and information systems to support the decision-making behaviour of people on quality

technological and managerial functions are essential for realising food quality (Luning and Marcelis, 2007).

The conceptual FQM functions model (Figure 1.5) shows the organisation in its environment, wherein technological and managerial functions interact, striving for food quality that meets or exceeds customer and consumer requirements. The model consists of four elements, i.e. consumer/customer quality requirements, environment, technological and managerial functions:

- The *'customer/consumer requirements'* element concerns the quality concept which is the starting point for FQM. Quality is described here in terms of physical and organisational quality dimensions (see also extended quality triangle, section 1.3.2).
- The *'environment'* element refers to social, political, economical, and technological circumstances in the environment that affect decision-making on food quality by the organisation, via interests and power.
- The *'technological functions'* element refers to the technology dependent activities needed to achieve a product with certain physical properties. Technology is defined as a set of processes, tools, methods, and equipment used to produce goods. The technology dependent activities include the actual *production activities* (e.g. heating, storing, transporting, etc.) as well as the *measuring activities* (e.g. taking samples, analysing, measuring, etc.) to provide information about the status of products and processes. These activities are executed by (process) equipment, machines, and tools alone or in combination with people.
- The *'managerial functions'* element refers to the necessary decision-making activities to activate the food production system as well as the management system, to give it the right direction, and to ensure that it meets consumer and customer requirements.

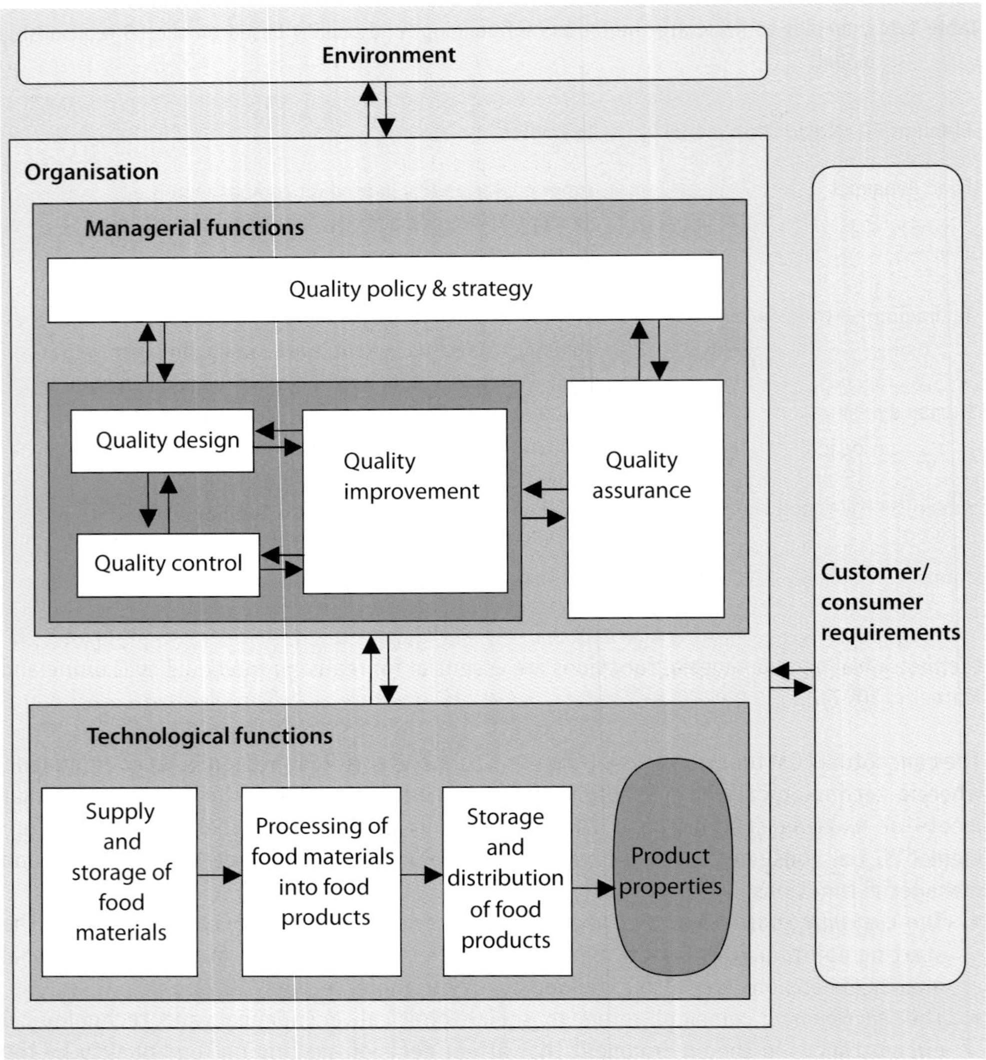

Figure 1.5. Food Quality Management functions model (from Luning and Marcelis, 2007).

Within the technological functions there are three with different objectives:

1. *Physical supply and storage* of incoming food materials. Technological activities are aimed at *obtaining* and *maintaining* the required physical properties of incoming food materials.
2. *Transformation* of food materials into processed food products. Activities are aimed at *transforming* incoming materials into food products with desired physical properties by applying appropriate technological conditions.
3. *Physical storage and distribution* of processed food products. Activities are aimed at *maintaining* the required food properties and preventing deterioration by applying appropriate storage and distribution conditions.

With respect to the managerial functions, five different ones were identified. In agribusiness and food industry, quality control is commonly accepted as a basic function in realising food quality. However, approaching food quality management only as a series of control processes is like reducing reality to cybernetic processes. This is too simplistic if we realise that the complexity of food quality management is determined by uncertainty due to lack of information, and by ambiguity due to lack of a full understanding of the food and human systems (Figure 1.2). Therefore, more quality-related activities are necessary, each with a different objective:

1. *Quality design* aims at incorporating quality in food product and process design. It starts with specifying consumer and or customer demands and translating them into product and process specifications. Requirements on resources are also specified, like suppliers, process and transport equipment, and people involved.
2. *Quality control* is the ongoing process of evaluating the performance of both technological and human processes and taking corrective actions when necessary; it is aimed at keeping processes within acceptable tolerances.
3. *Quality improvement* is aimed at changing people, processes, and resources in order to bring them to a higher level of quality performance and into 'new zones of control', which means working with lower tolerances.
4. *Quality assurance* activities deal with setting requirements for the quality system, evaluating its performance and organising any necessary changes. The objective is to control the quality system, which includes all activities and decisions to realise quality, and to provide confidence to customers and consumers that quality requirements will be met.
5. *Quality strategy and policy* are aimed at determining long-term food quality goals and objectives, and establishing how to achieve them using the quality system. This includes determining quality levels of both products, and technological and human resources, and the type of quality system (Luning and Marcelis, 2007).

The managerial functions of quality design, control, improvement, assurance, and policy & strategy have different objectives and include different activities. However, all managerial functions are decision-making activities that activate the food production system (including all technological functions) and the management system. So decision-making plays a central role in realising food quality as will be discussed in more detail in Chapter 3.

1.3 Quality

There is no unambiguous definition or concept for quality and more specifically food quality. Quality is not a scientific or technical word (Williams, 1993), but it is often used without defining the meaning (Bremner, 2000). There are many definitions of quality as applied to foodstuffs (Botta, 1995). For example, Bremner (2002) developed a nominal hierarchy of definitions related to quality, ranging from overall concept (not measurable), general definitions (such as ISO definition), defined quality (i.e. in terms of rigorously measured parameters determined by standard methods, such as in legislations and codes), to prescribed test methods to provide values related to quality criteria. It is therefore important to have some insight into quality definitions and concepts as a starting point. In this section, several common definitions for quality are given and a range of general and more food specific quality models are described. The different definitions and models illustrate that quality is not a physical entity or instance that has a fixed position in time and space but a concept.

1.3.1 Quality definitions

Definitions of quality first emerged with the quality philosophers in the eighties, who referred to the quality of products and services in general, not specifically to food quality. Juran (1990) defined quality as 'product performance that results in customer satisfaction and freedom from deficiencies, which avoids customer dissatisfaction'; in short fitness for use. Deming (1993) stated that 'a product or a service possesses quality if it helps somebody and enjoys a good and sustainable market'. Crosby (1979) described quality as 'complying with clear specifications, whereby the management is responsible for establishment of univocal specifications'. Chapter 3 provides more details about these quality philosophies and their role in developing quality management.

Several institutes also developed definitions for (food) quality. The International Organisation of Standardisation (ISO) defined a wide range of terms related to quality management, which are applicable for all products and services (http://www.praxiom.com/iso-definition.htm). In the ISO 9000 document quality is defined as a desirable characteristic that a product or service must have. For example, products must be reliable, useable, and repairable, whereas service should be courteous, efficient, and effective. However, not all qualities are equal. The most important qualities are the ones that customers want. So providing quality products and services is all about meeting the needs and expectations of customers.

The American Society for Quality Control (ASQC) (http://www.asq.org/) developed a glossary of terms related to quality. They defined quality as a subjective term for which everyone has his or her own definition. In technical usage, quality can have two meanings: the characteristics of a product or service that relate to its ability to satisfy stated or implied needs, and a product or service free of deficiencies.

The Institute of Food Science and Technology (IFST) (http://www.ifst.org/) has described more specifically the term food quality. When applied meaningfully to the character of a food, quality may refer to (a) the degree or standard of excellence, and/or (b) the suitability for purpose, and/or (c) the consistency of attainment of the specified properties of the food. In the context of food control, it is meaning (c) that applies.

In this book we use the commonly accepted definition:

> *'Quality is meeting or exceeding customer and consumer expectations'.*

Customers are defined here as those who receive a product (ranging from raw materials to finished products) or a service from a supplier within a food production chain. For example, the milk factory is the customer of farmers who supply the milk. This type of customer is called an external customer, whereas customers within a company are typified as internal customers (e.g. the cheese production is the customer of the quality control laboratory with respect to the quality results). Customers that are the ultimate purchasers or users of a food product or service are referred to as consumers. *The* consumer does not exist and there is no average consumer. There is a specific consumer, which in a specific situation and at a certain moment has a specific need to which the producer can respond.

1.3.2 Quality concepts

Defining quality is difficult but describing a model or concept for (food) quality is even more challenging. Several authors have attempted to describe such a concept to enable the analysis of quality and, more specifically, food quality.

Quality perspectives concept

According to Evans and Lindsay (2005) the concept of quality is often confusing because people consider quality from different perspectives depending on their role in the production-marketing value chain, and quality evolves as the quality profession grows and matures. They distinguished five different perspectives, i.e. judgmental, product-based, user-based, value-based, and manufacturing-based (Figure 1.6).

From a *judgmental* point of view, quality can be considered as synonymous with excellence or superiority. This view is referred to as the *transcendent* definition of quality, which rising or extending beyond ordinary limits. From this viewpoint quality is absolute and universally recognisable, a mark of uncompromising standards and high achievement. It is often loosely related to a real comparison of product features and characteristics and sometimes it is more of a quality image created by marketing departments. Typical examples of products attributed with this image are Rolex watches and Cadillac automobiles but also Coca-Cola

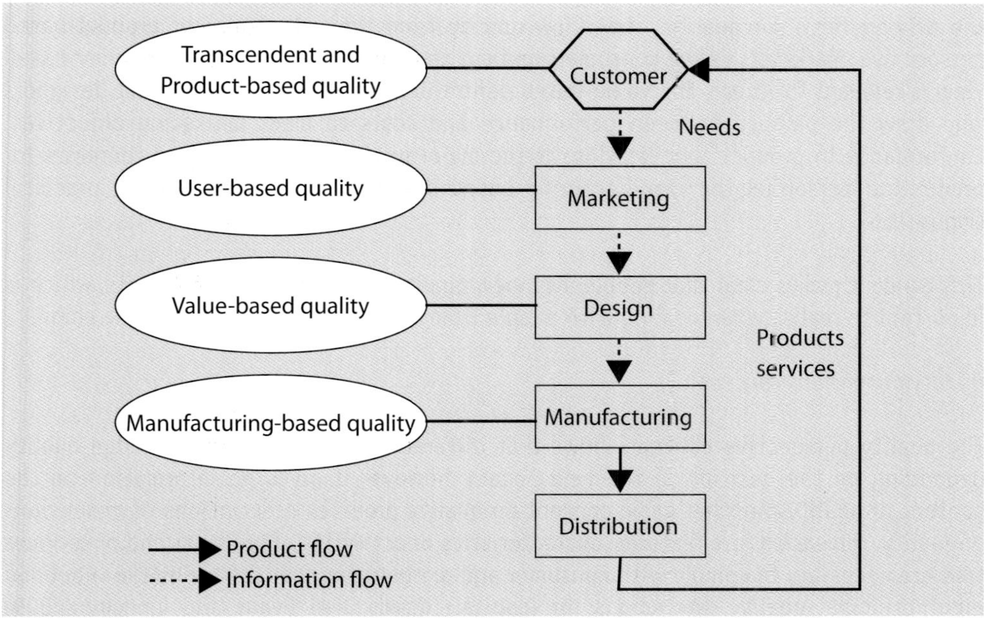

Figure 1.6. Quality perspectives in the value chain (from Evans and Lindsay, 2005).

and Nestlé coffee. These products obviously have to comply with consumer demands but the brand is a kind of guarantee for quality.

From a *product-based* view quality can be defined as a function of a specific measurable variable. From this viewpoint quality is reflected in a quantitative difference of a certain desired variable. This implies that higher levels or amounts are equivalent to higher quality. For example, higher concentrations of fresh fruit contribute to a better quality of fruit juice.

From the *user-based* perspective quality concerns the presumption that quality is determined by what a customer wants, in short 'fitness for intended use'. In other words, how well the product performs within its intended function or use. For example, instant coffee may be perceived as good quality when used during a trek in the mountains, but may be perceived as lower quality for use at home.

From the *value-based* perspective quality is defined as the relationship of usefulness or satisfaction to price. From this point of view a quality product is one that is as useful as competing products and is sold at a lower price or one that offers greater usefulness or satisfaction at a comparable price. For example, products with retail brands are often perceived as good quality because they have similar features to the A-brands at a lower price.

The *manufacturing-based* perspective on quality can be described as the desirable outcome of engineering and manufacturing practice; in short, conformance to specifications. These specifications include targets with tolerances, as specified by the designers of products and services.

Evans and Lindsay (2005) concluded that the criteria used for defining quality depend on one's situation in the production-distribution cycle. Consumers and customers are the driving force for quality. They view quality from the judgmental or product-based perspective. Marketers define consumers and customer needs and therefore the user-based view is relevant for them. The value-based definition is most useful for product designers who draw the balance between performance and costs to meet marketing objectives. Conformance to product specifications is the major goal in production, and therefore for production personnel the manufacturing-based definition of quality is most practical (Figure 1.6).

This concept makes clear that people may view quality from different viewpoints, which is important to realise when dealing with people from various positions in the value chain.

Quality dimensions concept

The quality perspective concept shows that different criteria are used to judge quality depending on the position in the value chain, however it gives no information on the content of quality. Another basic concept on quality provides descriptions of dimensions of quality. *Dimensions* are features, characteristics or activities of products and or services that are necessary to comply with consumer and or customer needs. Garvin (1984) defined eight principal quality dimensions for (capital) goods, and Evans and Lindsay (2005) mentioned typical dimensions for service quality, as summarised in Table 1.2. Services differ fundamentally from physical goods; for example, the output of service systems is intangible (manufactured foods are visible), services are produced and consumed simultaneously, customer needs and performance standards are difficult to identify, and it requires a high degree of customisation (whereas in manufacturing the goal is uniformity) (Evans and Lindsay, 2005). Although both authors use different terms, there are similarities in meaning, such as performance and accuracy, conformance and completeness, and serviceability and responsiveness (Table 1.2). Some dimensions are very typical for services, like accessibility, convenience and customisation, whereas aesthetics and perceived quality are typical quality dimensions for manufactured (capital) goods. This concept of describing quality in terms of dimensions clearly shows the multidimensionality of quality.

The above concepts are general concepts for quality and thus not specific for agriculture and food products. These products have typical characteristics, like the fact that they are living systems that decay over time, they may contain hidden safety risks that cannot be seen, etc. Moreover, consumers have a very critical and emotional attitude towards safety and quality of food. This means that dimensions as described for technical goods are not just applicable to agriculture products and food products. The following quality models and concepts deal specifically with (agri) food quality.

Table 1.2. Concept of quality dimensions.

(Capital) goods (Garvin, 1984)	Services (Evans and Lindsay, 2005)
Performance: primary operating characteristics	Accuracy: correct performance of the service
Conformance: degree to which physical and performance characteristics of a product comply with pre-established standards	Completeness: correctness of the delivery of items
Serviceability: speed, courtesy, and competence of repair	Timeliness: completion of the service in the agreed time
Features: additional properties, the 'bells and whistles'	Responsiveness: quick response of service personnel to unexpected problems
Reliability: probability of a product to survive over a specified period of time under stated conditions of use	Courtesy: friendliness and politeness of employees
Durability: amount of use one gets from a product before it physically deteriorates or needs repair	Expertise and customisation of service personnel
Aesthetics: how a products look, feels, or sounds	Accessibility and convenience: the easiness of obtaining the service or information
Perceived quality: subjective assessment of quality resulting from image, advertising, or brand names	Time: before service is delivered or completed

Analytical model of food quality

Peri (2006) described quality as the requirements necessary to satisfy the needs and expectations of the consumer. He developed an analytical model of food quality, represented as a set of requirements (Figure 1.7).

The analytical model of food quality consists of two parts: one containing requirements on the food product, involving us as a *'Homo edens'*, and one containing requirements on the product as market object, involving us as *'Homo oeconomicus'*. Peri (2006) described these requirements as follows:

- *Safety requirements,* which are expressed as absence of risk factors for consumer.
- *Commodity requirements,* which refer to the conformity of a product to its definition.
- *Nutritional requirements,* which refer to nutritional and health-giving product properties.
- *Sensory requirements,* which concern the sensory observable product properties. The fact that sensory properties are perceived make them an important means of interaction between products and consumers.

The set of safety, commodity, nutrition and sensory requirements constitutes the frame work of the *product quality itself.* However, consumer expectations are about more than

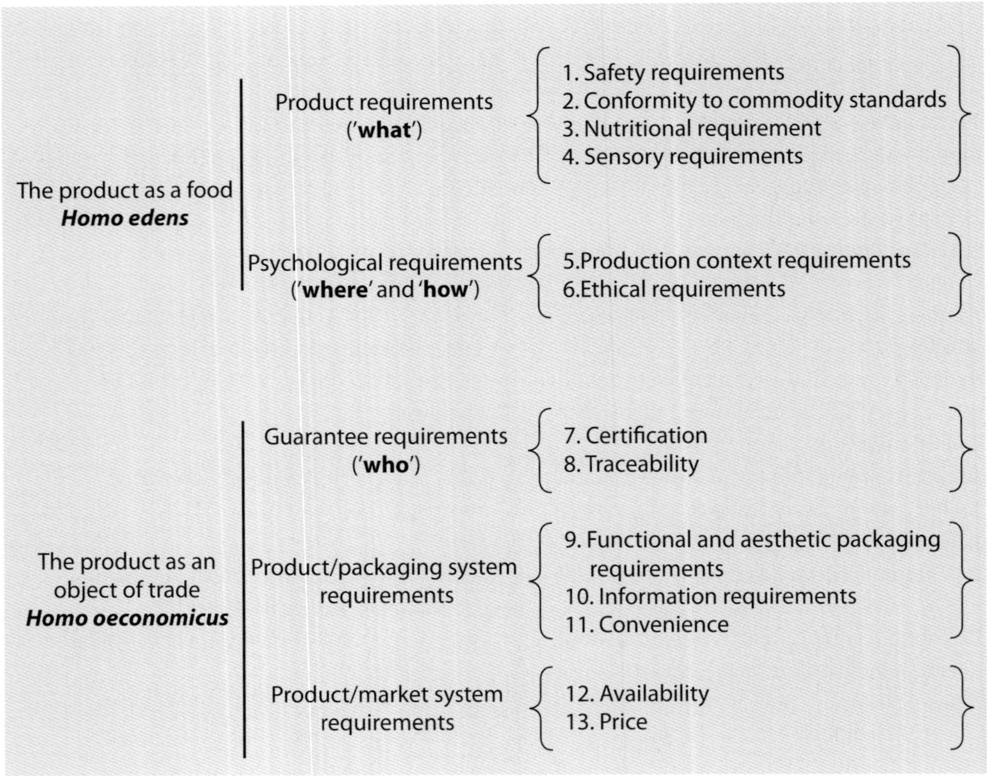

Figure 1.7. Analytical model of food quality (from Peri, 2006).

satisfying these requirements, moreover other requirements play a role in determining their choices, including:

- *Production context or immaterial requirements,* which concern the origin or tradition of a product, or the use of organic agriculture. These requirements satisfy psychological and cultural needs and their appeal does not depend on the 'what' of a product, but on the 'how', 'when' and 'where' it was produced.
- *Ethical requirements,* which relate to the system of values that conditions consumer behaviours. Typical ethical requirements include the defence of biodiversity against mass production, the well-being of animals, the protection of the environment, etc.

The requirements of production context and ethics cannot be verified or perceived when eating or analysing a food product; it is highly susceptible to fraud and deceit and therefore a third group of requirements is important:

- *Guarantee requirements,* which refer to certification and traceability procedures. So besides, 'what', 'how', 'when', and 'where' the product is produced, consumer perceptions of quality also involve requirements on 'who' produced it.

Food products are not offered as such but as a packaged product within a certain market context, which results in the final requirements:

- *Product/packaging system requirements,* which refer to the properties of packaging concepts (like convenience, aesthetics).
- *Product/market system requirements,* which concern the availability of the product at the right time in the right place and the desired amount, where price is also important.

This quality concept provides a detailed overview of the different quality requirements and their characteristics. It embraces not only what a consumer finds desirable but also what may guarantee the 'sustainable development' of the food production system (Peri, 2006).

Extended quality triangle

Within the philosophy of the techno-managerial approach, quality should not just be considered as physical product and production (e.g. environmentally sound) quality but should also concern quality performance of the organisation (e.g. flexibility and service).

Noori and Radford (1995) considered quality from a total quality perspective and identified six dimensions, i.e. product quality, cost, time, flexibility, dependability and service. This concept has been modified to the extended quality triangle, which includes a product-related triangle extended with an organisation-related one (Luning *et al.,* 2002; Luning and Marcelis, 2007), as described below and shown in Figure 1.8.

1. *Product quality* refers to the intrinsic and extrinsic quality attributes of the food product (linked to intrinsic-extrinsic quality attributes model, see Chapter 2). Quality attributes are the result of (various) product properties, which are noticeable by sensory observation or via communication, and thus contribute to quality perception.
2. *Cost* is primarily considered in a (consumer) customer-oriented net value sense. It is the basis for quality/price perception.

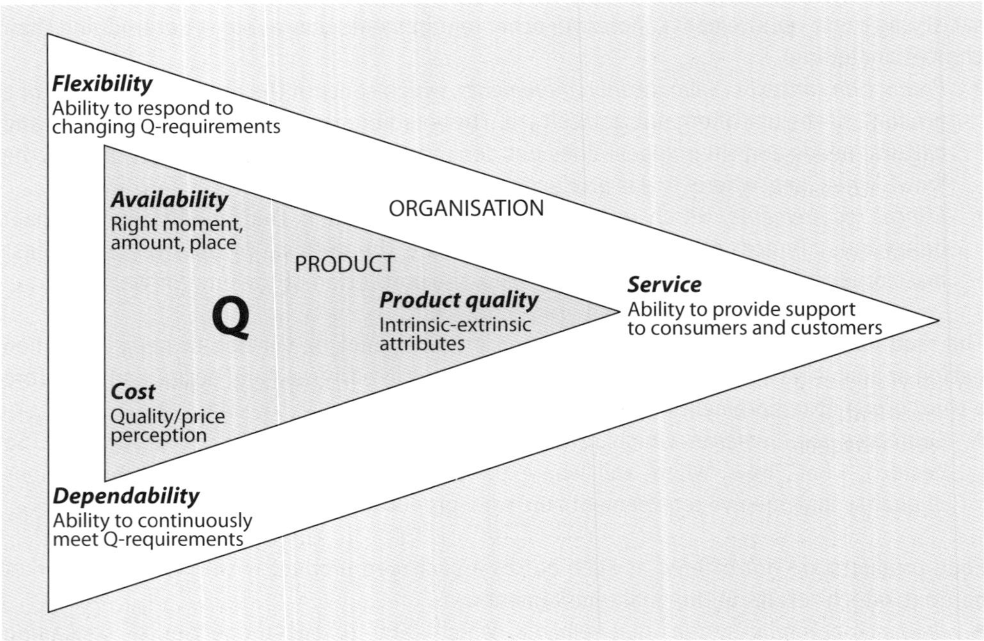

Figure 1.8. Extended quality triangle.

3. *Availability* concerns availability of products, ingredients and or raw materials, at the right moment, correct amount and in the right place. For food this is important due to the rapid decline in quality level of many food products, raw materials, and ingredients.
4. *Flexibility* is the ability of the organisation to respond quickly to changing requirements on products, processes and resources e.g. due to new customer wishes, legislative demands.
5. *Dependability* or *reliability* refers to the ability of the organisation to continuously meet requirements on quality, such as food safety level and other agreements, so providing confidence to customers and consumers.
6. *Service* involves providing support for consumers (e.g. complaint support) and customers (e.g. service in application of ingredients or in product design).

This model addresses quality for both the physical product (the first three dimensions) and the organisation (the last three dimensions).

In this book we use the extended quality triangle as the *quality concept combined with the intrinsic-extrinsic quality attributes model* for product quality, which is described in Chapter 2.

1.4 Structure of the book

The techno-managerial approach takes precedence in our book, which is aimed at supporting education and research on food quality management issues. This means that theories, models, concepts and techniques have been selected from technological and managerial sciences in order to provide some basic theories with which to understand and analyse food quality management issues.

The structure of the book is based on the need for a basic understanding of food quality and quality management and the essential food quality management functions, which are necessary to realise quality that satisfies or even exceeds customer and or consumer needs.

The book consists of nine chapters. Chapters 1-8 give an introduction to food quality management theories, models, principles, and concepts. The last chapter deals with future developments. The content of the chapters is briefly described below.

Chapter 1 gives an introduction to the basic philosophy and models beyond food quality management. The techno-managerial approach, food quality relationship model and food quality management functions model are introduced and explained.

Chapter 2 is aimed at providing a basic understanding of food behaviour from a technological perspective. This chapter gives an introduction to the complexity and dynamics of food systems and the technological conditions affecting these systems. It should provide people from a managerial background with some basic insight into the technological aspects of food quality management. For details and more background information we have attempted to refer to commonly acknowledged handbooks and review articles.

Similarly, Chapter 3 is aimed at providing a basic understanding of human behaviour from the perspective of managing food quality. This chapter gives an introduction to the complexity and dynamics of people working within the organisational and chain context, and how administrative conditions affect the behaviour of people striving to achieve food quality. It should provide people from a technological background some basic insight into the managerial aspects of food quality management. For details and more background information, we refer to commonly acknowledged handbooks and review articles wherever possible.

Chapters 4-8 deal with the food quality management functions, quality design (Chapter 4), quality control (Chapter 5), quality improvement (Chapter 6), quality assurance (Chapter 7), and quality policy and strategy (Chapter 8). In each chapter, the function (e.g. Q-design, Q-control, etc.) is described in relation to realising food quality and the process and principles are explained. Subsequently, we have considered both technological and managerial aspects that are relevant for analysing activities in each specific function, design, control, improvement, assurance and policy and strategy. The selected theories, models and concepts from technological and managerial sciences provide a *basis* for a techno-managerial analysis of food quality management issues and a *basic* understanding of its complexity.

The last chapter discusses future developments and prospects in food quality management in agribusiness and the food industry.

Chapter 2

Food quality

2.1 Introduction

Chapter 2 provides some basic knowledge about biological processes in quality perception, takes a look at the various quality attribute concepts, explains typical food system characteristics, typifies different technological conditions, and provides insight into how quality can be affected along the food supply chain, and how environmental requirements (especially legislation) also impact food quality. This essential knowledge is required to be able to recognise and consider the typical technological aspects when designing, controlling, improving, and assuring food quality.

The extended quality triangle in Chapter 1 shows that different dimensions contribute to the quality performance of products (i.e. physical product quality, costs, availability) and quality performance of organisations producing it (i.e. service, flexibility, and dependability) (Figure 1.8). Realising quality in food production systems along the whole supply chain requires consideration of these dimensions because they can affect the satisfaction of customers in the chain and consumers at the end of the chain. However, it is important to realise that the way in which consumers perceive and describe food quality is on a different level to that necessary for designing, controlling, improving, and assuring it.

Therefore, it is indispensable to have insight into how quality from a consumer perspective links to measurable physical product properties as a result of the food production process. Figure 2.1 shows that quality is the result of a quality perception process (section 2.2). A product can have different quality attributes (section 2.3), which result from the food system characteristics (section 2.4.1) and the technological conditions (section 2.4.2) that are applied. The different technological conditions along the chain have an impact on various quality attributes as perceived by final consumers (section 2.5), and food legislation too can affect food quality by influencing the decision-making of companies (section 2.6).

2.2 Quality perception

Before giving any further explanation about food quality, it is necessary to point out that agri-products and food products as such have no quality! These products have properties (such as concentration of sugars, level of spoilage bacteria) that influence product attributes (like taste, safety), which are perceived and judged as quality after use and or consumption by customers and/or consumers. Here, consumers are the end users and customers are actors in the food supply chain.

People have a special relationship with food as compared to other goods. We need to consume food everyday for our metabolism and immediate well-being, and quality is of major importance for a long-lasting acceptance and consumption of food products. The overall judgement of food quality is influenced by how people perceive the product through observation of sensory observable attributes (like taste, texture) and via communication

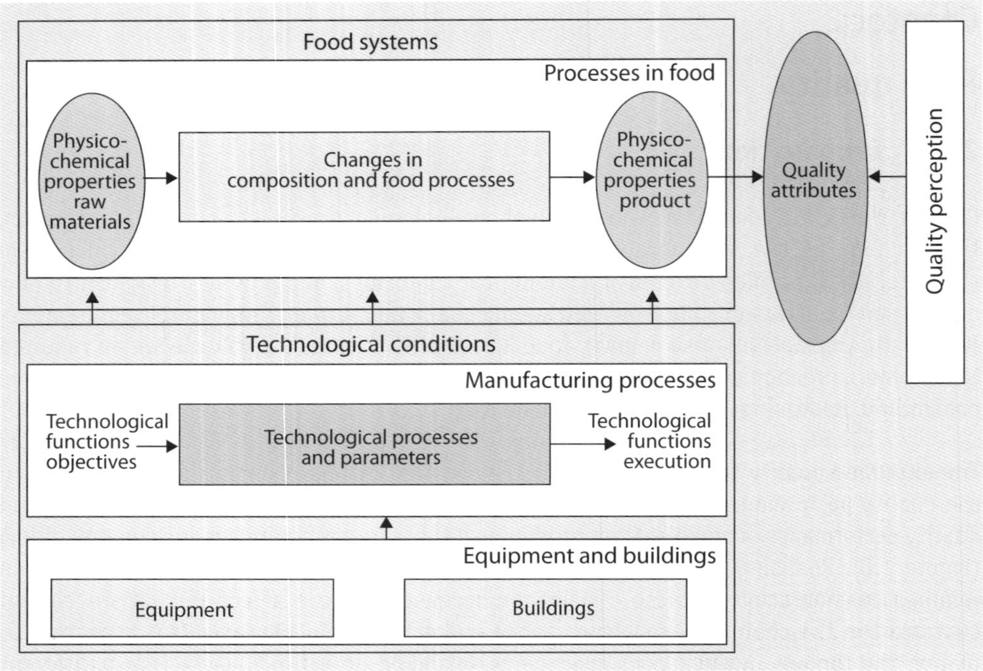

Figure 2.1. Food quality as a result of the food production process.

about non-observable attributes (e.g. information about health compounds, safety) (Steenkamp, 1989, 1990).

For food quality management, it is necessary to have a basic insight into the processes that constitute food quality perception. A commonly accepted model distinguishes physical and psychological processes contributing to food quality perception (e.g. Cardello, 1996; Dijksterhuis and Piggot, 2000). The physical processes occur when one puts food in the mouth, whereas the psychological processes start when the physical stimuli, initiated by the food, change into neural responses. Both are dynamic processes. Figure 2.2 provides a schematic presentation of these processes.

The *physical processes* take place while the food is in the mouth. Typical processes include:

- chewing to breakdown food structure;
- breathing to encourage release of volatile compounds to the olfactory epithelium, saliva flow to dilute or dissolve food and modify pH;
- swallowing to produce a pulse of air to increase transfer of volatiles to the nasal cavity;
- temperature changes to melt fatty foods and affect volatility of flavour compounds; and
- tongue movements to mix food and accelerate mass transfer (Dijksterhuis and Piggot, 2000).

The physicochemical properties of a food product concern the *physical* (product structure, viscosity) and *chemical* (e.g. composition of volatiles, non-volatile aroma compounds, and taste compounds) properties. The product composition determines which compounds are

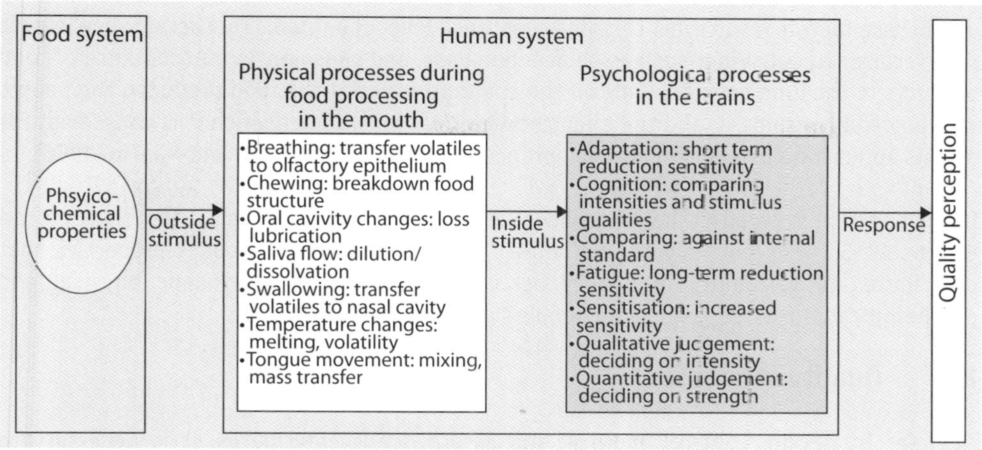

Figure 2.2. Schematic presentation of quality perception (modified from Dijksterhuis and Piggot, 2000).

produced and how they are released in the mouth. The physicochemical product properties stimulate the sensory systems via specific receptors for each sensory modality (like taste buds on the tongue, odour receptors in the nose), which result in sensory experiences in terms of taste or smell sensations. Five human senses can be distinguished, i.e. vision, hearing, taste, smell, and touch, although other sensory systems (e.g. for balance, body and limb position, and subcutaneous perception by pressure) can influence the perception of food quality as well. Each sensory modality can provide two different dimensions of sensory experience, i.e. quantitative and qualitative. For example, with respect to taste one can experience the four qualitative senses such as sweet, salt, bitter or sour. This qualitative experience is also called sensory quality. Moreover, one can experience the intensity of each of these sensations. Both the sensory quality and intensity contribute to our perception of food quality.

The psychological processes start after the physical stimulation (by the physicochemical properties) of the receptors into sensory experiences. These sensory attributes (e.g. hardness, sweetness, fried odour) turn into sensory preferences due to the interference of the central nervous system. Typical processes involved in the psychological processes are:

- adaptation, causing a short-term reduction in sensitivity during exposure to a stimulus;
- cognition, comparing intensities and stimulus qualities, involving memory;
- comparison with internal standard, probably involving memory;
- fatigue, causing a long-term reduction in sensitivity and ability to concentrate after prolonged exposure to stimuli;
- sensitisation, i.e. increased sensitivity during exposure, e.g. due to pungent compounds;
- qualitative and quantitative judgement of the sensory senses (Dijksterhuis and Piggott, 2000).

Learning and experience, expectations, and cognitive factors arising from e.g. information, price, brand, and labelling modify the overall sensory preferences. Physiological factors such as hunger, thirst and other internal factors may also interfere at this stage. The integration of all these factors results in a certain acceptance level of an item. This acceptance level is the *precursor* to behaviour such as choice, purchase, and consumption. Acceptance of food products in the long term depends on the characteristics of the food products, the people who buy and consume it, and the circumstances (context) under which it is consumed. For details about food perception and acceptance, we refer to Meiselman and MacFie (1996).

It is important to realise that the above processes are dynamic and differ from person to person. So, it is not only the variability in physicochemical product properties that contributes to variability in quality perception, but also the dynamic physical and psychological processes involved in people's perception of food.

2.3 Quality attributes

In order to design, control, improve, and assure quality attributes, it is necessary to understand what the different quality attributes comprise and how they relate to technological product and process properties. One should be aware that individuals have different concepts in mind, so it is important to make it explicit. This section aims at obtaining a clear understanding of the more detailed concepts beyond major quality attributes.

The intrinsic-extrinsic model is the basis for selection and classification of the different quality attributes (Figure 2.3). As previously stated a food product as such has no quality but consists of physicochemical properties. Quality attributes are the result of various product properties, which are noticeable by sensory observation or via communication, and in this way contribute to the quality perception and experience of consumers and customers (Steenkamp, 1990; Van Trijp and Steenkamp, 2005). The model distinguishes between two different types of quality attribute, i.e. intrinsic and extrinsic, both of which can influence the acceptance of food products.

Intrinsic attributes are *inherent* to the physical product. These attributes can be noticeable (e.g. texture and odour attributes) or communicated (e.g. safety, health). They are the result of physicochemical and other properties of a product (e.g. pH, composition, microbial contamination, etc.).

Extrinsic attributes do not necessarily have a direct relationship with the product properties but can affect consumers' quality perception. These attributes relate to the production and marketing aspects of a product. These extrinsic attributes can influence acceptance of products too (e.g. animal welfare, sustainability of production, brand name). Figure 2.3 shows that the product properties (food dynamics) and technological conditions that are applied (conditioning) influence the quality attributes.

> *In this book we use the intrinsic-extrinsic quality model as the* product quality *concept for food products, within the broader quality concept of the extended quality triangle.*

Moreover, those involved in food production affect these technological conditions by their behaviour (e.g. hygienic behaviour, quality control behaviour, etc.). Last but not least, legislation affects quality attributes by putting restrictions and requirements on the

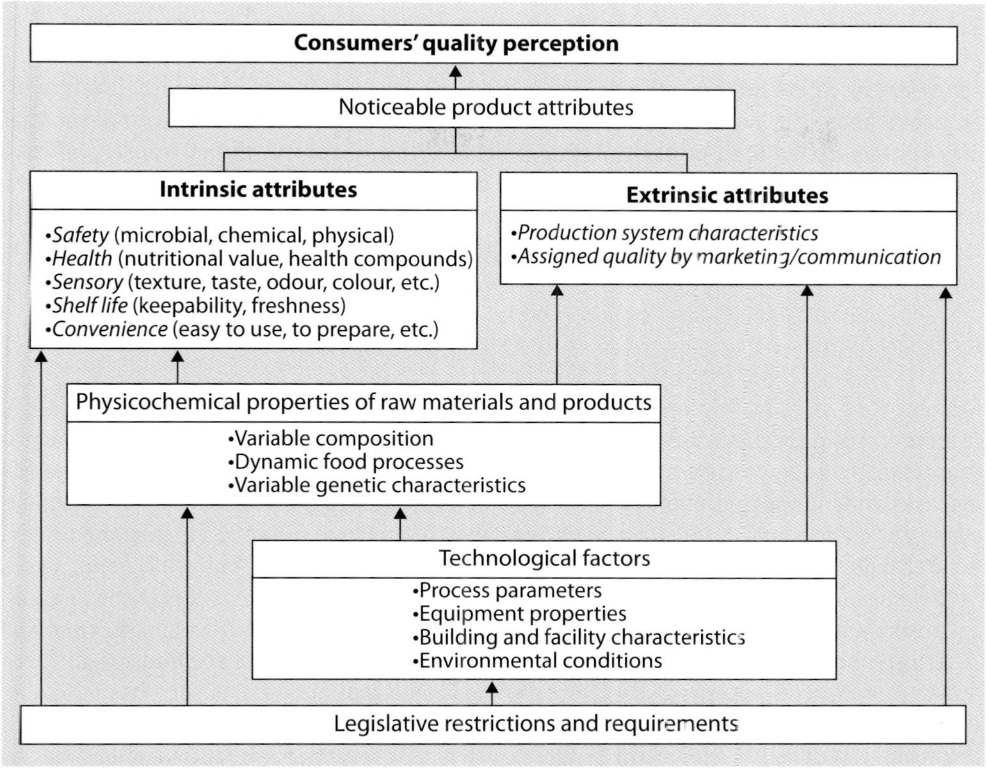

Figure 2.3. Intrinsic-extrinsic quality attributes model (modified from Luning *et al.*, 2002).

product properties (e.g. compositional limits), technological conditions (e.g. minimum T-t conditions) and managerial aspects (e.g. hygienic behaviour, procedures, quality systems), but also directly on the attributes (e.g. labelling requirements on health claims, or organic production).

For food quality management it is necessary to understand and to be able to unravel quality attributes for several reasons. Depending on the position in the food supply chain, actors may have different quality concepts and different interpretations of quality attributes in mind. Moreover, consumers judge product quality on this abstract level, e.g. they like the texture and flavour and prefer a certain packaging because it is convenient. Consumers do not usually judge quality in terms of product properties like the concentration of vitamins or minimum level of pathogens; they judge in terms of whether the food is healthy and safe. However, from a food production perspective these attributes must be translated into product properties and technological conditions that are necessary to obtain required levels and to set acceptable tolerances. The detailed quality concepts as provided below support the translation of consumer terms into technological requirements.

2.3.1 Safety

Food safety refers to the absence of hazards or the existence of hazard levels with an acceptable risk. From this perspective, a hazard is a potential source of danger that may contaminate a food product. A risk refers to the probability, during any step of food production, of a hazard contaminating food, which if ingested by a consumer can cause a health disorder; it is a measure of the probability and severity of harm to human health. Figure 2.4 shows a safety concept indicating three different classes of hazards, i.e. biological, chemical and physical.

Biological hazards

Biological hazards mainly involve living organisms, which are able to colonise foods and in some cases grow or survive in the food matrix or produce toxic metabolites. The microorganisms that may cause a health disorder are referred to as *foodborne pathogens*, and the diseases they produce are known as *foodborne illnesses*. Foodborne pathogens include a variety of gram-positive and gram-negative bacteria, moulds, parasites, viruses, and prions.

- Well-known gram-negative foodborne pathogens are species of following genus types *Salmonella*, *Shigella*, *Escherichia coli*, *Campylobacter* and *Yersinia*. Well-known gram-positive pathogens belong to species of genus types *Clostridium* (like *C. botulinum* and *C. perfringens*), *Staphylococcus aureus*, *Listeria monocytogenes*, and *Bacillus cereus*.

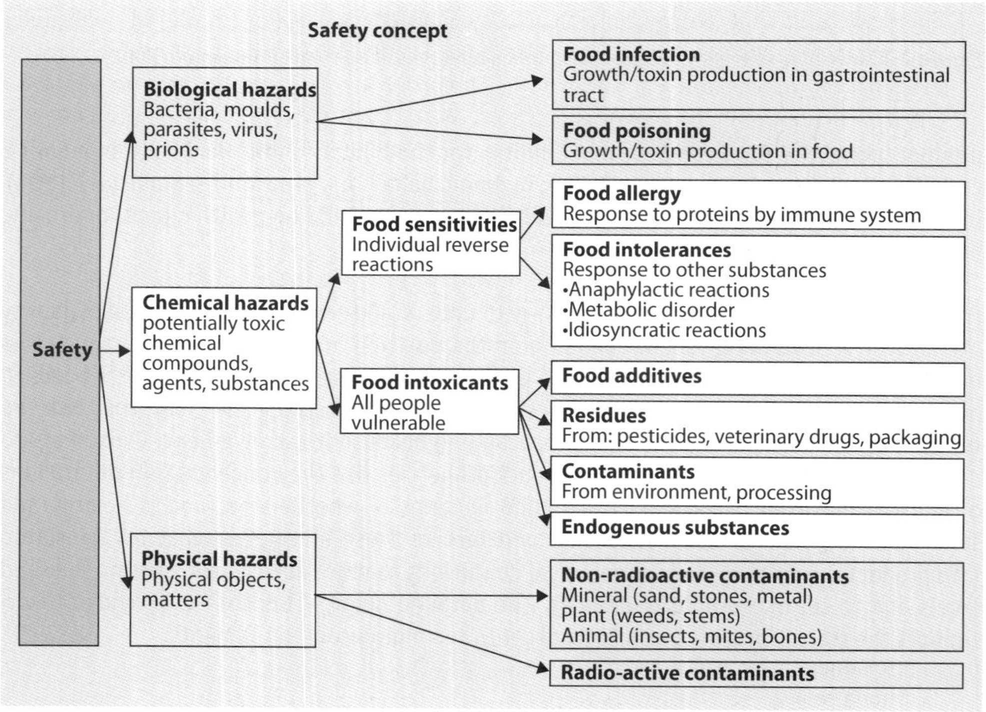

Figure 2.4. Food safety concept.

- Well-known moulds that produce mycotoxins (like aflatoxins, ochratoxin A, and patulin) are *Aspergillus flavus*, *A. parasiticus*, *A. ochraceus*, and *Penicillium expansum*. These mycotoxins are of greatest concern because they are highly toxic and potentially carcinogenic.
- Well-known groups of parasites are protozoa (e.g. cryptosporidium parvum, cyclospora cayetanensis) and helminths (e.g. tapeworms and roundworms). This group of biological hazards is often underestimated and/or not taken into account in food safety systems because the common assumption is that parasites are only present in particular areas of the world. However, the globalisation of world markets has caused them to become a real and worrying concern.
- Well-known viruses are poliovirus, hepatovirus, rotavirus and noroviruses.
- Two well-known cases of prion protein formation are bovine spongiform encephalopathy (BSE) and Creutzfeldt-Jakob disease (CJD). There are indications that ingestion of prion-containing meat may cause spongiform encephalopathy in human brains (Rovira *et al.*, 2006).

Foodborne pathogens are differentiated on the basis of two different modes of action (Figure 2.4):

1. *Foodborne toxico-infections* arise when a micro-organism, ingested in contaminated food, penetrates intestinal mucosa and grows in the gastrointestinal tract where it can produce a toxin(s) that damages tissues or interferes with normal tissue/organ functions, also called *food infection*. Symptoms include nausea, vomiting, abdominal pain, headache and diarrhoea.
   Major bacteria that are responsible for food infections are members of the genus *Salmonella*, *Shigella* spp. and certain strains of *Escherichia coli*, *Listeria monocytogenes*, but also *Campylobacter jejuni*. Beef, turkey, chicken, eggs, pork and raw milk products are typical vehicle foods for salmonellosis outbreaks. Poor personal hygiene is a common factor in foodborne Shigellosis, with shellfish, chicken, salads, fruits, and vegetables being prominent among vehicle foods. Specific strains of *Escherichia coli* often cause traveller's diarrhoea. Another source of food infection is foodborne listeriosis caused by *Listeria monocytogenes*. *Listeria* species are widely distributed in soils, animal faeces, sewage, silage and water. *Listeria* infections also occur in foods, especially raw milk, soft cheeses and other dairy products. Listeriosis is primarily a risk for pregnant women, young children and individuals with a suppressed immune system (for example, people suffering from AIDS, diabetes, etc.). *Campylobacter jejuni* is assumed to be the most common cause of acute bacterial diarrhoea in humans. It is not an environmental organism but the bacteria are carried through the animal intestinal tract. A large percentage of all major animals for meat production, especially poultry, contain *C. jejuni* in the faeces (Jay *et al.*, 2005).
2. *Foodborne microbial intoxications* occur by ingestion of a food containing harmful toxins or chemicals produced by the micro-organisms usually during their growth in food, also called *food poisoning*. Typical biotoxins include toxins produced by bacteria (i.e. *enterotoxins*), moulds (i.e. *mycotoxins*), and plants (i.e. *phytotoxins*). Consumption of the poisoned food can result in symptoms that range from acute abdominal pain and diarrhoea to long-term diseases such as cancer or histological changes in the liver.
   Clostridium botulinum and *Staphylococcus aureus* are well-known bacteria responsible for production of enterotoxins in food. *Clostridium botulinum* is an anaerobic spore forming

gram-positive bacteria and is widely spread in soil and water. Food packed under low oxygen concentrations containing low-acid foods is a particularly good as a vehicle food for Clostridium (e.g. canned foods, *'sous-vide'* processed or modified atmosphere packed foods). Staphylococci have been detected in a large number of vehicle foods (like meat, chicken, and bakery products). Moulds can cause food poisoning too. A well-known mycotoxin is aflatoxin, which can be produced by *Aspergillus flavus* and *A. parasiticus.* Major vehicle products are peas and cereals such as rice, corn and wheat, which are stored in warm and humid conditions. Depending on the causative micro-organisms, its mode of action, and the health status of the host, the symptoms of foodborne illness vary from acute symptoms such as vomiting and diarrhoea, fever, kidney failure, anaemia and even death to chronic diseases like dementia, inflammatory bowel disease.

From a food quality management (FQM) perspective, it is important to understand the different modes of action of pathogens because it may have consequences for safety control. In the case of micro-organisms causing food infection, safety can be controlled by a proper inactivation step reducing the number of these micro-organisms in food to a level that may not cause any harm (cooking at home may even be sufficient). However, in the case of micro-organisms that cause food poisoning, it is best to prevent the toxins being produced in the first place (e.g. by applying appropriate storage conditions, inactivation at an early stage), because the toxic compounds are often heat stable and cannot be removed in the food production process once they are formed. Heat treatment and temperature control are basic technological conditions affecting the behaviour of bacteria, moulds and parasites. Additional measures are hygienic handling to prevent contamination by soil, faeces, insects, or people, and control of water quality (parasites). Viruses and prions are difficult to control because they contaminate the food via infected animals, plant materials but also via infected people. Moreover, there are not yet quantitative methods for their detection. Infection by viruses and prions needs to be controlled by preventive measures, such as disease control, personal health, and no use of dangerous parts of animals like brains for food production. For specific analyses of microbial hazards, we refer to more detailed information in Jay and co-authors (2005), and Rovira and co-authors (2006).

Chemical hazards

Chemical hazards are chemicals in food that provoke adverse reactions in people. They can be classified as food intoxicants and food sensitivities (Figure 2.4). With respect to food intoxicants, each individual in a population is vulnerable to it as long as the exposure is high enough. However, food sensitivities refer to individual adverse reactions towards particular food components.

Food sensitivities can be divided in two categories, i.e. immunological (food allergy) and non-immunological (food intolerances) (De Meulenaer, 2006).

In case of *food allergy* the response of the immune system involves the formation of allergen-specific immunoglobulin E antibodies (IgE), which cause a range of reactions that can clinically manifest in various forms like anaphylaxis, eczema, vomiting, asthma, etc. In case of immediate hypersensitivity reaction begins within minutes after ingestion.

In delayed hypersensitivity reactions, symptoms begin to appear 24 hours or longer after ingestion (the most well-known example is Coeliac disease).

Food intolerances do not involve the immune system but other substances than proteins cause the adverse reactions. Typical reactions involve anaphylactic reactions, metabolic food disorder, and idiosyncratic reactions. Anaphylactic reactions are very similar to food allergy symptoms, since they are also provoked by a release of histamine; however, this release is not mediated via IgE antibodies (e.g. strawberry sensitivity). Metabolic disorders involve genetically-determined altered metabolic patterns resulting in deficient metabolism or increased sensitivity (e.g. lactose intolerance). Many adverse reactions that affect certain individuals have an unknown mechanism, and different ones may be involved in these idiosyncratic reactions. Some typical idiosyncratic reactions are associated with ingredients or food such as sulphites (asthma reaction), aspartame (urticaria), sugars (aggressive behaviour), and chocolate (migraine headache). There is often an incomplete understanding of the underlying mechanisms.

The other group of chemical hazards refers to the *food intoxicants*, which provoke undesired reactions depending on the concentrations but not on the sensitiveness of specific people. Food intoxicants are classified in different ways. A common classification distinguishes food additives, residues, contaminants, and endogenous toxicants (De Meulenaer, 2006).

The use of food additives is restricted and is highly dependent on scientifically underpinned analysis to assess the safety of the compounds and to determine acceptable concentrations. Residues refer to particular chemical compounds used during food processes but not intended to remain in the product (e.g. pesticides, cleaning agents). Contaminants may originate from different sources such as the environment or as a result of processing (e.g. PCB, dioxin, aromatic hydrocarbons) (see also section 2.4.1). Endogenous toxicants refer to toxic compounds occurring in nature and in our food. Most of these compounds are of plant origin; some well-known compounds are glycoalkaloids (e.g. solanine in potatoes), caffeine, and biogene amines (e.g. phenylethylamine). Another well-known endogenous toxin is tetrodotoxin, which occurs in various organs of the puffer fish. In Japan, this fish is a delicacy but needs to be prepared with great skill, because consumption of this toxin results to total paralysis and ultimately death.

In order to control and assure food safety, it is necessary to have insight into the possible sources of chemical hazards, and their impact on people. Some chemicals are hazardous to all people whereas others are specifically hazardous for sensitive individuals. For example, in the case of Coeliac disease just a few spores of gluten proteins can cause serious harm, which puts extremely high demands on control and assurance of the food production system (e.g. fully separated production lines) and appropriate labelling.

In contrast to biological hazards, most chemical hazards cannot be eliminated from the food processing system because the compounds are often heat stable. Therefore, the focus is more commonly on preventing the chemicals from entering the food system at unacceptable levels. Typical issues that need to be considered are:

- The *origin* of the compound, which could have consequences for the type of preventive measures that are needed. For example, with respect to residues adequate practices (cleaning practices, pesticide management practices, etc.) could help to prevent them from entering the food production system at excessive levels. In the case of process contamination, proper control of process conditions can prevent the formation of the undesired compounds, whereas in the case of environmental contamination control of incoming materials is needed to assess where the levels are below acceptable limits. In the case of endogenous compounds, genetic engineering and or selection of varieties may reduce the levels of unacceptable toxic compounds. These examples illustrate the impact of the origin of contaminants on food quality management.

- Knowledge about *properties* of the toxic compounds, which are necessary especially when one needs to assess the risk of the hazard (see also Chapter 7). Typical properties that should be considered are:
  The *fat solubility* of the toxic compound. Fat-soluble toxic compounds can accumulate in the food cycle production. For example, small amounts of polychlorinated biphenyls (PCBs) accumulate in the adipose tissue of fatty fish species. By far the biggest source of PCBs in the human diet is oily fish (Shapiro and Mercier, 1994).
  Another important property is the *degradability or inactivation*. Some toxic compounds are very stable and therefore remain in the food production cycle for a long period. DDT, a very stable pesticide used in the 40's and 50's, has accumulated in the food production cycle via milk fat, meat and breast milk. For many years, breast milk contained higher levels of DDT than cow's milk.
- The *toxicity* of the toxic compound itself and the toxicity of its degradation products also determine the risk. Compounds can be toxic at very low levels and they can have a direct or indirect toxic effect. For example, the enterotoxins produced by *Clostridium botulinum* are highly toxic and the fatality rate is much higher than the enterotoxins produced by *Clostridium perfringens* (Jay *et al.*, 2005). Oxalates and alkaloids, however, have an indirect effect; they can bind bivalent metals and so affect the bioavailability. For more details about chemical hazards the reader is referred to De Meulenaer (2006), Watson (1998, 2004, 2005a,b) and Hefle (1999).

Physical hazards

Physical hazards can be defined as additional matter or foreign objects normally not present in food, which could cause injury, disease or psychological trauma to the organisms and harm to human health (Figure 2.4). It can occur at any stage of the food chain and precautions need to be taken to minimise the chance of foreign bodies appearing in food. However, about 25% of consumer complaints are related to physical contaminants (Aladjadjiyan, 2006), about 1-5% of foreign objects ingested by people result in minor to serious injury (Olsen, 1998). Based on Aladjadjiyan (2006) physical contaminants can be classified as non-radioactive and radio-active ones. The latter are also considered as physical contaminants but are invisible to consumers and have no direct influence on the quality perception (unless there is information about it). However, radioactive contamination can cause serious diseases, but the chance on contamination is very low.

Here we focus on the common (non-radioactive) physical hazards, which are divided into three groups depending on their origins:

- Mineral physical hazards, e.g. soil, stones, dust, metals, class, fibre, plastic, jewellery, etc.
- Plant-derived hazards, e.g. weeds, leaves, stems, wheat-ears, etc.

- Animal-derived hazards, e.g. mites, insects, rodents, fowl, but also (fish) bones, etc.

The foreign objects can enter the food system via raw materials, water, floor covers and construction and building materials and via staff. Since physical contamination can occur at any stage and often cannot be removed by processing (like heating), these hazards should be controlled by preventive measures. In some cases, for instance metal pieces, measuring equipment is available to detect pieces above a certain size level. Another common way of control is visual inspection, which is often less accurate.

Insight into the characteristics of the different hazards is essential for developing effective safety control systems to realise safe products.

2.3.2 Health

Health-related refers to those properties of a food product that are necessary for the body functioning (e.g. respiration, circulation, physical work, and protein synthesis) and that positively contribute to people's well-being. Figure 2.5 shows a health concept, distinguishing three major determinants of health that relate to physicochemical properties of food, i.e. nutritional value, health improving agents, and satiation characteristics.

Nutritional value refers to the amounts and availability of macro- and micronutrients, which are necessary in the diet for growth and maintenance of the human body.

Health improving agents are compounds that are naturally present or added to food products, with the specific aim of stimulating, supporting or diminishing certain processes in the body that affect human health. The increased consumer interest in health and well-

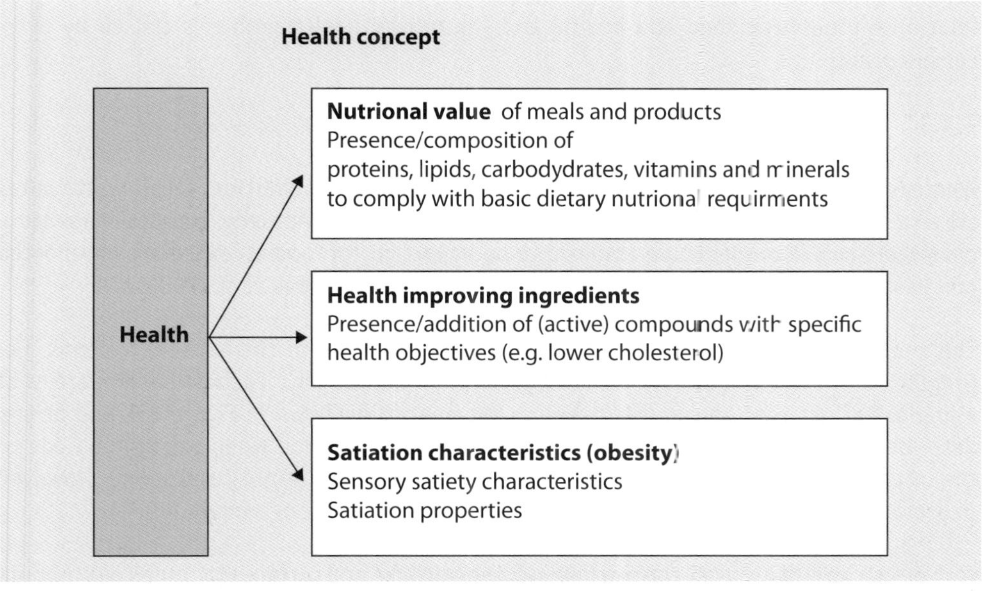

Figure 2.5. Food health concept.

being has stimulated the development and application of health improving agents in food. A major point of concern is to provide scientific evidence for the effectiveness of these compounds on improving human health and to maintain the (active) functions of these agents during food production. Typical examples of current health improving agents are omega-3 fatty acids (e.g. Nettleton, 1995; Simopoulos, 2002), glucosinolates in *Brassica* species (e.g. Dekker *et al.,* 2000; Cartea and Velasco, 2007) and flavonoids in various vegetable products (e.g. Evans, and Packer, 2003). Well-known health improving compounds are fibres to improve digestion (Food and Nutrition Board, 2005) and vitamin C to help maintain a healthy immune system (e.g. Davies *et al.,* 1991).

The third determinant in the health concept refers to the *satiation characteristics* of food products, i.e. how they affect meal termination (called satiation) and how they affect time intervals between meals (called satiety). Satiation for food comprises the physiological processes that result in the termination of eating and is evoked by physical and chemical qualities of ingested food, which trigger afferent signals to the brain from multiple sites in the body (including stomach, proximal small intestine, distal small intestine and colon).

Holt and co-authors (1995) developed a satiety index for foods, which refers to the ability of different food products to satisfy hunger. The chemical components of a food, is one of the factors that determines how it ranks on the index. For example, beans and lentils contain anti-nutrients, which delay their absorption so they make you feel full for longer. Roughly speaking, the more fibre, protein and water a food contains, the longer it will satisfy. Thus, the longer it will take before someone will start eating again. These characteristics of food products are attracting more and more attention due to the increasing number of people with obesity worldwide (e.g. Popkin, 1998; Hurk *et al.,* 2007).

Although food safety is still a major product attribute of concern, health attributes are becoming increasingly important for consumers due to an increased awareness of the relationship between food and health, and the typical health problems caused by over-consumption.

2.3.3 Sensory

Whereas safety is a typical dis-satisfier because it is a basic condition, sensory attributes are a satisfier and explicitly contribute to quality perception. Actually, appropriate sensory characteristics of products are assumed to be important for food acceptance and repeated purchase in the long term (e.g. Meiselman and MacFie, 1996; Lévy *et al.,* 2006).

The sensory attributes are observed via specific receptors in the sensory systems, i.e. tongue, nose, eyes, and ears (see also section 2.1). Various physicochemical properties of a product affect the sensory attributes, as schematically shown in Figure 2.6, and briefly described below. A definition of flavour is the overall sensation when eating food due to the interaction of taste, mouthfeel, odour/aroma, appearance (sight), and sound. However, in literature the term flavour is used synonymously with odour or aroma.

Texture can be defined as all the mechanical, geometrical, and surface attributes of a product perceptible by means of mechanical, tactile, visual and auditory receptors (definition ISO

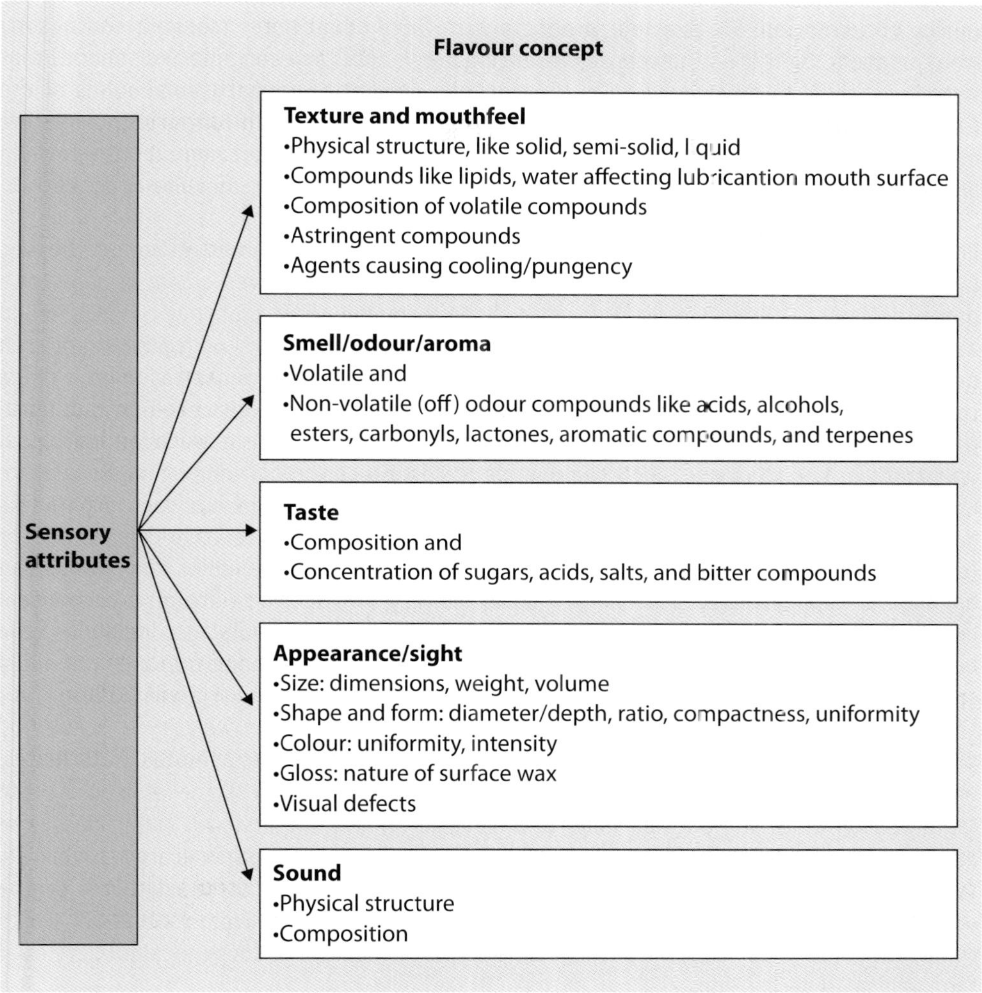

Figure 2.6. Sensory attributes concept.

5492). Food texture is about the perception that derives from the structure and how the product behaves when handled and eaten. Texture perception is a cyclical process, whereby information about texture is gathered during handling, chewing, and swallowing, which feeds back into our anticipation of the next portion. Information from handling comes from seeing and touching the product. During chewing information comes from (a) touching in the mouth, which gives feelings like elasticity and thickness, and (b) breaking down the food structure, which gives a range of textural perceptions, like crispiness and softness. When swallowing, impressions that arise from remnants of food give sensations like greasiness and chewiness (Rosenthal, 1999).

The physical structure of a product (e.g. liquid, semi-solid, solid), the presence of specific compounds (e.g. fat, lubricants, astringent compounds), and the interaction with volatile compounds, are product properties that can typically affect perceived texture or mouthfeel during consumption. For example, products with fatty odour notes (coconut and butter) were judged to be thicker than those presenting green notes (green apple and almond) that were considered to be smoother. In addition, the number of aromatic compounds of the flavouring agent appeared to affect texture perceptions. Products flavoured with a mixture of aroma compounds were perceived to be less thick and sticky, but more mouth-coating and smoother than those flavoured with a single aroma compound (Saint-Eve *et al.*, 2004).

Volatile and non-volatile (off) odour compounds like acids, alcohols, esters, carbonyls, lactones, aromatic compounds, and terpenes influence the perceived *aroma* or *odour* of a product (Maarse, 1991). One characteristic of these compounds is that they can be smelled at relatively low concentrations (part per million (ppm) to part per billion (ppb) range). Each odour compound has a characteristic threshold value; that is the concentration at which a compound can be observed. For example, sulphur compounds are noticeable at ppb level, whereas acids are typically observable at ppm level. However, there are large individual variations in the ability to smell compounds at certain concentrations; some people are much more sensitive then others. Moreover, aroma or odour compounds can become less or more attractive or repulsive depending on their absolute and relative concentrations (e.g. Swiegers and Pretorius, 2007). To make things even more complex, compounds can enhance or suppress each other, which makes it very difficult to relate physicochemical characteristics to perceived aroma or odour. For example, over 500 volatile compounds have been identified in coffee, but it is not known exactly how the composition and individual concentrations of volatiles contribute to the perceived coffee aroma (Flament, 2002).

Whereas there are numerous different odour attributes, there are only a few taste attributes, namely sweetness, saltiness, bitterness, sourness, and umani (savoury). The relation of these taste attributes to the physicochemical are relatively well understood. Major classes of compounds contributing to sweetness, saltiness, bitterness, and sourness are respectively sugars, salt, bitter compounds, and acids. However, the perceived taste attributes can be enhanced or suppressed by the presence of volatile compounds (e.g. Hollowood, 2002), which complicates the understanding of the relationships between physicochemical properties of food and taste perception.

Appearance and *sight* are important sensory attributes when purchasing a product. They greatly affect the initial quality perception of a food before consumption (Van Trijp and Steenkamp, 2005). Size, shape and form, gloss, visual defects, and colour are typical attributes, which one judges during purchase. For raw materials, these properties are highly dependent on seasonal differences, breeding practices, and cultivar differences, whereas for manufactured foods these attributes are a result of the recipe design and production process conditions.

Sound is of importance in crispy foods, and is often associated with freshness, for example in fruit, but also in snacks. Sound is affected by the physical structure of the product.

For more details about sensory attributes of food, we refer to Rosenthal (1999) and Voilley (2006).

2.3.4 Shelf life

Shelf life is an important quality attribute of all foods. It refers to the time between production and packaging of the product and the point at which it becomes unacceptable under defined environmental conditions (Ellis and Man, 2000). Shelf life is the time, during which the food product will (1) remain safe, (2) keep desired sensory, chemical, physiological and microbiological characteristics, and (3) comply with any label declaration of nutritional data (Kilcast and Subramaniam, 2000).

Major shelf life limiting food processes that deteriorate food products involve chemical, biochemical, microbiological, physical and physiological processes (Singh, 2000; Fennema, 1996; Kilcast and Subramaniam, 2000; Jay *et al.*, 2005), as shown in Figure 2.7.

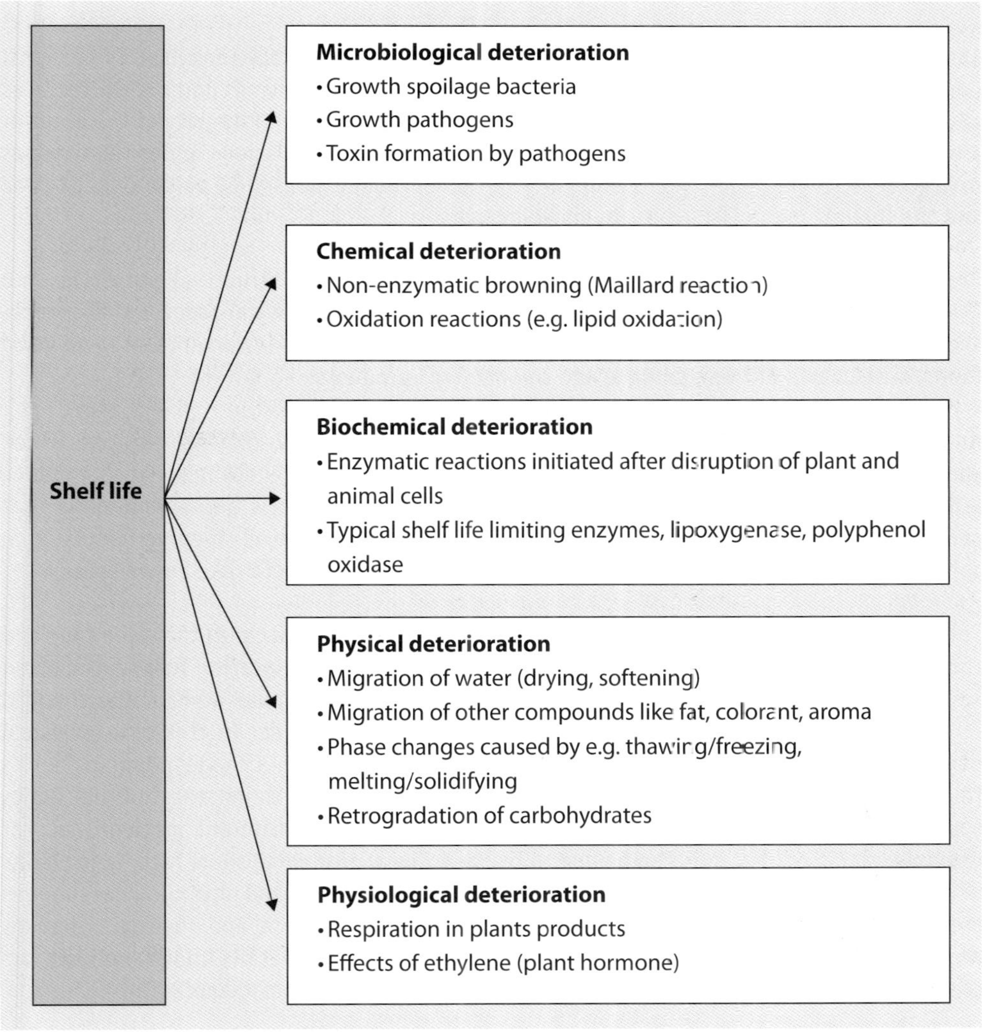

Figure 2.7. Food shelf life concept.

Microbiological deterioration is due to the growth of micro-organisms whereby a distinction must be made between spoilage bacteria and pathogens. Spoilage bacteria restrict shelf life because of their growth resulting in undesirable sensory characteristics, like loss of texture, development of off-odours and off-colours. Pathogenic bacteria restrict shelf life due to their growth and/or production of toxic compounds, resulting in negative health effects (Kilcast and Subramaniam, 2000; Jay *et al.*, 2005). Spoilage bacteria more commonly restrict shelf life than pathogens.

Typical *chemical reactions* that limit product shelf life are non-enzymatic browning and oxidative reactions. Non-enzymatic browning (also called Maillard reaction) primarily causes changes in appearance due to the formation of brown compounds, and the lowering of nutritional value by loss of essential amino acids. Oxidative reactions can alter various quality attributes like odour (e.g. due to lipid oxidation), colour (e.g. due to carotenoid oxidation), and nutritional loss (e.g. due to oxidation of vitamins).

Shelf life limitation by *biochemical reactions* is caused by undesirable enzymatic reactions. Generally, enzymes are compartmentalised, which means that enzymes and their substrates are separated at the sub-cellular or tissue level. Disruption of the integrity of the plant or animal tissue leads to enzymatic reactions, which may result in various undesired changes in quality attributes, such as browning of vegetable products caused by polyphenol oxidase and the formation of off-flavours by lipoxygenase.

Physical processes can also limit shelf life because they affect structural properties and thereby the texture and or mouthfeel of a product. A major physical process that affects shelf life is the migration of water resulting, amongst other things, in softening of dried foods, desiccation of fresh produce and smooth/soft products.

Another type of food process that affects shelf life is the *physiological reaction*, which commonly occurs during post-harvest storage of fruits and vegetables. Fresh produce still has a respiration rate and often produces ethylene, which is a plant hormone that can initiate various quality affecting reactions, such as accelerated ripening (resulting in softening, colouring), and formation of typical post harvest defects (like brown spots).

The shelf life affecting food processes, as described above, may occur at the same time and they may even interact. However, in practice, the fastest deterioration reaction is often responsible for the determination of product shelf life. For example, undesirable changes in texture of bread usually occur prior to the growth of moulds. In this case, physical degradation of the bread structure is faster than microbiological changes; degradation is thus the shelf life limiting factor. Although it is commonly one major reaction that limits shelf life, sometimes a typical quality defect may be due to different mechanisms. For example, rancid off-flavours can be due to lipase activity producing short fatty acid chains or by oxidation of the fatty acids. In general, the microbiological and physiological changes are of primary importance for short-life products, whereas chemical and physical processes are more relevant for medium- to long-life products. For more details about shelf life the reader is referred to Man and Jones (2000), and Kilcast and Subramaniam (2000).

To control and or improve shelf life, therefore, it is necessary to identify the deteriorative mechanisms responsible and find out how they are affected by intrinsic product properties (such as pH, a_w, etc.) and technological conditions (like temperature, relative humidity, etc.).

2.3.5 Convenience

Convenience is a product attribute that relates to the ease of use or preparation before consumption. Products are offered to consumers in such a way that purchase and preparation of a meal costs less money, and less physical and mental effort than when the original separate components are used and prepared. In current Western society this has become an important quality attribute of products because it fits with the changes in life-styles and demographic situations, such as the decreasing number of household members, a reduced desire for housekeeping, the increasing number of women in the workforce, and more welfare (e.g. Warde, 1999).

Convenience foods range from washed/sliced vegetables and fruits to complete ready-to-eat meals. The foods require minimum preparation, typically just heating or assembling (in case of salads), and are specifically packaged for the required shelf life with little loss of flavour and nutrients over time. Creating product convenience, however, often has considerable consequences for the shelf life of the product and puts great demands on the packaging concept. For example, sliced fruits easily decay within a few days, whereas fresh unsliced fruit can be stored for months under appropriate storage conditions. Due to cell disruption (after slicing) various deteriorating processes are initiated (e.g. enzymatic browning enzymes become active, spoilage bacteria can grow more easily) and decrease shelf life. Sliced fruits therefore need sophisticated packaging concepts with modified atmosphere tailored for the specific fruits inside to obtain a longer shelf life.

In contrast to the above quality attributes, convenience is not a direct result of certain physical and chemical properties but is due to certain preparations or handlings and packaging that create convenience in food preparation. However, these handlings directly influence various attributes like safety, flavour, shelf life, and health and are therefore closely interrelated. Creating convenience puts great demands on the control of packaging concepts, temperature as well as food handling (like prevention of misuse, hygiene).

2.3.6 Extrinsic attributes

In addition to intrinsic attributes, extrinsic attributes also contribute to quality perception (Figure 2.3). Extrinsic attributes do not necessarily relate to physicochemical product properties. Extrinsic attributes relate to *characteristics of the food production system* or they are *assigned* to the product by communication. Typical characteristics of food production systems that can affect consumers' quality perception are animal welfare, environmentally sound and sustainable, organically produced, and no use of genetically modified products (e.g. Torjusen *et al.*, 2001; Miele, 2001; Foster *et al.*, 2006; Kjærnes and Lavik, 2007; Poelman *et al.*, 2008).

However, the influence of *production system characteristics* on product acceptance is very complex. For example, there has been much concern about public acceptance of genetically modified food products and genetic engineering in general. Frewer and co-authors (1997) indicated in their study on consumer attitudes towards different food technologies used in cheese, that product benefits (such as product quality, animal welfare, environment) were a more important factor in their decision than process considerations. However, they did emphasise that the results could not be merely translated to other applications of genetic engineering in food production; each application should be considered on a case-by-case basis. Another study showed that the extent to which process characteristics affect quality perception is also dependent on individual attitudes and purchasing motives. In this study, subjects from the Netherlands and United Kingdom received pineapple pieces of three regularly grown varieties, in which they systematically varied the information about organic production and fair trade. Taken over all subjects, organic or fair-trade information did not seem to have much influence on liking or perception. However, subjects with a positive attitude towards organic or fair-trade information perceived the products to have an overall stronger sensory impact in the presence of such information than in its absence, whereas the opposite was true for subjects with a negative attitude towards organic or fair-trade information (Poelman *et al.*, 2008).

Moreover, the impact of process characteristics on quality perception is not always reflected in actual purchasing behaviour. Wandel and Brugge (1997) made a distinction between people as consumer and as citizen. As citizen, one may be concerned about environmental issues, animal welfare, etc., but that does not mean that as a consumer one behaves accordingly. This is particularly the case when the product represents a conflict between environmental soundness and other perceived benefits such as convenience, taste, flavour or price (Alwitt and Berger, 1993; Wandel and Brugge, 1997). In fact as citizens, people are concerned with social and environmental issues, while as customers the economic aspects are often prevalent in their choices. Nevertheless, the market for organically produced products has grown in the last decade in Western Europe, which means that these production concerns are increasingly reflected in changed purchasing behaviour (e.g. Jones *et al.*, 2001).

Attributed or *assigned quality* refers to quality that is ascribed to a product by communication (via labelling, product information) and advertisement (TV, internet, radio, papers, etc.). Perceived quality can be affected as well, although it is not necessarily directly related to the physicochemical properties of the product itself. Brand products often have a quality image, which can have a strong influence on the perceived quality. For example, in a study on different cola brands it appeared that participants who expressed a strong established preference for one cola over the other failed to label the two colas correctly. They concluded that pre-existing preferences for a particular cola were not really based on taste, but instead on previously acquired perceptions of image, quality or brand name. So, a brand name could influence consumers' quality perception as well.

Although these extrinsic attributes are often not directly related to the physicochemical properties of the food, they can impact consumers perception of food quality and therefore need to be considered when designing, controlling and assuring food quality.

2.4 Factors influencing food behaviour

The food quality relationship, as described in Chapter 1, shows that food quality is dependent on the behaviour of food systems and the behaviour of human systems involved in food production; FQ=f(FB, HB). The dynamic and variable characteristics of food systems (food dynamics) and the technological conditions that are applied affect food systems' behaviour over time.

Food dynamics refers to the dynamic character that is inherent to a specific food system. A food system is the whole range of variable product properties and dynamic food processes.
- Food systems refer to raw materials, ingredients, half fabricates, and (fully) processed products.
- Product properties concern the composition of components in food systems (i.e. which components and how much) and the intrinsic physical properties, like pH (acidity) and a_w (water activity) values.
- Food processes include the microbial, chemical, biochemical, physical and physiological reactions that can take place in food systems.

Food systems are complex and dynamic because of the wide range of components that can interact and have variable concentrations, and due also to the dynamic food processes that change the product properties over time. Food behaviour is the result of outcomes of the food system over time (Luning and Marcelis, 2006).

Technological conditions are applied to food systems to achieve and or maintain the required desired product properties within acceptable tolerances. By technological conditions, we also mean packaging conditions (section 2.4.3), equipment conditions (section 2.4.4) and food plant and environmental circumstances (section 2.4.5), which are described in more detail in the last three sections. To understand how technological conditions can influence food behaviour, we first describe major classes of components in food systems and their variability (2.4.1). In the subsequent section, we explain the mechanisms and dynamic character of major food processes (2.4.2).

2.4.1 Variable food composition

Figure 2.8 provides a schematic presentation of typical components and processes comprising a food system. Although food systems are rather complex, the components can be divided into a few major classes, i.e. nutrients, food additives, and undesirable contaminants and residues.

Nutrients consist of macronutrients, which are present at relatively high levels, and micronutrients (including trace elements), which are present at much lower levels (Food and Nutrition Board, 2005). The amounts and availability of macro- and micronutrients determine the nutritional value of a food system. Both macro and micronutrients are necessary in the diet for growth and or maintenance of the human body. Table 2.1 gives a concise overview of major macro- and micronutrients and their typical nutritional functions. In addition, the macronutrient composition contributes to the typical structure of a food system, which determines the textural characteristics of a food system. Some

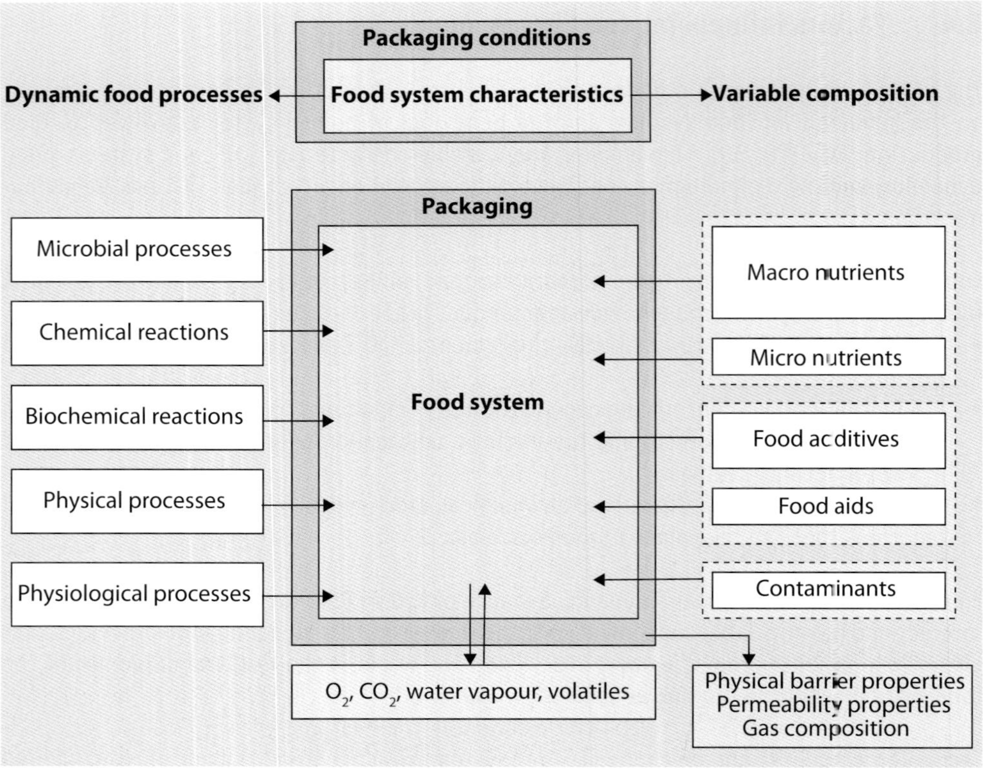

Figure 2.8. Schematic presentation of food system characteristics and within its packaging conditions.

macronutrients, like fat, affect various quality attributes. For example, fat influences the absorption of fat-soluble vitamins, colour components, and volatile aroma compounds, and thereby influence respectively health, colour and aroma attributes.

A wide range of factors, like seasonal variations, cultivar differences, cultivation and breeding conditions, affect the variation in composition of these nutritional compounds. This natural variation is inherent in a certain food system.

Food additives are a substance or a mixture of substances, other than a basic foodstuff, which are present in a food as a result of any aspect of production, processing, storage or packaging but do not include contaminants (Committee of the Food and Nutrition Board, cited in Branen *et al.*, 2002). There are different classes of food additives with different functions: i.e. preservatives, nutritional additives, colouring agents, flavouring additives, texturising agents, and miscellaneous agents (Table 2.1).

- Preservatives are aimed at preventing microbial growth and oxidative reactions (like lipid oxidation), and enzymatic and non-enzymatic browning, in order to extend shelf life.
- Nutritional additives include the primary nutritional additives (like vitamins and minerals), and additives with special purposes (e.g. alpha-omega lipid acid to reduce cholesterol, or fibres to reduce calories). The use of nutritional additives has risen over the last few decades due to the increased concern for affect diet has on human health.

Table 2.1. Main classes of components in food and their contribution to quality.

Components	Function	Contribution to quality
Macro-nutrients		
Fats Saturated fatty acids, mono and poly-unsaturated fatty acids, cholesterol	Major source of energy for body, aids absorption of fat soluble compounds, some polyunsaturated fatty acids cannot be synthesised in body and are essential in diet	Health, texture, colour, aroma, health
Carbohydrates Sugars and starches, dietary and functional fibres	Sugars and starch provide energy to cells in the body (particularly the brain) Dietary fibres are non-digestible carbohydrates intact in plant materials Functional fibres are isolated non-digestible carbohydrates that have beneficial physiological effects in humans	Health, taste
Proteins & amino acids	Proteins are major structural components of all cells of the body. Amino acids are building blocks of proteins; nine are not synthesised in the body and are essential in the diet Proteins and amino acids function as enzymes, membrane carriers and hormones	Health, aroma, texture
Micronutrients		
Various minerals Calcium (Ca), Chlorine (Cl), Iron (Fe), Zinc (Zn) Manganese (Mn, Molybdenum (Mo), Boron (B)	Minerals have various functions in the body Minerals come from soil and fertilisers; they enter food systems via vegetable produce grown on (artificial) soil	Health, texture
Food additives		
Preservatives	Antimicrobials to prevent growth of spoilage and pathogenic bacteria Anti-oxidants to prevent undesirable oxidative reactions Anti-browning agents to prevent enzymatic browning and non-enzymatic browning	Shelf life
Nutritional additives	Primary nutritional additives (vitamins, minerals) Special purpose additives, e.g. lowering cholesterol, fat replacement, cancer prevention	Health
Colouring agents	Natural and synthetic colour compounds	Colour
Flavouring agents	Sweeteners (low-caloric, non-caloric) to hide bitterness/sourness, to enhance sweetness Natural and synthetic aroma compounds to modify or reinforce aroma Enhancers to modify or enhance taste and aroma	Taste, aroma, smell

Table 2.1. Continued.

Components	Function	Contribution to quality
Food additives (Continued)		
Texturising agents	To add or modify overall texture or mouthfeel Emulsifiers and stabilisers affect texture properties Phosphates and dough conditioners affect texture of foods containing protein or starch	Texture, mouthfeel
Miscellaneous	Numerous other chemicals for specific but limited purposes, including various processing aids like chelating agents and enzymes	
Contaminants		
Environment contaminants	Organic and inorganic compounds mainly come from emissions from industrial processes via atmosphere and water	Safety
Processing contaminants	Processing contaminants are compounds formed during food manufacturing	Safety

Sources: Vlek (1985), Branen *et al.* (2002), Food and Nutrition Board (2005), De Meulenaer (2006), Soccaciu (2007).

- Colouring agents are synthetic or natural compounds, to enhance colour and/or to restore colour that is lost during processing (Socacciu, 2007).
- Flavouring agents are the most commonly used additives used in food. Three major types are sweeteners, natural and synthetic aroma compounds (>1,700), and aroma and taste enhancers. The latter enhance or modify flavour of foods without having a taste or aroma of their own.
- Texturising agents add or modify the overall texture and mouthfeel of food products. These agents are used in the greatest total quantity as compared to the other food additives.

The precision with which the right concentrations are added and the way in which the compounds are distributed in the food matrix affect the variation in composition of the additives applied. For further reading, see Branen (2002) and Watson (2005b).

Contaminants and *residues* are undesirable compounds that have entered the food system.
- Contaminants may originate from different sources such as the environment or due to processing. Environmental contaminants include inorganic and organic chemicals, which mainly come from emissions from industrial processes via the atmosphere and water. Typical environmental contaminants are aromatic hydrocarbons (e.g. benzene, toluene, etc.), dioxins and heavy metals (lead, mercury, cadmium, etc.). Processing contaminants are compounds formed during food manufacturing. Typical classes of process contaminants are polyaromatic hydrocarbons (produced during grilling, smoking, drying), heterocyclic amines (produced in the Maillard reaction during cooking of e.g.

meat and fish) and acryl amide (produced in the Maillard reaction in carbohydrate-rich products, like potatoes or cereal products).
- Residues refer to particular chemical compounds used during agricultural production, which, along with their degradation products, are not intended to remain in the agricultural product when processed or transformed to food. Pesticides and veterinary drugs are typical residues resulting from agriculture practices. Other important residues may originate from cleaning and disinfection practices. Remains of contact materials are also considered as residues, and typically derive from packaging materials as well as processing equipment.

The variability in the concentrations of contaminants and residues is greatly affected by the different circumstances along the food production chain (e.g. use of pesticides, amount of industrial emissions, effectiveness of sanitation programme, etc.). The entry of contaminants and residues should be prevented as early as possible, because they are not easily removed from the food systems due to their (often) stable character. Their presence is subject to strict legislative limits. For details about contaminants in food, we refer to Watson (2005a).

Acidity (pH) and *available water (a_w)* (Figure 2.8) are inherent in a certain product because they are determined by the types of components and their concentrations. The pH and a_w values have an impact on the reaction rates and behaviour of all food processes, as described in more detail below.

2.4.2 Dynamic food processes

It is not only the variability in composition that contributes to the dynamics of food systems. The typical characteristics of food processes also play a major role. This section concisely describes the microbial, chemical, biochemical, physical, and physiological processes in food systems (Figure 2.9) in order to obtain a basic insight into their dynamic characteristics.

Microbial processes

These processes involve living micro-organisms, which are able to colonise (multiply themselves) foods. They *grow* or *survive* in the food matrix and they may *produce* desirable compounds (like acids, volatile odour compounds) or undesirable compounds (like toxic metabolites). There are roughly three kinds of micro-organisms in food: the good, the bad, and the ugly ones (Figure 2.9). The good ones are those that produce a desirable change in the food product in order to extend shelf life or achieve the desired flavour attributes; examples include the micro-organisms used in fermentation processes, such as cheese-making, and beer and wine production. The ugly ones are the spoilage micro-organisms that reduce shelf life by changing the sensory properties, preventing consumers from eating them. The bad ones are to the micro-organisms that may cause a health disorder; these are referred to as pathogens. Micro-organisms include a variety of gram-positive and gram-negative bacteria, moulds, parasites, viruses, and prions, which are briefly typified below:
- Bacteria are prokaryotic micro-organisms, i.e. they do not have a true nucleus inside their protoplasm. Gram-positive and gram-negative bacteria have different cell wall structures and depending on their response to the gram-stain they are identified as

Type of food process	Major processes	Factors influencing process dynamics
Microbial processes	•Growth curves •Production of compounds and/or toxins •Inactivation curves	•Intrinsic product properties (aw, pH, Eh, nutrient content, antimicrobial agents) •Technological conditions (time-temperature, relative humidity, gas composition)
Chemical processes	•Maillard reaction •Oxidation reactions (of lipids, colour, vitamins) •Chemical modifications	•Intrinsic product properties (aw, pH, presence of oxygen, other precursors, reactants and catalysts) •Technological conditions (time-temperature, light)
Biochemical processes	•Enzymatic reactions (see Table)	•Intrinsic product properties (presence reactants, catalytic power enzyme) •Technological conditions (time-temperature)
Physical processes	•Transformation of phases •Mass and heat transfer •Diffusion/migration of compounds	•Intrinsic product properties (physical properties: aw, melting & boiling point, diffusion coefficients, viscosity & density) •Technological conditions (fluctuating time-temperature, forces e.g. by pressure, stirring)
Physiological processes	•Respiration (O_2/CO_2) •Transformation (evaporation of water) •Ripening (synthesis and degradation of compounds & affecting ripeness quality defects)	•Intrinsic product properties (genetic properties, (non)-climacteric, sensitivity to chilling injury) •Technological conditions (harvesting practices, storage conditions (T-t, RH, O_2/CO_2))

Figure 2.9. Overview of food processes and factors influencing their dynamics.

gram-positive or gram-negative. Bacteria can be involved in fermentation, spoilage processes (spoilage bacteria) or can cause health disorders (pathogens).
- Moulds have a double role because some are used in food processing (e.g. soybean fermentation to make soy sauces, ripening of blue cheese), but others can be dangerous for human health. It is not the moulds themselves that are hazardous but the secondary metabolites they produce, also called mycotoxins.
- Parasites need a host to complete their biological cycle. They can contaminate food by infecting animals, fish or shellfish or they can contaminate food via water or faeces. Parasites are pathogens and can cause health disorders.
- Viruses cannot live alone; they are very dependent on their hosts (bacteria, animals, and plants). Foodborne and waterborne viruses are increasingly recognised as the cause of disease in humans, which is attributed to the increasing trend in international food trade and travel.

A prion is not a micro-organism but an abnormal protein, which causes several fatal neurodegenerative diseases that affect animals and humans. The common characteristic of all diseases caused by prions is the deposits of protein aggregates in the brain that can be diagnosed *post-mortem* as sponge-like brains (Rovira *et al.*, 2006).

The *dynamic character* of micro-organisms is caused by their typical growth and inactivation behaviour. We will now focus on the bacteria, yeast, and moulds. The growth of these micro-organisms can be described by a lag phase λ (hour) (i.e. time that the micro-organism needs to adapt to its new environment), the growth speed μ (per hour) (correlated with slope of growth curve), the initial cell concentration N_0 (colony forming units (cfu)/ml), and the maximal cell concentration N_{max}. Each micro-organism has a characteristic growth profile with a specific *optimum* and *bandwidth* for certain growth circumstances.

Typical circumstances that affect growth and production of compounds or toxins are intrinsic product properties and external technological conditions. Major intrinsic properties include pH, water activity (a_w), oxidation-reduction potential (E_h), presence of antimicrobial agents and presence of nutrients necessary for growth. Important extrinsic technological conditions affecting growth are temperature (T), time (t), relative humidity (RH), and gas composition (Table 2.2).

The inactivation and inhibition of micro-organisms also encompasses a dynamic character. Depending on the type of micro-organism and the environmental circumstances, they can have various inactivation profiles (e.g. log-linear curves, sigmoidal curves, biphasic curves).

Considering the above dynamics, it is clear that in order to control growth, compounds and toxin production, and inactivation of desirable and undesirable micro-organisms, it is necessary to have a good understanding of their identity, their typical behaviour profile, and the major intrinsic and extrinsic factors that influence their behaviour. For more details, we refer to Jay and co-authors (2005), Rovira and co-authors (2006), and Devlieghere and co-authors (2006).

Chemical reactions

Chemical reactions involve reactions between different reactants (also called substrates, compounds, agents) resulting in new compounds, which may undergo further chemical reactions. These reactions require certain activation energy to initiate the reaction; heat often initiates chemical reactions. If enzymes initiate the reaction, they are called biochemical reactions (see next section for details). Some major chemical reactions that affect the quality of food are described below (Figure 2.9):

- *Non-enzymatic browning (also called the Maillard reaction)*: this includes very complex reaction mechanisms, involving proteins or amines and carbohydrates. This reaction influences both desirable and undesirable quality effects. It typically influences the brown coloration of food during heating (as in cooking, frying of meat, baking of bread), but it also occurs at room temperature (e.g. browning of dried milk powder). The Maillard reaction results, amongst other things, in a wide range of volatile odour compounds, and it therefore contributes to the typical aroma and smell of heated products. It negatively affects quality due to loss of essential amino acids (like lysine) and formation of toxic compounds (Fennema, 1996).
  The *dynamic character* of the Maillard reaction is affected by the reaction rate (i.e. how fast the reaction is) and which chemical pathway is followed (there are different reaction mechanisms). The Maillard reaction is dependent on several process and product variables like temperature, pH, moisture content, presence or absence of metal ions and effect of sugar structure. Some of these variables mainly affect the rate of the reaction (like temperature and metal ions, which can function as catalysts), whereas

Table 2.2. Important intrinsic properties and extrinsic technological conditions that influence the growth of micro-organisms (Jay, 1996; Jay *et al.*, 2005).

Growth factors	Impact on micro-organism
Intrinsic properties	
Acidity (pH)	Well established that most micro-organisms grow best at pH values around 7 (6.6-7.5), whereas few grow below 4. Bacteria tend to have smaller pH ranges, especially pathogens, whereas moulds (pH 0-11), and yeast (pH 2-8) can grow at a broader pH spectrum Fruits and vegetables typical range is pH 3-6; meat, poultry, seafood range is pH 5-7, and dairy 4.5-6.5
Water activity (a_w)	Most spoilage bacteria do not grow below a_w =0.9, whereas spoilage moulds can grow as low as 0.80; most spoilage yeast have a minim a_w of 0.88. With respect to pathogens, *Staphylococcus aureus* grows as low as 0.86, whereas *Clostridium botulinum* does not grow below 0.94. The pH can change due to growth of the micro-organisms Most fresh foods have an a_w above 0.99
Oxidation-reduction potential (E_h)	The presence or absence of appropriate quantities of oxidising-reducing agents (e.g. ascorbic acid, reducing sugars) affects growth and activities of all micro-organisms
Nutrient content (water, source of energy & nitrogen, vitamins, growth factors and minerals)	Moulds have the lowest requirements with respect to the nutritional compounds followed by yeasts, gram-negative bacteria, and gram-positive bacteria
Antimicrobial constituents	Typical substances that protect food against attacks of micro-organisms are eugenol (cloves), allicin (garlic), thymol (sage), (iso) thymol (oregano, lactoferrin, conglutinin and lactoperoxidase (milk)), glucosinolates (cabbage, broccoli, etc.)
Extrinsic technological conditions	
Temperature-time conditions (T-t)	Micro-organisms can be placed roughly in three groups: *psychotrophs* (grow between 7-30 °C, optimum ca 20 °C), *mesophiles* (grow between 20-45 °C, optimum 30-40 °C), *thermophiles* (grow >45 °C, optimum 55-65 °C). Under ideal conditions bacteria can multiply by dividing into two every 20 minutes; in this way a single bacterium could increase to 2,097,152 within 7 hours
Relative humidity (RH)	The RH affects the a_w of foods and thereby the growth of micro-organisms, and it influences the growth of micro-organisms at surfaces. Products that suffer surface spoilage due to moulds, yeast and certain bacteria should be stored at a relatively low RH
Concentration of carbon dioxide (CO_2) oxygen (O_2) in environment	Gram-negative bacteria are more sensitive to CO_2 than gram-positive bacteria. CO_2 increase affects fungi grow in fruit so retarding spoilage. The ratio impact of the modified CO_2/O_2 is highly dependent on the product and bacteria present

others determine the pathway through which the reaction occurs, and so affect the types of melanoidin pigments (from yellow to dark brown) and volatile aroma compounds formed. At times the Maillard reaction is desirable but in other situations not. In either case, it is necessary to understand the impact of the product and process variables on the nature and extent of the reaction (Fennema, 1996).

- *Oxidation reactions* involve a group of chemical reactions in which oxygen is one of the reactants. Typical oxidation reactions include:
  Lipid oxidation (or autoxidation) is one of the major causes of food spoilage in edible oils and fat-containing foods. Lipid oxidation is a process that proceeds via a radical mechanism, which means that after initiation it goes very fast. It consists of three phases: (a) initiation whereby free radicals are formed (e.g. initiated by light, or free radical producing substances); (b) propagation where oxygen addition results in the formation of hydroperoxides, which can enter into numerous and complex breakdown and interaction mechanisms resulting in a myriad of compounds (often with distinct odours, i.e. when rancid); (c) termination when radicals react together to form non-radicals.
  The lipid oxidation reaction is a very *dynamic reaction* process that, once initiated, rapidly leads to a wide range of volatile compounds with distinct odours (both desirable and undesirable). The rate of lipid oxidation is affected by various product (e.g. fatty acid composition, free fatty acids, moisture, surface area, presence of pro-oxidants) and process (temperature, oxygen concentration) variables, which makes it very difficult to obtain a full insight into the kinetics (dynamics) of the reaction.
  Chemical *oxidation of colour pigments* occurs in typical plant pigments (e.g. carotenoids, chlorophyll) but oxygen also plays a crucial role in the complex reactions involved in fresh and cured meat colour (Fennema, 1996; Soccaciu, 2007). Oxidation reactions are very *dynamic* in nature and their reaction rates are affected by, amongst others, the presence of oxygen and temperature.
  Chemical *oxidation of vitamins* is an important reaction responsible for the breakdown of vitamins. The dynamics of these reactions are dependent on diverse variables. For example, oxygen and heavy metals affect the oxidation rate of vitamin C, which follows approximately a first order reaction. This means that the reaction follows an exponential increase depending on the reactants. In the case of vitamin C, the presence of metal catalysts considerably increases the reaction rate as compared to spontaneous oxidation, which means that a few parts per million of these metals may cause serious loss of vitamin C in products.
- *Chemical modifications* typically occur in proteins. Application of heat affects the three-dimensional structure of proteins and initiates various chemical reactions, which result in a wide range of new compounds like volatiles compounds (e.g. typical sulphur compounds in meats, fish, milk). Temperature plays a crucial role in the dynamics of these reactions and influences the extent and nature of the reactions.

Above we have described some of the major chemical reactions in food; for more information about chemical reactions in food, we refer the reader to Fennema (1996).

Biochemical reactions

The main characteristic of biochemical reactions is the involvement of enzymes (Figure 2.9). Enzymes are proteins that catalyse *specific reactions* with extraordinary *catalytic* power. They catalyse (initiate) a reaction between substrates and are released again after the reaction is finished. In contrast to chemical reactions, enzyme reactions occur at much lower temperatures (20-50 °C), and they drive the reaction through very specific reaction pathways, which means that specific compounds are produced (in contrast to chemical reactions that may lead to a wide range of different compounds depending on diverse variables). Enzymes can be naturally present in food or added in order to initiate specific reactions resulting in desirable quality attributes such as a smooth texture, certain aroma attributes, etc.

Table 2.3. Overview of biochemical processes which can influence food quality (Compiled from Wong, 1995).

Enzymes	Description	Application/impact on product quality
Amolytic enzymes	Enzymes which degrade starch (e.g. amylase)	Play important role in brewery industry, distilleries, and baking processes
Cellulolytic enzymes	Enzymes that hydrolyse cellulose or its chemically modified polymers	These enzymes play a role in converting cellulose into smaller units (often together with pectinases). They can be used as an alternative for chemical peeling of vegetable products
Protolytic enzymes	All enzymes that hydrolyse peptide bonds of proteins	Major use of proteases occur in meat tenderising, cheese manufacturing and for texture improvements
Lipolytic enzymes	Consists of two groups, i.e. the triacylglycerol acylhydrolyases and phospholipases; they typically act on the lipid/water interface	
Pectic enzymes	Enzymes that break down plant cell walls, classified in polygalacturonase, pectin esterase and pectate lyase	Contribute to softening of plant/ vegetable tissues
Lipoxygenase	Catalyses the oxidation of poly-unsaturated fatty acids containing ci, cis, 1,4 pentadiene systems into hydroperoxides (ROOH)	Contributes to the formation of volatiles aroma compounds (as well desired as undesirable aroma compounds)
Polyphenol oxidase	Enzymes that oxidise polyphenols resulting in brown pigments (melanoids)	Contributes to undesirable browning (e.g. in peaches, bananas, grapes, tropical fruits) or desirable browning like in tea, coffee, cocoa, black raisins, and figs

Enzymes are highly valuable in the food industry because of their catalytic power (reactions occur at much lower temperatures then in chemical reactions) and their specificity. Table 2.3 gives a concise overview of important enzymes used in the food industry.

Their *kinetics* (or dynamics) is typified by the so-called Michaelis-Menten equation. In other words, they follow a typical behaviour, with an initial velocity and maximum velocity, which is affected by the substrate (reactant) concentration that is converted into a new compound(s), the catalytic power of the specific enzyme, and temperature conditions. These factors should be well-understood to control enzyme reactions in such a way that they result in desirable physicochemical properties and related quality attributes (taste, texture). For further reading we refer the reader to Tucker and Woods (1995), and Whitehurst and Law (2002).

Physical processes

Physical processes are processes that affect the physical structure of a food system, which often results in changes in texture, mouthfeel, and or appearance (Figure 2.9). Each food system has specific physical properties, which influence the extent to which physical processes may occur during processing and or storage. Important physical properties of food systems are:

- their phase transition properties, like melting and boiling point (i.e. when does the product become solid, fluid or volatile, e.g. ice, water, water vapour);
- viscosity and density characteristics that also determine physical structure (solid, semi-solid, fluid);
- the rates of various physical processes, like mass and heat transfer rates, which affect, amongst other things, the velocity of heating and cooling of products;
- the diffusion coefficient of water and other compounds, which influence how fast compounds migrate within the food system.

Water plays an important role in physical processes affecting the quality of products. Typical examples of moisture migration are wilting of leafy vegetables, loss of crispiness of dry products (like cereals, biscuits, chips), and drying of smooth products. In addition, in complex composite foods, the migration of other compounds like fat, colorants, and aromas contribute to undesirable changes in texture, aroma, taste, etc., thereby restricting shelf life.

Physical processes are dynamic in time like the other food processes. The inherent physical properties of the product (e.g. diffusion coefficient of water and product viscosity) largely affect the rates of these physical processes. Although, higher temperature increases the rate of physical processes, it often has a bigger impact on microbial, chemical, and biochemical reactions (see section on factors influencing processing). Application of force (e.g. stirring, applying pressure, etc.) can affect physical processes too. However, *fluctuating temperatures* are often destructive. For example, temperature fluctuations cause re-crystallisation of ice - leading to undesirable sandy texture but also formation of large ice crystals, which damage the cell structure of frozen vegetables. Temperature fluctuations may also initiate phase changes involving melting and solidifying fats and are detrimental to the quality of candies, chocolate and other lipid containing confectionary products.

Understanding the impact of physical properties and maintaining constant temperature are thus important factors for controlling physical processes in food. For more detailed reading about physical food processes we refer the reader to Karel and Lund (2003), and Figura and Teixeira, (2007).

Physiological processes

Physiological processes only occur in fresh produce like vegetables and fruit (Figure 2.9). Fresh produce *after* harvesting are still living systems with a continuing metabolism. Major physiological processes that can occur after harvesting of fresh produce from plant or tree include:

- *respiration*, whereby oxygen is consumed and carbon dioxide is formed;
- *transpiration*, whereby water evaporates;
- and *ripening* whereby synthesis and degradation of compounds takes place, like formation of ethylene, breakdown of cell walls resulting in tissue softening, formation of taste and odour compounds can occur.

Basically, there are two distinct patterns of ripening which are termed *climacteric* and *non-climacteric*. In non-climacteric fruits (e.g. strawberry, cherry, cucumber, orange, lemon, pineapple) the process of maturation and ripening is a continuous but gradual process. In contrast, the climacteric fruits (e.g. apple, pear, peach, apricot, melon, tomato, banana, avocado) undergo a rapid ripening phase when triggered by changes in hormonal composition. The onset of climacteric ripening is a well-defined event marked by a rapid increase in the rate of respiration and the natural evolution of ethylene gas by the fruit, which is called respiratory climacteric.

The volatile compound 'ethylene' plays a major role in (post-harvest) ripening of fresh produce. Ethylene is present in all fruit and is recognised as the central fruit ripening hormone which, in climacteric fruits, can actually initiate ripening when present at concentrations as low as 0.1 to 10 parts per million (ppm). Non-climacteric fruits also respond to ethylene application by increasing their respiration rate but the actual ripening process is only triggered by the fruit itself (Harris, 1988).

Respiration, transpiration, and ripening processes after harvesting are *dynamic processes* and continue over time. The rates of these processes are affected by factors such as genetic properties of fresh produce, whether they are (non)climacteric, harvesting practices, and storage conditions (e.g. Harris, 1988; Crisosto *et al.*, 1995). To illustrate the dynamics, the importance of the respiratory climacteric is that fruits such as bananas may be held at a reasonable temperature when in the green state, but as they begin to ripen they will rapidly increase their respiration and generate much more heat. The consequence may be that this heating cannot be controlled and even more respiration will occur in an inflationary spiral rapidly leading to spoilage of the fruit in a very short time. Once climacteric fruits start to ripen, there is very little that can be done except to market them for immediate consumption.

An understanding of the respiration, transpiration, and ripening characteristics of fresh produce (e.g. climacteric versus non-climacteric fruit, sensitivity to chilling injury) and an understanding of the impact of product handling and storage conditions are required to maintain the quality characteristics of fresh produce. For more detailed information about this topic, we refer to Kays (1991), Bartz and Brecht (2002), and Thomson (2003).

2.4.3 Packaging conditions

In this section, we briefly describe typical conditions of packaging, which can affect the behaviour of food systems. Packaging is an integral part of a food system (Figure 2.8). However, packaging concepts can condition the physicochemical properties of fresh produce and processed products. Therefore, we discuss packaging from the viewpoint of technological conditions.

The basic function of packaging is containment in order to protect the environment from a myriad of products, which are moved from one place to the other on numerous occasions each day in modern society. Packaging thus plays an important role in the maintenance of quality attributes of raw materials, fresh produce, and manufactured food; the protection function. Packaging plays an important role in convenience too. Packaging enables the right portions to be created or can create convenience. Moreover, packaging is an important marketing tool for communication about quality (product information and labelling). Depending on the characteristics of a food system, certain packaging conditions should be selected to optimally maintain the initial characteristics over time.

Packaging conditions refer to the properties of the packaging material itself (i.e. their mechanical barrier and permeation properties) *and* the air conditions within the packaging. Packaging forms an integral part of the (final) product from the perspective of customers or consumers. The mechanical barrier properties imply how well the packaging material is able to prevent the product from mechanical injuries during its distribution and storage life. For example, preventing cookies from breaking or fruit from bruising during transport. The permeation properties refer to the extent to which components can migrate through the packaging materials. Major migrating compounds through packaging materials are water vapour, oxygen (O_2), carbon dioxide (CO_2), nitrogen (N_2) and volatile aroma compounds, but also chemical compounds (Tehrany and Desobry, 2004).

Air conditions refer to the gas composition in the packaging concept. There are approximately three conditions: (a) normal air, (b) modified gas, and (c) vacuum (oxygen is removed). The basic idea beyond changing the initial gas composition is to influence the metabolic activity of fresh produce (typically fresh fruits and vegetables) or influence typical chemical deterioration reactions (like oxidative reactions), by removing oxygen and or replacing it with other gases in order to maintain microbial safety and extend shelf life. Note that the fine-tuning of packaging properties and modified gas compositions is very crucial for product quality and safety, because changing gas composition may inhibit the growth of some bacteria (the ones that need oxygen to grow) but favour others (the ones that do not need oxygen for their growth and or toxin production) (Riquelme *et al.*, 1999). Appendix 2. provides an overview of typical packaging materials, their barrier properties and application for packing of agri-products and food.

So, it is very important that packaging materials have the right barrier properties and gas composition to prevent products from e.g. drying or changes in texture (due to migration of water vapour), discoloration and/or off-flavour formation due to oxygen permeation, growth of pathogens due to certain modified gas compositions. Moreover, some other product-packaging interactions may lead to undesirable product properties too. For example, migration of packaging components into the food product, and scalping or absorption of flavour compounds by the packaging materials. Migration of packaging materials may lead to unsafe products and/or the alteration of product flavour. All packaging concepts must therefore comply with strict regulations on food contact materials (EU food contact materials regulation 1935/2004; Food contact substance notification, US Food and Drug Administration (FDA)) (Baner *et al.*, 1999). Scalping or absorption of aroma compounds may lead to unacceptable changes in the smell/odour (Linssen and Roozen, 1999). Packaging conditions are thus essential for maintaining desired product properties over time.

2.4.4 Food processing equipment conditions

A food processing system is an assemblage of equipment linked by some form of interaction or interdependence. Processing equipment constitutes the food processing system, and manufactures or transforms raw materials into products. *Food process equipments* are interconnected by means of transport and handling systems, e.g. conveyor belts, hydraulic transport channels, pipes, liquid pumping systems, etc. (i.e. *material handling systems*). Additional facilities such as energy supply, hot water or steam, cold water or cold air, etc., are necessary for equipment operation (i.e. *energy handling system*). All processing systems require devices to monitor and correct possible deviations in the established process conditions (i.e. *process control systems*). The material and energy handling systems, and process control systems should support the food process equipment (López Gómez and Barbosa Cánovas, 2005). Although conditions of material and energy handling systems could also influence product properties, we focus primarily on the typical process parameters of major food processing equipment and their design.

Process parameters

In the handling and processing of raw materials into food products, various equipment with different process parameters are used. The type of process techniques (e.g. hot air dryer or spray dryer) and the process parameters (e.g. short-time high temperature) that are applied affect the ultimate physicochemical properties of products. Although a wide range of food processes play a role in the formation of these properties and associated quality attributes (section 2.5.4), relatively few of the parameters can be conditioned during food processing depending on the technique that is applied. Firstly, we describe some major process parameters for typical processing techniques in food. Secondly, we briefly explain how equipment design can condition certain aspects of food systems.

- *Heating and cooling conditions.* The application of certain *time-temperature* combinations is probably one of the most dominating ways to condition quality parameters of food systems. Time-temperature conditions play an important role in heat treatments and are aimed at creating desirable safety and or sensory properties within acceptable limits. Time-temperature conditions in heat treatments affect the rate of microbial processes

(e.g. inactivation of spoilage bacteria, and pathogens), the kinetics of biochemical and chemical processes, but also the rate of physical processes (e.g. evaporation of water). However, one should realise that the same T-t conditions can influence various food processes in different ways and at different rates. Chemical reactions often reveal an exponential character, which means that once it starts it goes very fast, whereas enzymes have an optimal temperature at which they operate (too high or too low temperatures will result in suboptimal conditions).
Time-temperature conditions are relevant in cooling and freezing applications too. Cooling is commonly used to stabilise certain quality attributes (taste, smell, safety, etc.), whereas freezing conditions are used to stabilise quality attributes, and/or create conditions for physical processes (like formation of crystals or evaporation of water, i.e. freeze-drying). As with heat treatments, the low temperature conditions can influence various food processes in different ways and to different extents.

- *Dielectric heating conditions.* There has been a considerable evolution in the development of heating techniques, because traditional heating applications often have negative effects on desired sensory properties as well. Various new ways of heating have been explored (Boom *et al.,* 2005). Relatively new ways of heating include use of microwave (i.e. dielectric heating), lower-frequency electromagnetic radiations (referred to as radio frequency (RF) heating), and ohmic heating (application of low frequency alternating current). The basic process parameters that can be conditioned in these processes are *application time* and *frequency* (Herz) (Galema, 1997; Boom *et al.,* 2005).
- *Non-thermal heating strategies.* In addition to the above developments, other strategies have been developed to inactivate micro-organisms in order to preserve foods without applying heat, like ultra-high pressure processing (UHPP), and pulsed electric field treatment (PEF). In UHPP, the basic parameter is the combination of time-pressure combinations (e.g. Erkman and Dogan, 2004), whereas in PEF the action relies on the creation of an electric potential difference over the cell membrane by electric pulses, which disrupts membranes of plant cells and micro-organisms. Both techniques are often accompanied by a certain degree of heating (which is much lower than in traditional heat treatments) (Boom *et al.,* 2005).
- *Extraction conditions.* Extraction processes have been important in the food industry for years. Examples are removal of oil from olives and seed oils usually extracted with the help of (apolar) solvents. Herbs and medical plants are commonly extracted with an alcohol-water mixture. Other well-known extraction processes are decaffeination of coffee and tea, removal of fat in low-fat foods, and removal of alcohol in light alcohol drinks (Boom *et al.,* 2005). Characteristics of the *solvent* for extraction (e.g. more polar or apolar), *extraction time*, and *extraction surface* are essential extraction conditions.
- *Membrane conditions.* A trend in food processing is the application of techniques for fractioning and separation. For example, in the production of butter, cream must be separated from milk, in apple juice production particles must be removed to obtain clear liquids, or for demineralisation of liquid foods. For these purposes they use membranes in various applications (like micro-, ultra-, and nano filtration), whereby typical process parameters are *pore size* and *shape, size* and *sequences* of membranes (Boom *et al.,* 2005).
- *Extrusion and high-shear conditions.* Extrusion is a relatively new technique in food processing that appeared in the early 20th century. Nowadays it is an important process for structuring and cooking foods. The extruder exerts a strong shear stress accompanied with heat transfer, which can lead to various modifications in the product, for example,

starch can gelatinise and partially degrade, proteins can denature, and porous materials can be formed after the product comes out of the extruder as a result of expansion (Boom *et al.*, 2005). Important process parameters that can condition product properties are *shear rate, residence time, temperature, screw rotation velocity, flow rate*, and *extruder geometry* (e.g. Rayas-Duarte *et al.*, 1998; Ficarella *et al.*, 2006).

Food process equipment design

Not only process parameters but also the design of equipment itself can condition certain product properties to keep it within acceptable tolerances. A major point of interest in food industry nowadays is the hygienic design level of equipment. Hygienic design refers to the use of appropriate food contact materials, specific geometry and lay-out criteria of equipment to avoid cross contamination, and certain installation criteria (Lelieveld *et al.*, 2003; 2005). The extent of hygienic design can have a considerable impact on the risk of cross-contamination, and thereby provide strict conditions for desirable safety levels of products. See Chapter 4 for more details.

2.4.5 Food plant and environmental conditions

The food plant (or food processing plant) comprises the food processing equipment and supporting systems, and buildings. Processing equipment and supporting systems must be placed in logical order in different buildings of the food plant, in which proper working conditions (like comfort, hygiene, reliability, and safety) are established (López Gómez and Barbosa Cánovas, 2005). Typical food plant conditions that require attention are deviation of temperature zones (to prevent condensation, high temperature fluctuations, etc.), product routings to prevent crossing of raw and processed materials, high-care zones with restricted access, special waste disposal units, use of filters for air quality, and effective design of storage facilities for raw and processed materials. Control and assurance of these conditions are of major importance for quality (e.g. in the area of safety, shelf life) of the manufactured or fresh produce.

Environmental conditions relate to the conditions outside a food plant, which mainly relate to weather and climate conditions. They have a considerable impact on outside breeding and cultivation of fresh produce, but may also influence certain conditions in a food plant. Weather conditions can affect these parameters particularly when areas are not temperature and humidity controlled. For more details about food processing systems and food plants, we refer to Saravacos and co-authors (2003), Richardson, (2004), Boom and co-authors (2005), López Gómez and Barbosa Cánovas, (2005), and Lelieveld and co-authors (2007).

2.5 Food quality from a chain perspective

It is important to realise that the physicochemical properties of products for consumption are the result of a complex food production chain where various technological conditions and many quality decisions contribute to final food quality. It is necessary to look at this from a chain perspective because it gives some insight into *where* and *how* to take design, control, improvement, and or assurance measures to realise physicochemical properties and related attributes that comply with consumer demands. For example, with respect to food safety control and assurance, processing can often eliminate many pathogens and spoilage micro-organisms but heat-stable toxins, many environmental contaminants, and various residues cannot be removed by processing. Their occurrence should be kept below certain limits by preventive measures as early in the food production chain as possible.

This section focuses on the description of major technological factors along the food supply chain that can influence final product quality in primary production (sections 2.5.1-2.5.3), processing and distribution (section 2.5.4), and retail and consumption (section 2.5.5). Managerial considerations from a chain perspective are explained in Chapter 3.

2.5.1 Primary production

Primary production consists of animal and vegetable production. Animal production refers to animal breeding (i.e. animal feeding and keeping, and production of animal products like milk and eggs), fishing (from sea, lakes, rivers), and primary processing of animal products resulting in fresh meat and fish. Vegetable production concerns cultivation, harvesting and initial storage of fresh fruits, vegetables, nuts, and spices. Both production chains consist of complex networks of chain actors executing various technology-dependent and managerial activities that can influence final food safety. The following sections describe both animal and vegetable production in more detail.

2.5.2 Animal production

The primary animal production chain consists of the following major chain actors: farmers, transporters, and slaughterhouses (sometimes combined with processors). Shell-fish and fish production differs from other types of food production in a number of ways. Most seafood is still extracted from a 'wild' population, and fishermen are hunters with no influence on the handling of their prey before it is caught. Thus, it is not possible to imitate the situation for animals that are slaughtered, selecting only the most suitable specimens for slaughter and resting and feeding them well before killing. The seafood processor is limited to what is available in the choice of raw materials as regards size, condition, and fish species landed by the fishermen. However, nowadays some seafood is processed in a modern fish industry under well-controlled conditions. Figure 2.10 provides an overview of influencing factors at farms, in the transport sector and in the slaughterhouse.

Factors in farming practices

The production conditions at farms are very important for the physicochemical properties and related quality attributes of fresh and processed meat, and fish (e.g. safety, texture, taste, health). However, the characteristics of the productions systems themselves are extrinsic quality attributes (e.g. animal friendly, environmentally sound, etc.). Major factors in animal production (Figure 2.10) that can influence final quality are animal breed, feeding, living conditions, and animal health (e.g. Knura *et al.*, 2006; S.M. Osés, personal communication 2009).

The type of animal *breed* can have a significant influence on quality attributes of fresh and processed meat (e.g. Arsenos *et al.*, 2002; Sans *et al.*, 2004; Hopkins *et al.*, 2005). For

ANIMAL PRODUCTION

Farming practices
Breed choice, feeding (amount & composition), animal health, housing conditions (hygiene & stocking ration)

Transport & slaughtering
Stress factors (e.g. loading density, transport duration), hygienic conditions

VEGETABLE PRODUCTION

Cultivation practices
Crop varieties, cultural practices, environmental

Harvesting practices
Harvest time, product handling (mechanical injuries)

Post harvest storage
T-t, relative humidity, air composition, ventilation, treatments, sun exposure

DISTRIBUTION (See storage)

FOOD PROCESSING

Storage & preparation materials	Transformation materials	Packaging products	Storage products
•T-t, RH, packaging integrity raw materials, hygiene conditions •Adequate tools (sieves, knives, etc), appropriate chemicals for pre-treatments, hygienic equipment design	•Choice of preservation technique and type of processing equipment (e.g. batch or continuous) •Selected process parameters (like, T-t, Herz) •Hygienic equipment design	•Adequacy of barrier properties (permeation O_2, CO_2, water) •Integrity (no leaks) •Gas composition •Appropriate labelling	•T-t of storage areas •Relative humidity •Hygiene conditions

DISTRIBUTION

RETAIL & FINAL FOOD HANDLING

Retail practices	Final food preparation
•T-t, RH, and hygiene conditions in storage area and •Cooling and freezing cabinets, packaging integrity	•T-t and hygiene conditions in domestic/catering storage •Hygiene conditions at food preparation •Appropriateness of cooking conditions (T-t) & recipes

Figure 2.10. Overview of factors along the food supply chain, which can influence final product quality.

example, some pig breeds have a genetic predisposition for typical quality parameters of meat. For instance, the Duroc breed which is associated with darker, redder muscle colour, increased fat firmness and increased tenderness, although carcass conformation may be poorer (e.g. Edwards and Casabianca, 1997; Edwards *et al.,* 2003). These breeds are often crossed with other breeds to obtain optimal characteristics including meat quality. Another typical example is the cow breed Wagyu; the high degree of marbling makes this meat a culinary delicacy. This type of meat is a delicacy in Japan but is now attracting attention in Europe as well (www.wagyu.net/home.html). Nowadays, besides traditional breeding programmes, advances in molecular biology and gene technologies make it possible to select for quality attributes, such as eating quality, meat appearance (colour), flavour, juiciness and fat content (Cunningham, 2005).

Animal *feeding* can also affect the quality of animal products. Both the *amount* and the *composition* of feed can influence yield and/or the composition of animal products, such as milk, meat, and eggs (e.g. Harding, 1996; Vestergaard *et al.,* 2000a,b; Andersen *et al.,* 2005, Sahin *et al.,* 2007). For example, recent advances in animal nutrition have shown that essential fatty acid profiles in meat can be altered through feeding (e.g. Wachira *et al.,* 2002; Dymnicka *et al.,* 2004). Other studies have shown that feeds from plants that contain tannins produce leaner carcasses and therefore leaner meat (e.g. Vasta *et al.,* 2007).

Moreover, environmental contaminants (like dioxin) and pathogens (like *Salmonella, Aspergillus* spp.) in feed can affect the safety of animal products (Knura *et al.*, 2006). More specifically, forage containing aflatoxin (a mycotoxin produced by *Aspergillus* spp.) fed to dairy cows can be hydroxylated to a metabolite of aflatoxin, which is partly carried over to the milk (Harding, 1996).

Additionally, there is concern about the addition of antibiotics to feed to accelerate growth of the animals. Animal pathogens may become resistant to the antibiotics and resistant pathogens might transfer from animals to human. There is a potential risk that these resistant pathogens cannot be treated with human antibiotics, which could have major consequences for human health. The quality of animal feed is thus an important factor influencing the quality and safety of animal food products.

Animal health and the use of veterinary drugs can also influence product quality. For example, the occurrence of mastitis (i.e. inflammation of the mammary gland and disturbance of milk secretion) leads to alterations in the composition and chemical and physical properties of the milk (Auldist and Hubble, 1998). The number of somatic cells is an indicator of the qualitative and hygienic properties of the milk and reflects the mastitis situation of the given animal (Heeschen and Harding, 1995). Moreover, animal diseases can be treated with antibiotics and their residues in animal products (milk, meat) are assumed to be hazardous for human health. For the protection of consumers, the European Union and the Codex Alimentarius Commission have set maximum residue limits of veterinary drugs in foods of animal origin. Whereas residues of antibiotics can have a direct effect on food safety, the use of veterinary drugs as such is an important extrinsic quality attribute. There is some consumer concern about the effects on human health of using antibiotics in animal production (Kaufman, 2000).

Housing conditions of animals refer to hygienic conditions and movement circumstances. *Hygienic housing conditions* can have a considerable impact on the bacterial load of the exterior surfaces of animals. Large numbers and a great variety of micro-organisms are found on exterior surfaces (e.g. skin or teats) and in the intestinal tract of cattle, sheep and pigs; pigs generally have higher counts of micro-organisms than cattle (Forsythe and Hayes, 1998; Oses, 2009). For meat production, both exterior and interior load are important factors for food safety. Although the underlying tissue of slaughter animals is assumed to be sterile, high bacterial loads on interior and or exterior surfaces can lead to the contamination of other animals during transport or can lead to contamination of sterile meat during slaughter (S.M. Osés, personal communication 2009). In general, the cleaner the housing conditions the lower the load.

In milk production, preventive hygienic measures, like cleaning of teats, and cleaning and sterilising of milk equipment, prevent bacterial contamination of milk (e.g. Schreiner and Ruegg, 2003).

For fish, variations in marine environments (uncontrollable) and water quality at fish farms, affect the bacterial flora on skin surfaces. Psychrotrophic gram-negative bacteria dominate in cooler seas, whereas in warmer seas mesophiles and micrococci are the major spoilage bacteria (Forsythe and Haynes, 1998; Luten *et al.*, 2005).

The other aspect of housing, *movement circumstances* or *stocking ratio*, refers to the space in which animals move and live. Two distinct conditions are intensive versus extensive free-range or outdoor housing. Whilst some aspects of animal health and hygiene may be improved in more extensive conditions, exposure to parasites and contact with wildlife may increase zoonotic infection risk (Edwards, 2005). The movement circumstance can have an impact on both safety and sensory attributes. Various studies have reported improved meat characteristics among outdoor reared animals (e.g. Dufey, 1995; Pugliese *et al.*, 2005). Major influences on sensory quality derive from the choice of breeds most appropriate to outdoor systems and the increased proportion of forage in the diet. The indirect consequences of outdoor rearing on meat quality may result from influences on physiological stress responses in the pre-slaughter period. However, in some sensory evaluation studies it was shown that when panellists were aware of the origin of the animals, they rated free-range pork as juicier, less bland, more tender and pleasant. The extent to which real and perceived differences in product quality are reflected in changed economic value is more a function of marketing than the production system (Edwards, 2005).

Factors in transport and slaughtering

A critical point in the primary production of animal products is the transport of living animals to the slaughterhouse. *Stress* during transport and before slaughtering has a significant impact on the sensory properties (texture), safety (contamination by pathogens), and microbial shelf life (initial contamination) of fresh and processed meat products (Figure 2.10).

Too much exercise, fear, and fluctuations in hot and cold temperatures are factors that cause stress in animals. Stress can result in different quality defects depending on the type of animal. For example, meat of pigs and poultry exposed to stress has a soft texture, poor water-holding capacity and a pale colour also called pale-soft-exudative meat (PSE) (e.g. Barbut *et al.*, 2005). Stress causes various chemical reactions to occur, which results in excessively rapid pH decrease and in turn affects protein degradation during the post-mortem process. For cattle, excessively high post-mortem pH due to *ante-mortem* stress can result in beef with dark, firm and dry characteristics (DFD meat) (e.g. Viljoen *et al.*, 2002).

Factors influencing stress during transport and handling in slaughterhouses include (e.g. Warriss *et al.*, 1984; Tarrant and Grandin, 1993; Troeger, 1996; Gade and Christensen, 1998; Terlouw, 2005):

- *Loading density*, excessively high density results in formation of PSE meat for pigs, DFD meat for cattle and serious transport-induced hematomas in musculature.
- *Loading and unloading facilities*, i.e. angle of inclination of ramps. Steep slopes result in high heart rates, thus stress. Moreover, pitchforks are often used as driving agents which leads to carcass rind or even bleeding in the back fat, which eventually has a negative effect on meat quality (Troeger, 1996).
- *Duration of transport* can also negatively effect meat quality. Excessively short transportation times of pigs can increase the number of carcasses showing PSE characteristics. Longer transportation times probably lead to habituation and calming, so normalising the metabolic disturbances due to loading.

- *Mixing of animals* unknown to each other at the slaughterhouse has been shown to be particular stressful and results in DFD or PSE meat. It has proved worthwhile in practice to get the animals into individual passageways with limited inclination.

Hygienic conditions also play an important role during slaughtering of animals and or fish preparation (e.g. Abu-Ruwaida, 1994; Forsythe and Hayes, 1998; Azanza *et al.*, 2001; Sarli *et al.*, 2004; Osés, personal communication 2009). The slaughter procedure includes many steps like killing, bleeding, scalding, skinning and evisceration during which the sterile underlying muscle tissue can be contaminated by the intestinal tract, the exterior surface, by hands, knives and other utensils used. Total bacterial counts for freshly cut meat surfaces are likely to vary between 10^3 and 10^5 organisms per cm^2. Reduction of the microbial load of freshly killed animals can be achieved to a certain extent by spraying the carcasses with hot water containing chlorine, lactic acid or other chemicals. Moreover, strict hygienic conditions in the abattoir such as proper cleaning of walls, floors, knives and other utensils reduce the risk of contamination.

2.5.3 Vegetable production

Major actors in the primary production of vegetables and fruits (i.e. fresh produce) are plant breeders (seeds/crops), cultivating farms, and post-harvest stores, in addition to suppliers of supporting agents like fertilisers and pesticides. In this section, we focus on the cultivating farms including their suppliers and next customers (post-harvest stores). The two major factors at farms are *cultivation practices* and *harvest conditions*. These greatly influence properties of fresh produce and processed products, including nutritional composition, sensory properties (taste, odour, texture and colour), and levels of natural toxins, anti-microbial agents and anti-oxidants.

Factors in cultivation practices

Important quality-affecting factors in cultivation practices (Figure 2.10) include the selection of suitable varieties, cultivation practices, and environmental conditions (Ramdane and Mohan, 2004). Cultivation can occur outdoors or in sophisticated greenhouses, which has an impact on the extent of controllability of the influencing factors.

- Among *crop varieties,* there may be major differences in various quality attributes like health, colour, taste, texture, and odour (e.g. Gonzalez *et al.*, 1992; Miller and Mcnew, 2005). However, the selection of suitable varieties is commonly based on yield potential, disease resistance, or suitability for local conditions, often without sufficient consideration of market preferences and post-harvest behaviour.
- *Cultivation practices* that could have an impact on the quality of fresh produce include sowing date, nutrient supply (e.g. use of nitrogenous fertilisers), irrigation, and plant protection (i.e. use of pesticides and herbicides). Soil fertility has a direct effect on all aspects of crop growth and development. In some cases, post-harvest disorders are directly related to deficiency of a particular mineral. Irrigation increases crop turgidity at harvest but, depending on commodity, stage of maturity and other factors, may make the crop more susceptible to bruising and splitting. Over-irrigation and under-irrigation are a double-edged sword and can reduce yield, quality and marketability of the crop. All chemicals used on growing fruits and vegetables have an effect on quality, and in some

cases are applied specifically for their effect on post-harvest behaviour. The impact of these factors on the quality and safety of fresh fruits and vegetables has been discussed in various studies (e.g. Jongen, 2005; Nestby *et al.*, 2006; Anttonen *et al.*, 2006).
- *Environmental factors* that affect quality are environmental temperature, day length, and amount of rainfall (Feil and Stamp, 1993); these are uncontrollable, given factors.

Factors in harvest practices

With respect to harvesting practices, harvest time and mechanical injuries due to poor harvesting practices are major factors influencing the quality of fresh produce (Figure 2.10).
- *Harvest time* has an impact on the course of processes that occur during growth and ripening of fruits and vegetables, typical processes are:
  - changes in cell wall constituents, e.g. softening of fruits;
  - starch-sugar transformations. For example, degradation of starch into simple sugars results in a sweet taste in bananas. In other commodities like potatoes, corn, and peas the synthesis of starch predominates;
  - pigment metabolism. The green chlorophyll pigments often disappear upon ripening while other pigments like carotenoids and flavonoids are synthesised;
  - formation of aroma compounds and their precursors during ripening.

Two characteristics are normally used in deciding crop maturity, both very dependent on the experience and personal opinion of the grower, i.e. *observation*, or looking for characteristic signs of readiness (such as, size and shape, colour, texture, hardness and softness), and *sampling* of characteristic specimens and testing by smell and or taste. Some crops, such as cabbage and yam, are acceptable for consumption over a wide range of development and selection for harvest is often more dependent upon price and the size preferences of the market. Other crops must be harvested at a specific maturity or they will be unmarketable for reasons such as poor flavour, high fibre content, and/or rapid post-harvest deterioration. Several studies have demonstrated or underpinned the importance of understanding the relationship between harvest time and final quality attributes of fresh vegetables and fruits (Haard, 1985; Salunkhe *et al.*, 1991; Hagg *et al.*, 1999; Ramdane and Mohan, 2004).
- *Mechanical injury* can occur upon harvesting and transport. Damaging of plant organs can result in:
  - Wound healing, i.e. formation of physical barriers such as a waxy barrier or lignin deposition in asparagus.
  - Formation of stress metabolites. These metabolites appear to have a protective function for the plant but can have negative effects on food quality. For example, the stress metabolites (coumarins in carrots and glycoalkaloids in white potato tubers) have a bitter taste. Moreover, some stress metabolites exhibit toxic properties or are precursors for undesirable enzymatic reactions.
  - Enzymatic-catalysed browning occurs upon mechanical injury and leads to undesirable brown coloration of fresh produce.
  - Ethylene production upon wound healing stimulates respiration and accelerates the ripening and senescence process leading to a shorter shelf life (Haard, 1985; Salunkhe *et al.*, 1991).

Injury to fresh produce after harvesting will accelerate its deterioration, and reduces product shelf life. Injuries take many forms, including cuts, punctures, scraping of outer surfaces, internal and surface bruising, sunburn, heat damage and cold damage. Injuries are caused, amongst other things, by careless harvesting practices, rough field handling of produce, using unsuitable containers, overpacking containers, and by people walking or sitting on produce in containers (Piñeiro and Diaz Rios, 2004).

Factors in post-harvest storage

After farming, the fresh produce are transported to store houses for sorting, grading, pre-treatments and packing. Major conditions to which produce are exposed after harvesting are the transport and storage conditions, such as exposure to sun, product handling, ventilation, storage time and temperature, relative humidity, gas compositions, and treatments against fungi and bacteria (Figure 2.10).

Harvested produce should be protected from the sun at all times, because exposure results in defects due to sunburn. Produce must be carefully loaded and unloaded to prevent physical damage. Products should not be stacked in compact piles or masses, to prevent a rise in temperature at the centre of the mass, due to respiration of the fresh produce. Forced air ventilation, refrigeration, appropriate humidity, and modified atmosphere conditions can be used to lengthen post-harvest life (see also section 2.5.4).

For further reading on factors influencing quality of fresh produce along the supply chain, we refer to Bartz and Brecht (2002), Thompson (2003), Jongen (2002, 2005), and Prasanna and co-authors (2007).

2.5.4 Processing and distribution

Distribution conditions are important in the transport of products from one actor to another (e.g. by ship, truck, plane). The status of products (e.g. fresh produce, which are still respiring or packaged processed products) determines the factors of influence. The influencing factors for distribution are similar to those for storage of the specific products (Figure 2.10).

Fresh produce or animal products are directly sold to the market or further processed into ingredients (e.g. vegetable oils from seeds, flour from grains), food additives (e.g. emulsifiers, taste compounds), or manufactured food (ranging from sauces, bread, ready-to-eat meals, etc.). The motives for food processing are, amongst others, preservation to extend shelf life, improvement of sensory properties and digestibility, and enhancement of the health effects of food. The main approaches in preservation used to be heat treatment, addition of preservatives (e.g. salt and sugar to lower aw), and the use of micro-organisms for fermentation of food (e.g. to lower pH). However, changing consumer demands have forced the industry to use alternative processing techniques, like mild preservation techniques using less heat (Boom *et al.*, 2005) and hurdle technology where a range of mild treatments are applied to achieve food stability safety and quality (Leistner and Gould, 2002; Chawla and Chander, 2004).

Although each food product consists of a specific sequence of handling and process steps, the following main activities can be distinguished: storage and subsequent preparation of raw materials; transformation of raw materials and packing of final products; and storage and distribution of finished products (Luning and Marcelis, 2007). Figure 2.10 shows common influencing factors at this stage of the food production chain. They are briefly described below.

Factors in storage and subsequent preparation of raw materials

A dominant factor influencing the quality and safety of processed products is the initial quality of raw materials, for instance, the presence of initial microbial contamination, concentration of (un)desirable compounds, etc. (e.g. Bover-Cid, 2001; Lechevalier, 2005; Pons-Sanchez-Cascado *et al.*, 2005). Storage conditions, like temperature and relative humidity conditions, integrity of packaging, and hygiene, are common factors that can (negatively) affect the quality of raw materials during storage (Figure 2.10).

Subsequently, raw materials are prepared for processing. Common preparation activities are sorting, cleaning, washing, removal of physical agents, peeling, evisceration, dicing, slicing, and mixing of materials (Luning and Marcelis, 2005). The tools (e.g. sieves, knives, etc.) that are used, appropriate application of chemicals for pre-treatments (e.g. the use of chemicals to treat washing water or for spraying of carcasses), and hygienic design of equipment are important factors influencing product quality at this stage of food production. Besides these technological factors, the quality behaviour of food handlers (i.e. appropriate personal hygiene, accurate working) can considerably influence quality as well.

Factors in the transformation of raw materials and packing of final products

The product can then undergo various process steps, such as heating, drying, evaporating, extruding, fermenting, etc., to extend shelf life, improve safety, create certain sensory properties (e.g. by frying), and or improve digestibility (e.g. by fermenting). At this stage, the choice of preservation technique (e.g. severe heat treatment versus hurdle technology), type of processing equipment (continuous or batch, big or small), appropriateness of selected process parameters, and process equipment conditions (e.g. hygienic design level) are important factors influencing the final quality (see also section 2.4.2).

At this transformation stage, it is important to have a good understanding of how certain product properties (like a_w, pH) and applied technological conditions (like T-t conditions) influence different processes in the food systems. For example, water activity is an important factor in regulating the growth of micro-organisms, enzyme activity and the occurrence of chemical reactions. Typical reaction rate-a_w relationships for several important reactions that occur in food systems at a temperature range of 25-45 °C are shown in Figure 2.11 (Fennema, 1996). It is obvious that no microbial growth occurs at a water activity below 0.6, whereas lipid oxidation rate increases both at very low and at higher a_w values. It is common knowledge that foods protected against microbial spoilage by low water content are still susceptible to enzymatic spoilage. The exact reaction rate, position and shape of the curve, can change depending on composition, physical state and food structure. Another illustration, the rate of temperature change (dT/dt) in a process determines the

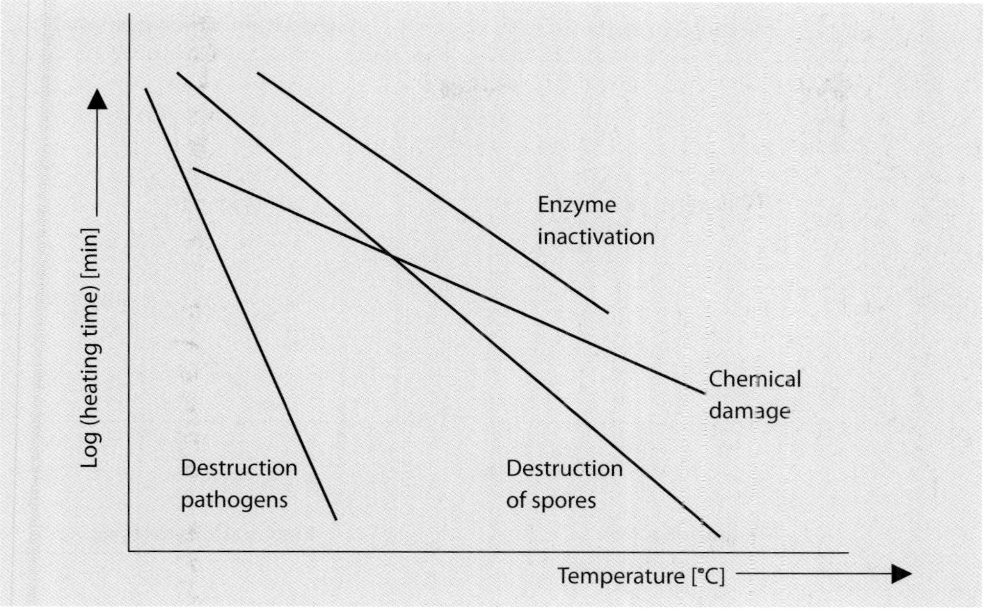

Figure 2.11. Schematic overview of destruction/damage lines for different reactions (microbial and(bio)chemical) at specific time-temperature combinations.

relative rates of competitive chemical reactions and the rate at which micro-organisms are destroyed as illustrated in Figure 2.12. In food manufacturing there is a sustained effort in optimising time-temperature combinations to produce safe and stable foods with the least heat damage possible (Van Boekel, 1998; Richardson, 2004).

Last but not least, packaging of final products is a very crucial step in food production systems. Poor packaging integrity, inappropriate packaging materials, and improper labelling (e.g. of allergens) can negatively influence the shelf life and safety of final products.

Factors in the storage and distribution of final products

Subsequent storage and distribution conditions aim at maintaining the properties of finished products. The main factors for packaged manufactured foods are storage temperature and duration, relative humidity, and appropriateness (correct barrier properties) and integrity of packaging materials. Comparable factors are relevant for bulk storage as summarised in Table 2.4.

2.5.5 Retail and final food handling

Retailers are commonly the last actor before purchase and food preparation by catering establishments, restaurants, and consumers. Critical factors that can influence final quality at this stage are storage conditions of supplies, conditions in the shop (e.g. temperature in freezers and cooling cabinets), hygienic handling of personal when slicing and packing fresh

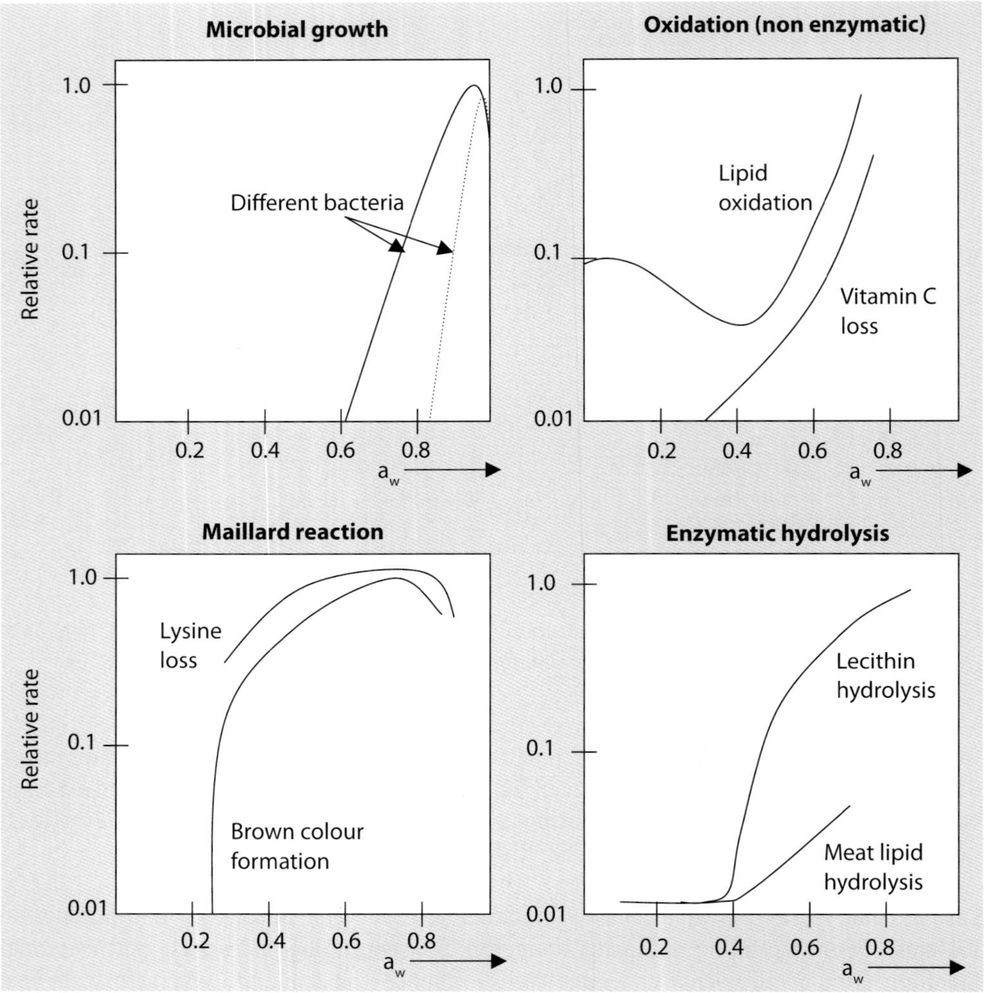

Figure 2.12. Schematic presentation of examples of relationships between a_w and different reactions.

products, and packaging integrity (Figure 2.10). Consumers are the last in the food supply chain, or the first if considered from a reversed chain viewpoint. Consumers buy their food products in supermarkets, specialised shops (e.g. bakery, butcher), and/or at fresh markets. Furthermore, they consume food outdoors from food outlets, catering establishments and restaurants. These outdoor food establishments have become increasingly important in the preparation of foods with appropriate quality and safety characteristics. Common factors influencing quality at home are related to conditions of domestic storage equipment (e.g. conditions and capacity of refrigerator and freezer), hygienic circumstances at food preparation (e.g. cleanliness of cutting boards), and cooking conditions (e.g. T-t, stirring). Similar factors are important in food preparation in restaurants and catering establishments (e.g. Little *et al.*, 2003, Ramirez *et al.*, 2006; Jones and Angulo, 2006). At this stage, the actual behaviour of food handlers as concerns, for example, following recipes, complying with cleaning rules, etc., plays an important role in final product quality and safety. In fact,

Table 2.4. Factors affecting product features during storage and distribution.

Relevant factors	Fresh produce (bulk storage)	Manufactured food
Initial bacterial load	High bacterial load reduces shelf life	Hygienic conditions during manufacturing determine initial bacterial load
Temperature-time [T-t]	Low temperature decreases respiration Too low a temperature may cause chilling injury	Low temperature delays (bio) chemical and microbial reactions
Relative humidity [RH]	Too high a RH favours growth of moulds and bacteria	RH in combination with packaging barrier properties determine the effects
Controlled atmosphere	Oxygen reduction combined with increased carbon dioxide tension slows respiration and subsequent decay	Effect of modified atmosphere packaging conditions depends on product properties
Germination inhibitors	Delay in e.g. sprouting of potatoes	Not applicable yet in packaging
Infestation control	Prevention of product loss by insects and mites; use of pesticides	Prevention of product loss by insects and mites by proper barrier properties of packaging
Poor handling	Mechanical injury (see harvesting practices)	Broken packages. Cross contamination due to mixed product storage

poor food handling practices in catering establishments and households has contributed to various food-borne outbreaks (Bryan, 2002; Much *et al.*, 2007). The role of final food handlers in affecting final quality at consumption should thus not be underestimated.

The above description of factors that may affect the quality of food products provides insight into the complexity of realising food quality. A profound understanding of which factors significantly contribute to variability in quality attributes of final products will support the assessment of critical locations for design, control, improvement, and or assurance measures.

2.6 Environmental factors

The environment of the organisation has an impact on food quality too (Figure 1.4). Environment refers to social, political, economical, and technological circumstances that affect decision-making on food quality by the organisation, via interests and power. Interests from social ethics provide sets of rules that define right or wrong behaviour, while interests from the firm's direct environment (e.g. stakeholders) have a direct influence on decision-making. Typical examples of interests are human health, animal welfare (social, political), biotechnology (technological, political), and supplies from developing countries (social, economical, political). Depending on the power position these interests contribute to the judgement on decision alternatives for food quality. Moreover, the government is quite dominant with a strong power position and clear interests in food safety. The government influences food quality along two lines, via rules and procedures and via inspection (Luning and Marcelis, 2007). In this last section of Chapter 2, we provide some basic information about laws and regulations (e.g. Codex Alimentarius, European Food Law, and Food and Drugs Administration (FDA)) that impact decision-making on food quality.

Codex Alimentarius

Codex Alimentarius means in Latin 'food code'. The Codex Alimentarius Commission was created in 1963 by FAO and WHO to develop food standards, guidelines and related texts such as codes of practice under the Joint FAO/WHO Food Standards Program. The main purposes of this programme are protecting the health of consumers, ensuring fair trade practices in food trade, and promoting coordination of all food standards work undertaken by international governmental and non-governmental organisations (www.codexalimentarius.net).

The Codex Alimentarius consists of a collection of internationally established laws and rules about procedures, guidelines, and recommendations related to food, production of foodstuffs, and food safety. These food standards are established through an elaborate procedure of international negotiations. About 173 countries collaborate in the formulation of these regulations. A major goal is that from 2010 each food transported and/or traded product will comply with the Codex regulations (http://nl.wikipedia.org/wiki/Codex_Alimentarius).

The Codex Alimentarius is a Napoleonic type of law, which means that issues not explicitly acknowledged by Codex regulations are automatically forbidden. The regulations treat foodstuffs and food supplements as toxicological compounds, which are subject to maximum levels. They make risk calculations from a toxicological disciplinary viewpoint, whereas opponents plead for a judgement based on biochemical methods. The Codex Alimentarius develops food regulations about:

- *food hygiene*, i.e. general rules, regulations about hygiene practices in specific industries, and guidelines for uses of hazard analysis according to HACCP;
- *contaminants* in foodstuff, i.e. general norm, tolerance for specific contaminants including radiated products, aflatoxins and other mycotoxins;
- *additives* for foodstuffs, i.e. general norm about accepted use, specifications for chemical compounds in food production;
- *labelling* of foodstuffs, i.e. general norm, guidelines about declaration of nutritional value, guidelines about labelling requirements;
- *pesticide* and *veterinary contaminants* in food (and maximum residue levels);
- procedures for risk assessment of products made by *biotechnology* (genetically modified plants and micro-organisms);
- *Specific norms* for specific products. Codex comprises more then 240 commodity standards, over 40 food hygiene and technological codes of practice, over 1000 food additives and contamination evaluations, and over 32,000 maximum residue limits for pesticides and veterinary drugs. It also includes requirements on labelling and presentation, and on methods of analysis and sampling (Van der Meulen and Van der Velde, 2007).

The Codex standards do not represent legally binding norms, but slightly resemble the directives in European law. They present models for national legislation, without the obligation to implement them. Member States undertake to transform the Codex Standards into national legislation but no sanctions are applied if they do not honour this undertaking. The strength, however, of the food standards of the Codex is that they function as a reference and so create a common language during negotiations between countries about food quality and safety requirements. Moreover, the national specialists on food law that enter the discussions on these standards will affect them in their work at home (Van der Meulen and Van der Velde, 2007).

European General Food Law

The objective of the European food policy is to obtain a single market in foodstuffs without trade barriers, while maintaining a high level of food safety and quality, public health and consumer protection. The EU regulations are very complex, but the most commonly used instrument for harmonisation are the Directives. A Directive is a measure addressed to the Member States. The latter must incorporate the Directives in their national legislation within a set time-schedule (De Sitter, 1998). For example, the Council Directive 93/43/EC on Food Hygiene is implemented in Dutch law via the Commodity Act regulation 'Hygiene for Food'. In this way, all EU directives are implemented in the national laws of the Member States.

With respect to control and enforcement of EU directives, each Member State is responsible for it themselves. The general principles for inspection, sampling and control of foodstuffs have been established in a framework Directive as well. The Member States must only inform the EU Commission about their control activities. The European food policy tends to shift responsibility for the control of foods towards the industry. Preventive food safety systems are being developed and implemented to identify and control potential hazards. This system aims at safety control prevention instead of inspection during food production. As a consequence, the function of national authorities is merely to check that these systems are well established and controlled instead of performing product inspection only.

However, due to various serious food crises in the agri-food sector in the last decade (e.g. BSE, dioxin issue, foot and mouth disease, rise in number of food poisoning cases, etc.) food safety has become a priority issue. On January 12th in 2000 the Commission published its White Paper on food safety. The White Paper aimed to restore and maintain consumer confidence. They scheduled eighty-four laws and policy initiatives in a very ambitious legislative programme. They have also taken other initiatives to support food safety besides legislative measures.

The Commission founded the European Food Safety Authority (EFSA), some of whose main tasks include monitoring (*quality assurance*), risk assessment, scientific advice (*scientific research*) and dissemination of food safety information (*communication*). The actual responsibility for legislation, control and enforcement is at the European parliament and Council level. Nevertheless, the European Food Safety Authority will be the supplier of scientific and technical information for the European Commission to develop policy and legislation for food safety issues in the agri-food chain. Two other initiatives on a European level are the European Food Safety Network (EFSN) and Safe Food in Europe (SAFE). The

EFSN and SAFE consist of 16 different governmental research institutes and Universities whose objectives are to identify and collect relevant knowledge and expertise with respect to food safety, to develop food safety research programmes and communicate with the public about food safety issues. Both EFSN and SAFE may support the European Food Safety Authority in fulfilling the scientific basis and communication role.

The popular name for the general principles and requirements of food law laid down in procedures and matters of food law is 'the General Food Law' (GFL). The GFL is not a code encompassing all food legislation. It is a basis for a general part of food law. The three major objectives of the GFL are:
1. to lay down the principles on which modern food legislation should be based in the European Union as well as in the Member States;
2. to establish the European Food Safety authority;
3. to establish procedures for reactions to food safety crises including the so-called Rapid Alert System.

Next to the GFL many other European and national rules and regulations play their role (Van der Meulen and Van der Velden, 2007).

Table 2.5 shows the subjects that form the basis of European food law, i.e. general principles, substantive, and procedural. The *'general principles'* on which modern legislation should be based address four food law principles (e.g. consumer protection, precautionary principle, science-based), two transparency principles (i.e. public consultation and information), three general obligations of food trade (e.g. about international standards), and eight articles containing general requirements on food law. The subject *'substantive'* includes guidelines and norms for content issues (e.g. limits for contaminants, allowed use of ingredients, etc.), product handling issues (e.g. production of HACCP), communication (e.g. about labelling), and miscellaneous (e.g. food contact materials). The subject *'procedural'* refers to procedures and guidelines for authorisation agencies, enforcement, and legal protection.

More specifically, the *precautionary principle* is an important addition to the risk analysis principle in the food law. In specific circumstances, following an assessment of available information, where the possibility of harmful effects on health is identified but scientific uncertainty persists, provisional risk management measures necessary to ensure the high level of health protection chosen in the Community may be adopted, pending further scientific information for a more comprehensive risk assessment (article 7 GFL). This precautionary principle ensures health protection by setting a uniform basis for the use of precautionary measures within the Community, in order to prevent arbitrary decisions giving rise to unjustified barriers to the free movement of food and feed (Van der Meulen and Van der Velde, 2007).

European Food Law thus affects food quality and safety by providing norms (e.g. residual limits on additives), by setting safety requirements for production, processing and distribution of food (e.g. HACCP, and T&T requirements), and by setting requirements on communication (e.g. labelling). These measures affect the decision-making of companies on food quality and safety. The other line of influence is via inspection by official authorities;

Table 2.5. Elements of the European Food Law (modified from Van der Meulen and Van der Velde, 2007).

General principles		
	General principles	General objectives (e.g. protection of human life and health, consumer interests) Consumer protection (e.g. prevention of misleading practices) Science-based (e.g. based on risk analysis) Precautionary principle
	Principles of transparency	Public consultation Public information
	General obligations of food trade	Import Export International standards
	General requirements	Food safety (no placing of unsafe foods on markets) Feed safety Presentation Responsibilities (e.g. business operators from farm to fork) Traceability (e.g. from farm to fork) Responsibilities for food (e.g. operator who has reason to doubt should withdraw product) Idem for feed Liability (i.e. product liability)
Substantive		
	Content	Ingredients Premarket approval (i.e. additives, novel foods, GMO's) Contaminants (i.e. residues)
	Handling	Production (i.e. HACCP) Trade (i.e. T&T, withdrawal & recall)
	Communication	Labelling Risk communication
	Miscellaneous	Food contact materials
Procedural		
	Authorisation procedures	
	Enforcement	Inspections Sanctions
	Legal protection	

this is addressed in the subject 'procedural' (e.g. guidelines for enforcement, authorisation procedures).

For details about the contents and background of the European food law we refer to Van der Meulen and Van der Velden (2004, 2008).

Food and Drug Administration (FDA)

The Food and Drug Administration (FDA) is an agency of the United States Department of Health and Human Services and is responsible for the safety regulation of most types of foods, dietary supplements, drugs, vaccines, biological medical products, blood products, medical devices, radiation-emitting devices, veterinary products, and cosmetics. The FDA also enforces section 361 of the Public Health Service Act and the associated regulations, including sanitation requirements on interstate travel as well as specific rules for control of disease on products ranging from pet turtles to semen donations for assisted reproductive medicine techniques.

Most federal laws administered through the FDA are codified into the Food, Drug and Cosmetic Act, also called Title 21, Chapter 9 of the United States Code (21 USC 9). The programmes for FDA safety regulation vary widely by the type of product, its potential risks, and the regulatory powers granted to the agency. It regulates products with a set of published standards enforced with a modest number of facilities inspections.

The agency is currently organised into the following major subdivisions, each focused on a major area of regulatory responsibility:
- The Center for Food Safety and Applied Nutrition (CFSAN)
- The Center for Drug Evaluation and Research (CDER)
- The Center for Biologics Evaluation and Research (CBER)
- The Center for Devices and Radiological Health (CDRH)
- The Center for Veterinary Medicine (CVM)
- The National Center for Toxicological Research (NCTR)
- The Office of Regulatory Affairs (ORA)
- The Office of the Commissioner (OC)

The *Center for Food Safety and Applied Nutrition* (CFSAN) is the branch of the FDA, which is responsible for ensuring the safety and accurate labelling of nearly all food products in the United States. One exception is products derived from traditional domesticated animals, such as cattle and chickens, which fall under the jurisdiction of the United States Department of Agriculture Food Safety and Inspection Service. FDA does regulate products that contain minimal amounts of meat. The exact boundaries are listed in a memorandum of understanding between the two agencies. However, medicines and other products given to all domesticated animals are regulated by FDA through a different branch, the Center for Veterinary Medicine (CVM). Other consumables which are not regulated by the FDA include beverages containing more than 7% alcohol (regulated by the Bureau of Alcohol, Tobacco, Firearms and Explosives in the Department of Justice), and non-bottled drinking water (regulated by the United States Environmental Protection Agency (EPA)).

The Dietary Supplement Health and Education Act of 1994 mandated that the FDA regulate dietary supplements as foods, rather than as drugs. Therefore, dietary supplements are not subject to safety and efficacy testing and there are no approval requirements. FDA can take action against dietary supplements only after they are proven to be unsafe. Manufacturers of dietary supplements are permitted to make specific claims of health benefits, referred to as 'structure or function claims' on the labels of these products. They may not claim to treat, diagnose, cure, or prevent disease and must include a disclaimer on the label (http://en.wikipedia.org/wiki/Food_and_Drug_Administration). For more details about the FDA we refer to their official website http://www.fda.gov/.

The globalisation of markets has increased the importance of internationally acknowledged legislation on food quality and safety. A good understanding of the different legislative requirements on food safety and food production systems is necessary in order to be able to take appropriate decisions and realise safe food with a high quality standard.

Appendix 2. Overview of common packaging materials for agri-products and food

Types	Typical properties
Plastic packaging materials	
Polyethylenes (low/ high density LDPE, HDPE)	LDPE: relatively chemically inert, moderately permeable to water vapour; highly permeable to oxygen; poor odour barrier. Contains relatively few additives, good processability HDPE: stiffer then LDPE, less transparent, more resistant to oil/grease, lower water permeability
Polypropylene (PP)	Permeability to gases and vapour between LDPE and HDPE
Polyvinyl chloride (PVC)	Excellent resistance to moisture, low gas permeability, good resistance to oils and fats. PVC is hard, stiff and clear, and has high impact strength.
Ethylene vinyl alcohol (EVOH)	Good gas barrier properties, but at high humidity levels oxygen barrier is severely reduced; EVOH tends to be soluble in water
Polystyrene (PS)	Good barriers to gases except water vapour, polymer is transparent
Polyamides (nylons)	Low gas permeability and they may absorb moisture. Nylons are strong, with high melting and softening points
Polyesters (like polyethylene terephthalate PET)	Good mechanical barriers, excellent transparency and relatively low permeability to gases
Polycarbonates (PC)	Poor gas barrier properties and they absorb moisture, resulting in deterioration of their mechanical properties and processability. Most commercial PC materials are stiff, hard and transparent
Metal packaging	
Tin plates/steel	Total barrier to water vapour and all gases. Not inert but sensitive towards corrosion, needs to be coated. Not transparent. Total package integrity affected by closures and seals
Aluminium	Total barrier to water vapour and all gases. Often combined with other materials like paper and plastics. Not transparent, easily damaged by physical handling
Glass	
Glass	Total barrier to moisture vapour and all other gases, relatively inert (i.e. not affected by other agents). It is a transparent and rigid material, which is well resistant to high temperatures. Critical area affecting total pack integrity is closures
Paper/card board	
Paper/cardboard	Poor barrier properties against moisture vapour and gases, only against light, and good protection properties (mechanical barrier). Often combined with plastics (laminates)

Compiled from Stöllman *et al.* (1994); Robertson (1998) and Emblem (2000).

Applications
LDPE accounts for largest portion of plastics Wide range of applications especially in combination with other materials (like paper/aluminium)
Wrapping of confectionery bars, potato crisps
Commonly used as packaging material because of good gas barrier properties
PET bottles for sweet drinks Commonly applied in sterilisable pouches, or boil-in-bag applications
Restricted applications
Canned foods
Used for trays to contain products Widely applied as coating
Ambient stable products like sauces, pickles
Specific applications: fresh bread, fresh fruit/vegetables. Combined with plastics, aluminium then wide application (e.g. dried soups, fruit juices, milk)

Chapter 3

Quality management

Besides technological factors, management activities are essential for the actual performance of product and production quality. To understand the influence of management activities it is necessary to have a sound knowledge of the basic principles. This chapter gives a brief introduction to general principles of quality management.

After a comprehensive overview of important focus-points in food quality management (section 3.1), management activities are described in general from the viewpoint of decision-making (section 3.2). A concise description of the theory of decision-making results in four administrative concepts, which indicate how decision-making activities can be allocated in the organisation.

In section 3.3 an overview is given of the historical development of quality management. Moreover, the contribution of the major quality gurus to the evolution of quality thinking is also described.

The subsequent sections focus on quality management in food production. The basic food quality management functions are described regarding their specific goals and activities. Decision-making activities within these functions are typified (section 3.4). Then, attention is paid to people's quality behaviour and the factors determining decision-making behaviour (section 3.5). The chapter closes with the topic of chain management (section 3.6), including aspects such as customer-supplier relationships, supply chain management and partnerships.

Whereas this chapter deals with general principles of quality management, the following chapters contain a more detailed analysis of specific aspects of quality management activities within the different food quality management functions.

3.1 Introduction

An organisation can be defined as a collection of people working together to achieve a common goal. Within an organisation, people are able to accomplish tasks, which are not possible when acting alone. The basic purpose of an organisation is to produce goods and/or services that satisfy the needs of customers.

Organisations are generally considered as open systems that interact with their environments in the continuous process of transforming resource inputs into outputs in the form of products and/or services. Here, the environment plays a vital role, on the one hand as an input of resources and on the other hand as a user of products and services. The organisation has to justify its existence by realising added value in the eyes of customers and other stakeholders (Schermerhorn, 2007).

Within organisations, managerial processes take place (Figure 3.1). The basic activities in these processes are aimed at 'getting things done'. Management can be defined as:

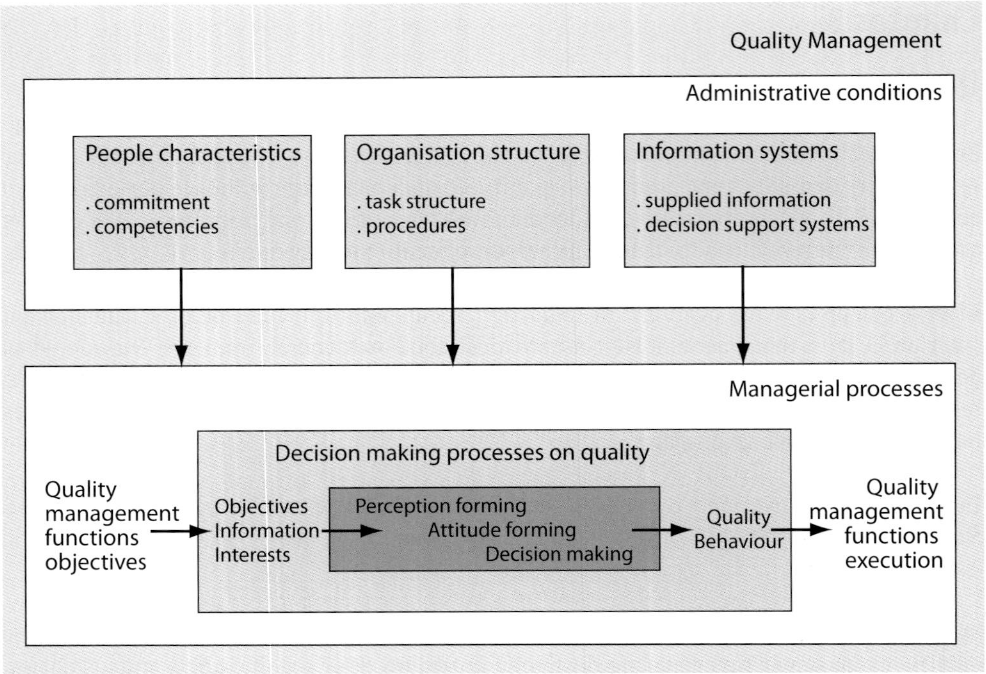

Figure 3.1. Quality management: basic processes and conditioning factors.

'Initiating, directing and controlling goal-oriented activities' (Kampfraath and Marcelis, 1981), or – more related to people –: 'Coordinating and overseeing the work activities of others so that their activities are completed efficiently and effectively' (Robbins and Coulter, 2007).

Figure 3.1 shows that managerial processes are aimed at realising quality management functions. Quality management functions are groups of necessary activities performed to reach certain quality objectives. Later in this chapter we explain in more detail the *food* quality management functions.

For all managerial activities, decision-making is the core activity. Decision-making is the process of defining problems, generating alternative solutions and deciding which alternative is the best one. Decision-making processes are complex, due to people having their own objectives, information and interests, and to the different sub-processes in people's mind, like perception and attitude forming, which are different for different people (Robbins and Coulter, 2007). These complicate prediction of the outcome of decision-making processes. The outcome of decion-making processes over time is the quality behaviour of people. It is showing a certain consistency over time, but is not always exactly predictable.

It should be noticed that decision-making is not exclusively the task of managers. In fact, everyone in the organisation is making decisions and contributing to management. For example, in self-managing teams, operators are able to do much of the managerial work themselves.

Kampfraath and Marcelis (Kampfraath and Marcelis, 1981; Marcelis, 1984) proposed that the organisational structure, the characteristics of people in the organisation, and the available information, should be considered as conditions for managerial processes. These factors create a mid-term or long-term administrative infrastructure having its own dynamics and life cycle. Under various administrative conditions, managerial processes and therefore management decisions will be different. For instance a quality department, reporting to the production managers, will hardly be able to influence quality decision-making in the purchasing or marketing department. However, a quality department, reporting to the chief executive officer and being a member of the management team, will not meet this problem. It is like perceiving management processes as the flow of a river between its banks. The banks are the conditions, which determine how the river flows. At the same time the flow (management processes) is changing its banks, so creating new conditions for itself. Therefore, effective managers not only make decisions themselves but they also influence the decision-making of people in the organisation by creating appropriate administrative conditions.

3.2 Management

In this section, managerial objectives are described as management concerns. Further, the process of decision-making and factors influencing decision-making processes are described. Finally, theory of decision-making is summarised in important decision-making models and administrative concepts.

3.2.1 Management concerns

According to Kampfraath and Marcelis (Kampfraath and Marcelis, 1981; Marcelis, 1984) a sharp distinction should be made (1) between long, mid- and short-term concerns, and (2) between concern for products and concern for resources. They stressed that decisions on resources are made long before actual manufacturing or service processes take place. In the short term, these processes and their outcomes are dependent on which resources are available at that moment. Therefore, management activities are quite different dependent on being:
1. Long-term or short-term oriented.
2. Resource- or product-oriented.

They are described now in three basic groups, ranging from long term to short term: strategic (long term), innovation (mid term) and operations management (short term). They are showed in Figure 3.2, where also three main resources are distinguished, i.e. suppliers, technical infrastructure and markets; here following the basic input-transformation-output model of organisations. The basic point concerning resources is that large investments are often required, which take time to be implemented and used. With respect to buildings and machinery (technical infrastructure) this will be clear, but also investments and time are needed for suppliers and markets. For instance, marketing is aimed at creating markets by investments in market channels, promotion and customer-relations, whereas purchasing requires, among other things, building supplier relationships.

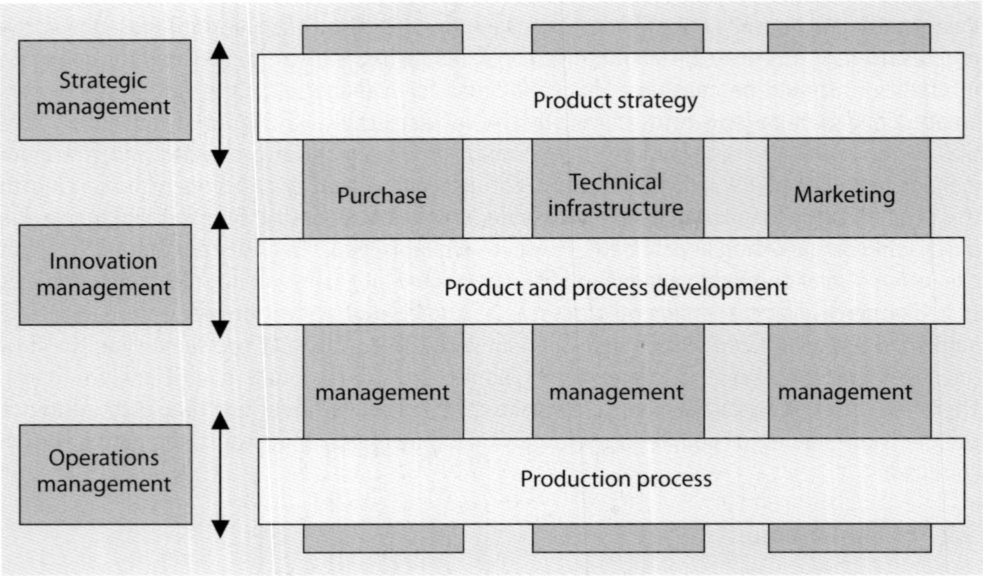

Figure 3.2. Management concerns.

A short description of management concerns, paying attention to product and resource-oriented activities:

- *Strategic management* is the process of deciding on the organisation's mission, overall objectives, product/market combinations and major resource allocations.
  Decisions on *products* deal with, e.g. products and product-groups, segmentation of the market, and the quality level of products.
  Decisions on *resources* are of course related to the product decisions. They include decisions such as supplier selection, degree of co-operation with suppliers, investments in machinery and buildings, degree of automation in manufacturing, and investments in market development and customer relations.
  In strategic decision-making, it is very important to look ahead and to imagine future market and production conditions. However, the final decisions will be mainly on resources, because of the need for investment here. From the moment of investment these decisions are more or less irreversible and have long-term consequences.
- *Innovation management* is the process of deciding on product and/or process innovations, which are necessary to reach strategic goals and objectives, and which provide conditions and knowledge for future manufacturing.
  Product decisions concern, for example, new product development, defining (new) process requirements and conditions, product testing, and the development of product and material specifications.
  Resource decisions concern, for example, technological process development, machinery modification, maintenance programmes, detailing supplier relations, assuring purchasing channels, detailing customer relations and market development.
  Innovation management is both product and resource oriented and functions as the natural link between strategic and operations management activities.

- *Operations management* is the process of deciding on customer orders and processes of supply, production, and distribution of products or services.
  Product decisions involve, for example, quality, quantity, delivery time, production schedules and storage and distribution schedules.
  Resource decisions include, for example, the way resources are used in production processes, including necessary maintenance, hygiene in the manufacturing part of the process, and co-operation with suppliers and customers in the remaining part of the process.
  Operations management is more product-oriented than resource-oriented due to the fact that customer satisfaction is mainly dependent on a product or service. Additionally, resources hardly change in the short term.

In most companies in the sixties and seventies, management paid much attention to operations management. In the eighties, strategic management increased strongly due to globalisation and the concerns of stakeholder. Later on, innovation management took its place among them due to increasing customer-orientation and growing intensity of competition.

Quality management is managing a part of the total business performance, but it addresses all areas of management concern and their interrelations. Within the strategic management area, quality strategy and policy are a logical part of the strategy and policy of the company. In the innovation management area quality design is perhaps the biggest part of total management's concern. In operations management, quality control is an important part of management activity in ensuring customer satisfaction. The broader the definition of quality (for instance quality, delivery time, price, service, flexibility and dependability), the greater the impact of quality management on total business management.

3.2.2 Decision-making

Decision-making is generally considered as the core activity in management processes. It is often described as a process consisting of the essential steps (Schermerhorn, 2007) shown in Figure 3.3.

1. *Identify and define the problem*
   Identifying and defining the problem starts with collecting, processing and reflecting on the information. It often begins with the appearance of problem symptoms. A problem is a difference between an actual situation and a desired situation. Problem analysis is a critical part of decision-making, because its determines the direction of the decision-making process and ultimately the decision that is made. Special care must be taken not just to address a symptom while ignoring the true problem.
2. *Generate and evaluate possible solutions*
   This step involves identifying courses of action that can lead to a solution of the problem. At this stage more information is gathered, data are analysed and the advantages and disadvantages of possible solutions are identified. The better the pool of alternatives, the more likely a good solution will be achieved. Common errors at this stage include selecting a particular solution too quickly (i.e. jumping to a conclusion), and choosing an alternative that, although convenient, has too many side effects.

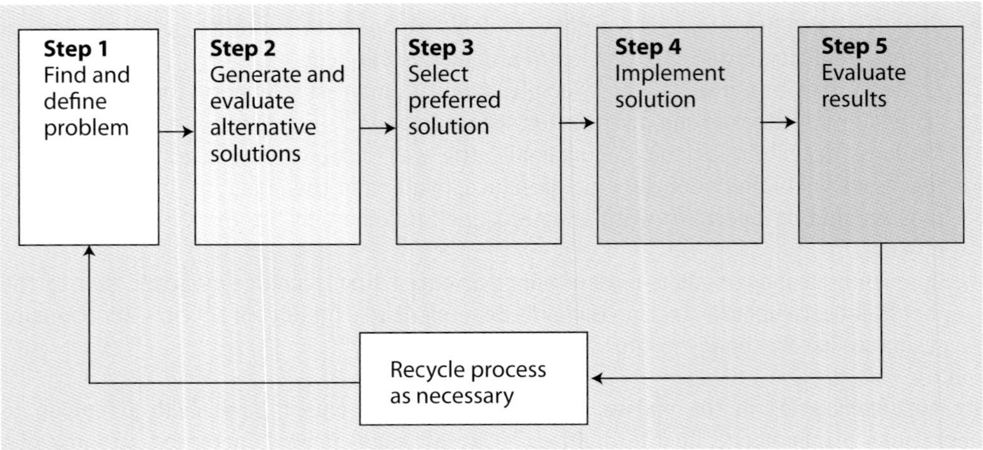

Figure 3.3. Steps in decision-making.

3. *Choose a solution*
   At this point a decision is made to select a particular course of action. How this decision is made, and by whom, must be properly resolved for each problem situation. In some cases, the best alternative may be selected based upon cost-benefit criteria, whereas in other situations additional criteria maybe relevant, such as stakeholders interests or ethics. However, once alternatives are generated and evaluated, a final choice must be made. This is the point of ultimate decision-making.
4. *Implement the solution*
   Appropriate action plans must now be established and fully implemented, based upon the preferred solution. This is the stage at which directions are finally set and problem-solving actions are initiated. People need here the ability and willingness of employees to implement the decision. Difficulties at this stage are often related to lack of participation or a failure to involve the right people.
5. *Evaluate results*
   The decision-making process is not complete until results are evaluated. If desired results have not been achieved, then the process must be restarted to allow for corrective action. Moreover, in any evaluation both the positive and negative consequences of the chosen course of action should be examined. If the original solution appeared inadequate, a return to earlier steps in the problem solving process may be required to generate a modified or new solution.

Decisions can be divided into either programmed or non-programmed decisions (Schermerhorn, 2007). *Programmed decisions* are decisions which are made in response to situations that are routine. They are more or less structured and fairly repetitive. These kinds of decision are suitable for establishment in some type of procedure. Examples of programmed decisions include deciding to re-order inventory when qualities fall below a certain level, or deciding to reduce temperature when temperature has exceeded a certain level in a production process. Many business decisions on routine activities and basic operations are programmed decisions. *Non-programmed decisions* are made in response to situations that are unique. They are relatively unstructured, undefined and/or have major consequences for the organisation. No standardised procedures exist to resolve these situations because the problem has never arisen before. The problems are complex or uncertain, or they are of such significance to the organisation that it requires tailor-made decisions. Examples of non-programmed decisions include deciding to build a new plant, to enter into a strategic alliance with another company or to develop a new product.

3.2.3 Factors influencing decision-making

Management can also be defined as goal-oriented decision-making. Decisions have to be made in order to stimulate actions towards more or less specified objectives. There are two main factors having great influence on decision-making, i.e. the availability of information and the existence of interests (Figure 3.4).

Starting with a theoretically infinite number of alternatives, lack of information results in uncertainty and existing interests result in constraints. Both uncertainty and interests reduce the scope for decision, because some alternatives must be excluded, while others may become preferable.

Decision-making processes could be characterised on a scale ranging from the extreme 'fully depending on information availability and information analysis' (problem solving) to the other extreme 'fully depending on interests and the way conflicts are managed' (conflict resolution). Most decision-making processes show a mix of both elements, information and interests. The two major influencing factors, information and interest, are discussed in more detail below.

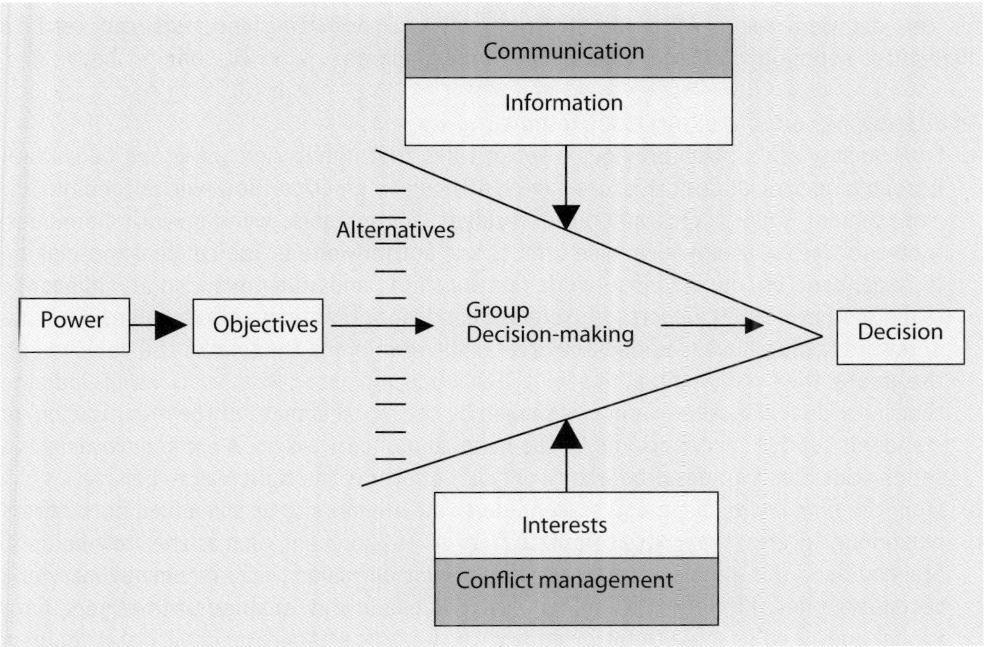

Figure 3.4. Factors influencing decision-making.

1. Information

The availability of information is an important condition affecting decision-making. In an ideal situation one would have all the information needed to make decisions with certainty. However, most business situations are characterised by incomplete or ambiguous information, which affects the level of certainty and so reduces the room for decision. With respect to the availability of information, three major situations can be distinguished, i.e. certainty, risk and uncertainty (Hellriegel *et al.*, 2005):

- Certainty is the situation that exists when decisions are made in the context of being fully informed about a problem, its alternative solutions and their respective outcomes. Under this condition one can anticipate and even exercise some control over events and their outcome. However this situation almost never occurs.
- Risk is the situation that exists when decisions are made in the context of incomplete, yet reliable information. Under a state of risk the future outcomes associated with alternative courses of action are not known for certain. The results are subject to chance. However, the decision-maker has enough information to determine the probabilities associated with each alternative.
- Uncertainty is the situation that exists when decisions are made while being incompletely and improperly informed about a problem, its alternative solutions and their respective outcomes.

2. Interests

Interests are a second condition affecting decision-making which reduce the scope for decisions. By making a decision, interests of some people or groups will be served, while interests of others will be neglected. Conflicting interests may result in stressful situations for the decision-maker. In group decision-making, conflicting interests can lead to differences in opinion and a struggle between participants.

An interest may originate from three resources, i.e.:

a. From *society ethics*: this provides a set of rules that define right or wrong behaviour. Recognising special ethical issues can be difficult in practice. However, in food quality management some topics are rather evident, for instance working with hormones, food additives with unknown side effects and animal welfare. Ethical decision-making is dependent on opportunity, work relationships, and individual moral philosophy (Schermerhorn, 2007). *Opportunity* refers to conditions that limit unfavourable behaviour or reward favourable behaviour. The greater the reward and the smaller the punishment, the greater the likelihood that unethical behaviour will occur. *Work relationships* indicate that individual ethical behaviour is shaped by the moral climate of the organisation, or standards set by supervisors and/or the behaviour of co-workers. A *moral philosophy* is a set of principles that describe what a person believes is the right way to behave.
b. From the *firm environment*, e.g. stakeholders who are more or less involved in the firm's behaviour. Interests may also originate from an environment such as the stakeholders. Stakeholders who are affected by the firm's decision-making may include primary and secondary ones. Primary stakeholders, such as employees, suppliers and buyers, have formal and/or contractual relationships with the firm, whereas secondary stakeholders

have less formal connections. Secondary stakeholders are for instance animal rights groups, consumer groups and environmentalists.
Relationships with stakeholders are rather complex because of their interrelationship. The firm is at the centre of a network of actors, which the organisation should identify and manage as far as possible (Schermerhorn, 2007).

c. From *inside the organisation*: the organisation's culture can influence decision-making behaviour. The organisational culture is the system of shared beliefs and values that is developed within an organisation and guides the behaviour of its members (Schein, 1990). Strong cultures are clear, well-defined and widely shared among members. In strong cultures there is commitment to do things within the interest of the organisation.

Information and interests are discussed for individual decision-making. In group decision-making however, the immediate interaction between participants plays a complicating role. For decision-making in a group, additional information is used which is exchanged between group members in *communication* processes, in addition to the regular information input. With respect to interests, in a group decision-making process, people often have contradictory interests and also the possibility to influence each other, sometimes causing conflicts. How groups or managers deal with conflicts is called *conflict management*.

In addition, in a group situation there are differences in power. *Power* is the ability to drive others in a direction you prefer and the way you want (Schermerhorn, 2007). Following that definition, power is perhaps the most important factor in decision-making. It will direct choice among alternatives. Communication, conflict management and power will be briefly discussed below.

Communication is the process through which information and meanings are transferred from one person to another (Schermerhorn, 2007). The process starts with a sender, which encodes a message into symbols. Then the message arrives at the receiver, who then decodes and/or interprets its meaning (Figure 3.5). Feedback is the reverse process, whereby the receiver responds to the sender.

The receiver's interpretation almost never matches the sender's original intention. The reason for this is noise in the communication process. Noise is anything that interferes with the message being communicated effectively. For instance, the process of encoding and decoding, or inadequate channels. Disturbances in communication arise through filtering (only information is sent that is important for the sender) and selective perception (only information is used that is meaningful for the receiver).

Formal communication represents the flow of communication within the formal organisational structure and the groups and teams within it. However, also very important is the informal communication channel (the grapevine). Individuals and groups spread information to other people when they feel the need for it. Informal communication depends strongly on personal relations and the existence of informal groups.

In the past informal communication was not of great concern to managers. Today managers view the quality of communication through formal channels as being poor. Quality of

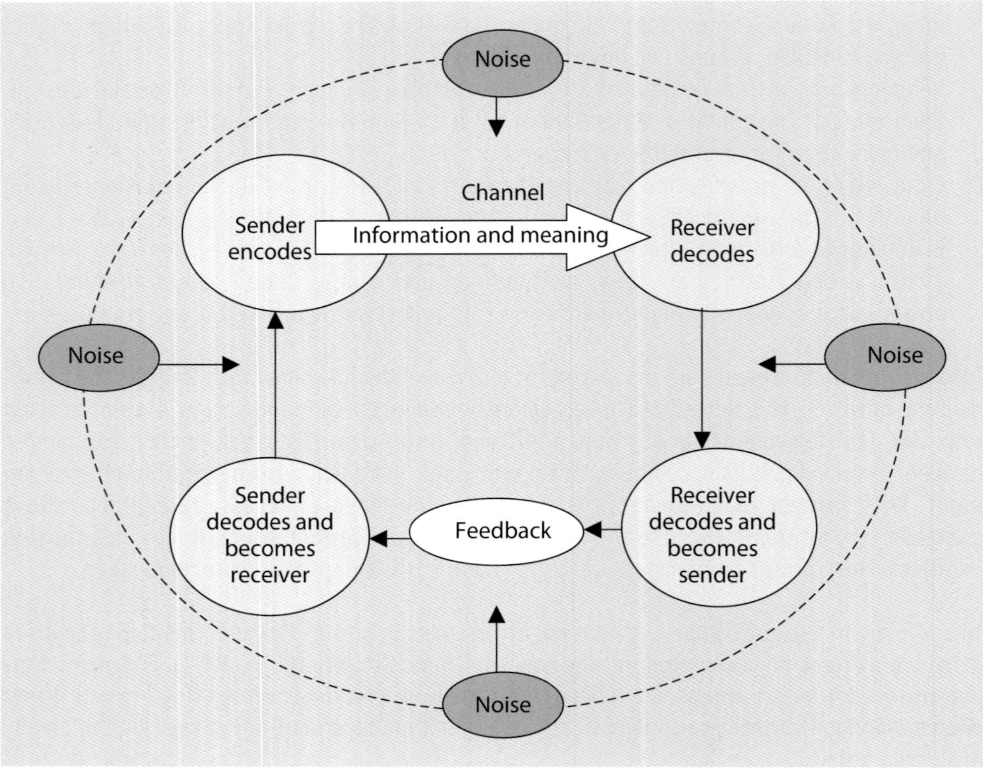

Figure 3.5. Communication process.

informal communication also is not perfect, but managers agree that these channels are often a much better source of good and accurate information than formal channels.

Conflict management involves the way managers handle conflict situations. Conflicts can arise when different conflicting interests are at stake. There is a real conflict when someone blocks the goals, intentions or behaviour of someone else.

Conflict has traditionally been viewed as negative; however, the modern view is that conflict is both inevitable and in many cases desirable, if properly managed. Conflict can stimulate people toward greater efforts, co-operation and creativity. Conflict does not always take the form of an open fight between two parties (hot conflict). In fact, there is a conflict process in which the conflict, in the course of time, grows and builds up to a climax (Robbins and Coulter, 2007).

Kenneth W. Thomas (1976) distinguished five conflict management styles. These styles evolved from different levels of co-operation and assertiveness (Figure 3.6). Co-operation is attempting to satisfy the other party's concerns and goals. Assertiveness is attempting to satisfy one's own concerns and goals.

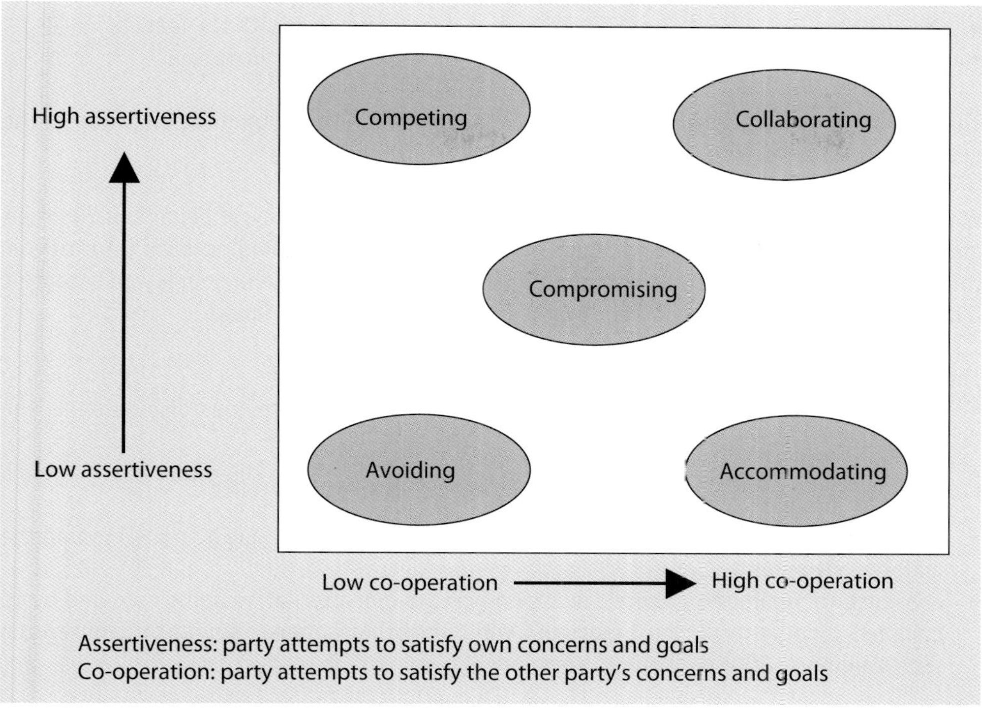

Figure 3.6. Conflict management styles.

Briefly stated, these conflict management styles involve the following behaviours (Schermerhorn, 2007):

- Avoidance, i.e. denying the existence of the conflict and hiding one's true feelings.
- Accommodation, i.e. playing down the conflict and seeking harmony among parties.
- Competition, i.e. forcing a solution to impose one's will on the other party.
- Compromise, i.e. bargaining for gains and losses to each party.
- Collaboration, i.e. searching for solutions that meet each other's needs.

Of course various conflict management styles have quite different outcomes. Collaboration or problem solving tries to reconcile underlying differences and is often the most effective style. It is a form of win-win situation whereby issues are solved to the mutual benefit of all conflicting parties.

Power is the ability to make things happen the way one wants them to happen. Power is more than just influence; it gives the possibility to get someone to do something even when he/she does not agree with it.

One of the most used frameworks for understanding power has been developed by French and Raven (1960). They identified five types of power:

- Legitimate power is based on the formal position in the organisational hierarchy.
- Reward power stems from the ability to reward followers.
- Coercive power results from the ability to obtain compliance through fear or punishment.

- Referent power is based on followers' personal identification with the leader.
- Expert power is based on having specialised knowledge and information.

The first three bases of power are related to the position in the organisation, whereas the last two are linked with personal characteristics (Figure 3.7).

When power is actually turned into action, one can speak of political behaviour (Robbins and Coulter, 2007). Political behaviour involves activities like holding back information, forming coalitions, threatening with punishment. Often political behaviour is well accepted, but there is also political behaviour that is perceived as non-legitimate.

The use of power in group decision-making can be influenced by:
- authority rule, where the leader decides most times with previous discussion in the group;
- minority rule, where two or three people are able to dominate the group into making a mutually agreeable decision;
- majority rule, where formal voting may take place or members may be polled to find the majority viewpoint;
- consensus or unanimity, where discussion leads to one alternative being favoured by all members (unanimity) or most members, while the others agree to support it (consensus) (Schermerhorn, 2007).

3.2.4 Decision-making models

Management theory offers a range of models for decision-making. From this broad range, three major models are described here, i.e. the rational model, the administrative model and the political model. These are combined in the Mintzberg process model.

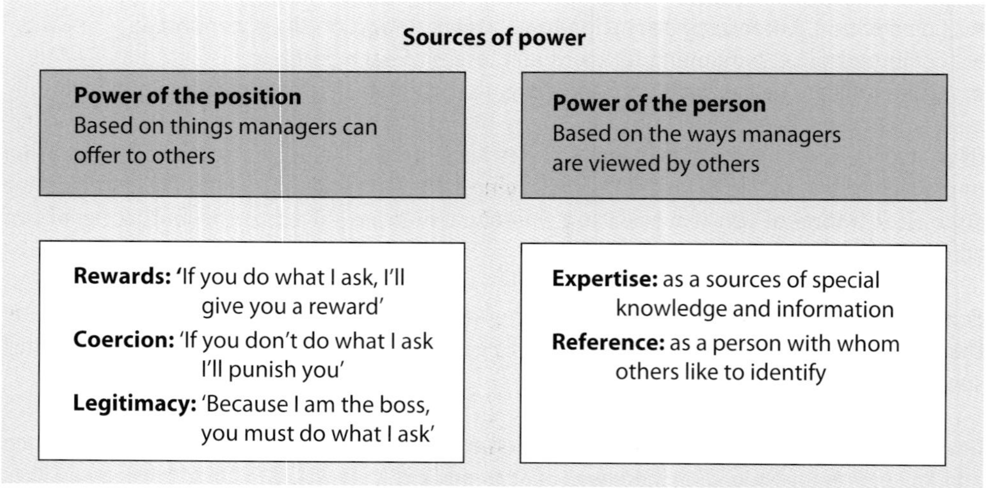

Figure 3.7. Sources of position power and personal power used by managers.

The *rational* model is an approach that outlines how managers should make decisions. It is also called the *classical model*. The rational model is based on economic assumptions and asserts that managers are logical, rational individuals, who make decisions that are in the best interests of the organisation. The model is characterised by the following assumptions.

- The manager has complete information about the decision situation and operates under a condition of certainty.
- The problem is clearly defined and the decision maker has knowledge of all possible alternatives and their outcomes.
- Through rationality and logic, the decision-maker evaluates the alternatives and selects the optimum alternative. That is the one that will maximise the decision situation by offering the best solution (Schermerhorn, 2007).

The description of the decision-making process in Figure 3.3 is derived from the rational model. The main hypothesis is that analysis leads to solution of the problem.

The *administrative* model of decision-making is an approach that outlines how managers actually make decisions. It is also called the *behavioural model*. The administrative model is based on the work of Simon (1960), psychologist and economist, whose research and findings resulted in his winning the Nobel Prize for economy. Simon recognised that people do not always make decisions with logic and rationality and introduced two additional concepts into his administrative model, i.e. bounded rationality and satisfying behaviour. The basic assumptions of the model of Simon include:

- The manager has incomplete information about the decision situation and operates under a condition of risk and uncertainty.
- The problem is not clearly defined and the decision-maker has limited knowledge of possible alternatives and their outcomes.
- The decision-maker shows satisfying behaviour by choosing the first satisfactory alternative; the one that will resolve the problem situation by offering a good solution.

In fact, in this approach one is accepting that people have their own way of dealing with problems, judging more or less subjectively. Implicitly they rely on so called judgmental heuristics. They simplify search and choice procedures based on readily available information, representation (as with previous occurrences) and/or propositions (ideas about causal relationships) (Hogarth *et al.*, 1994).

The *political* model of decision-making describes the process in terms of particular interests and objectives of powerful stakeholders (Hellriegel *et al.*, 2005). The model considers uncertainty (information is limited, managers have their limitations) and potential conflict (decision-makers differ in goals and objectives, values and experiences).

The organisation consists of different coalitions striving for the accomplishment of goals that are based on their own particular interests. The goals are consistent within a coalition, but are inconsistent within the organisation as a whole.

The composition of the coalitions depends on the content and circumstances of the problem or conflict. The same groups can be enemies in one conflict, whereas they can be allies in

another. The decision-making process, according to the political approach, consists of three steps:
- forming coalitions;
- gathering information and making rules and procedures;
- choice as a result of bargaining and negotiation.

Mintzberg and co-authors (1976) developed a decision-making model that offers the possibility of integrating the previously described models (Figure 3.8). Mintzberg divided the decision-making process in three phases, i.e. identification, development and selection.
- In the *identification* phase something is recognised as a stimulus requiring a decision. This is followed by a diagnosis, i.e. an attempt to understand the situation in terms of cause-effect relationships.
- The *development* phase starts with searching, which is trying to find existing solutions adequate for the actual problem. Searching is followed by screening, which is reducing the number of potential solutions by neglecting 'non-fitting' solutions. Finally, in the design step, a new solution for the particular problem is developed.
- In the *selection* phase the actual decision is made, which contains three parallel decision activities, i.e. analysis, judgement and bargaining. By analysis, choice is a logical outcome of rational activities. By judgement, choice is made on the basis of vague criteria. By bargaining, choice is a result of decision-making activities of several decision-makers having different interest.

These activities analysis, judgement and bargaining, largely correspond with the basic ideas of respectively the rational (i.e. all information available, selection of optimal alternative), the administrative (i.e. restricted information, choosing a satisfactory alternative) and the political model (i.e. using criteria and decision rules giving all parties satisfying solutions).

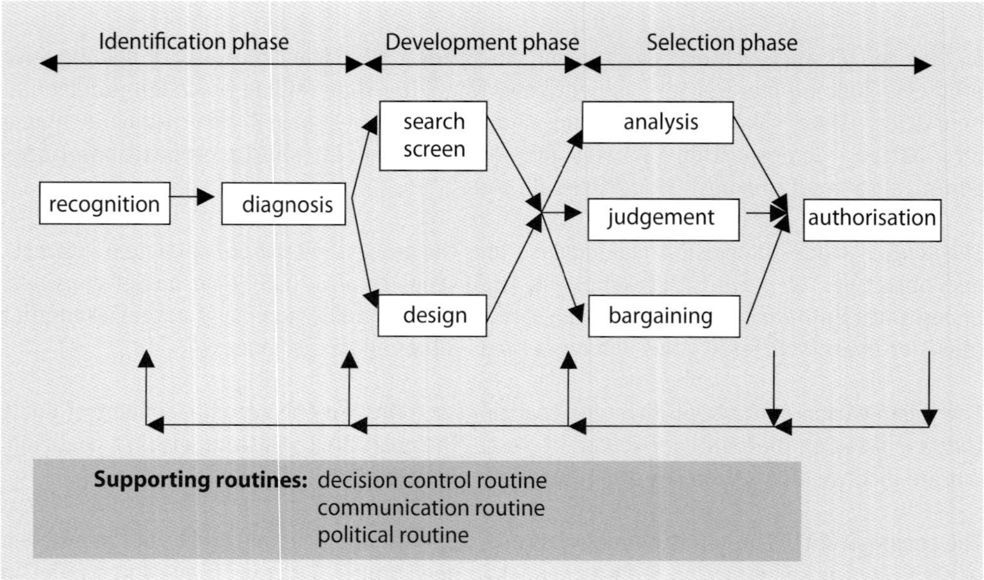

Figure 3.8. Mintzberg process model of decision-making.

Mintzberg paid much attention to the dynamics of decision-making. Several disturbing factors can influence decision-making in such a way that phases must be repeated, phases are delayed, conflicts have to be resolved and so on.

Mintzberg *et al.* (1976) defined three supporting routines, i.e. communication, political and decision control routines. These are rules or procedures that structure the decision-making processes. Communication routines ensure information availability. Political routines prescribe how to deal with (conflicting) interests. Decision control routines bring structure in making the actual decision.

3.2.5 Administrative concepts

An administrative concept is a broad view on how decisions in an organisation should be made. On the one hand, managers make decisions themselves, whereas on the other hand they create organisational conditions in order to influence the decision-making of others (centralisation resp. decentralisation). In small organisations, managers influence people more directly. In larger organisations, rules, procedures and structures are frequently needed in order to direct people's behaviour (low and high formalisation). Figure 3.9 illustrates how a manager can select for these possibilities resulting in four administrative options, i.e. the authority, bureaucratic, business and flexible concept.

The authority concept is the basic concept in small firms. Formalisation is not needed because a strong leader is making decisions, granting subordinates to make decisions 'like he would do it'.

The bureaucratic concept relies on structure, rules and procedures, giving top management the guarantee that people in the organisation behave according to central goals and objectives. The top management makes all the important decisions.

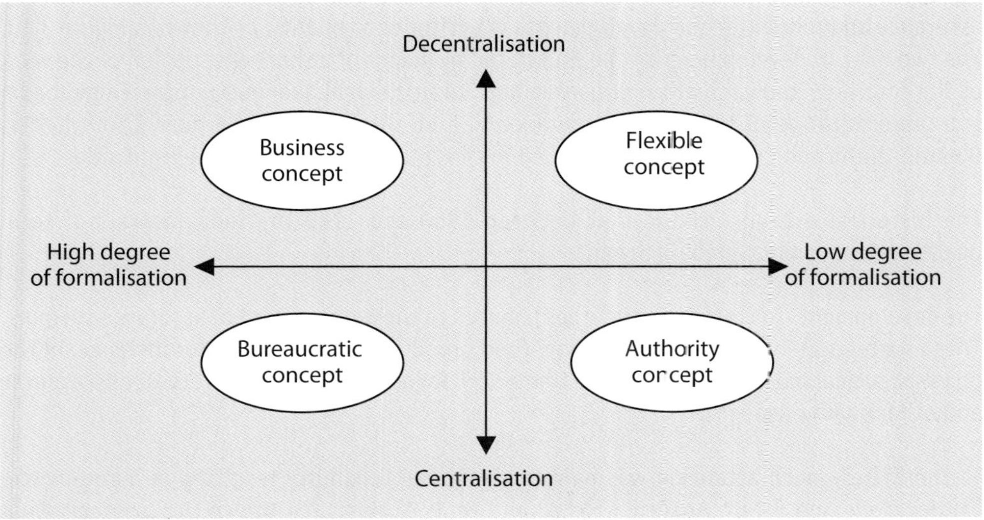

Figure 3.9. Administrative concepts.

The business concept allows decision-making to be dependent on market and product circumstances. Top management has influence by means of rules and procedures and by making policy statements.

The flexible concept accepts less influence from top management. Top management in this case clearly defines policy and communicates it in a broad sense to all organisation members. The members, however, evaluate and weigh policy statements against other factors important at that moment and decide for themselves.

3.3 Quality management

In this section a short overview of quality management history is given, with special attention to quality management as an assurance function, compared to quality management from the prevention point of view. The philosophies of the main quality gurus are described in more detail. Finally, the main activities in quality management are described.

3.3.1 Quality management history

In the first decades of the last century, statistical explanatory models had begun to be used in science. In 1931 Shewart (Shewart, 1980) published his book 'Economic control of quality of manufactured products'. The central topic in his publication is how to take care of data and draw conclusions from them in order to supervise and reduce variation in the production process. In those years he was already focused on customer needs. He stated that 'the first step of the engineer in trying to satisfy these wants is to translate them into the physical characteristics of the thing manufactured to satisfy these wants'. Shewart has had a profound influence on the modern view of quality control. If one single individual could be called the father of modern quality philosophy, that person would probably be Walter A. Shewart (Bergman and Klefsjö, 1994).

Quality management developments can be summarised along two lines: developments in assurance and reliability and developments in attitudes to quality, as shown in Figure 3.10. The two lines of development will be illustrated by means of a short description of the work of the four best known quality gurus Deming, Juran, Feigenbaum and Crosby. Feigenbaum is a representative of the quality assurance line on the left side of Figure 3.10, whereas Deming, Juran and Crosby should be placed in the total quality line on the right side.

The historical overview shows that between 1950 and 1980 the fundamentals of total quality management were laid down.

The developments in statistics surely contributed to the development of assurance systems. These systems are very important in agri-food production (as they are, for instance, in the pharmaceutical industry), forming the framework for controlling food safety, health aspects and other quality aspects.

In the fifties, much attention was paid to statistical reliability resulting in methods like Failure Mode and Effect Analysis (FMEA) and Fault Analysis. In the sixties, systems were developed like the Good Practice Codes (e.g. Good Manufacturing Practice) and AQAP (Allied

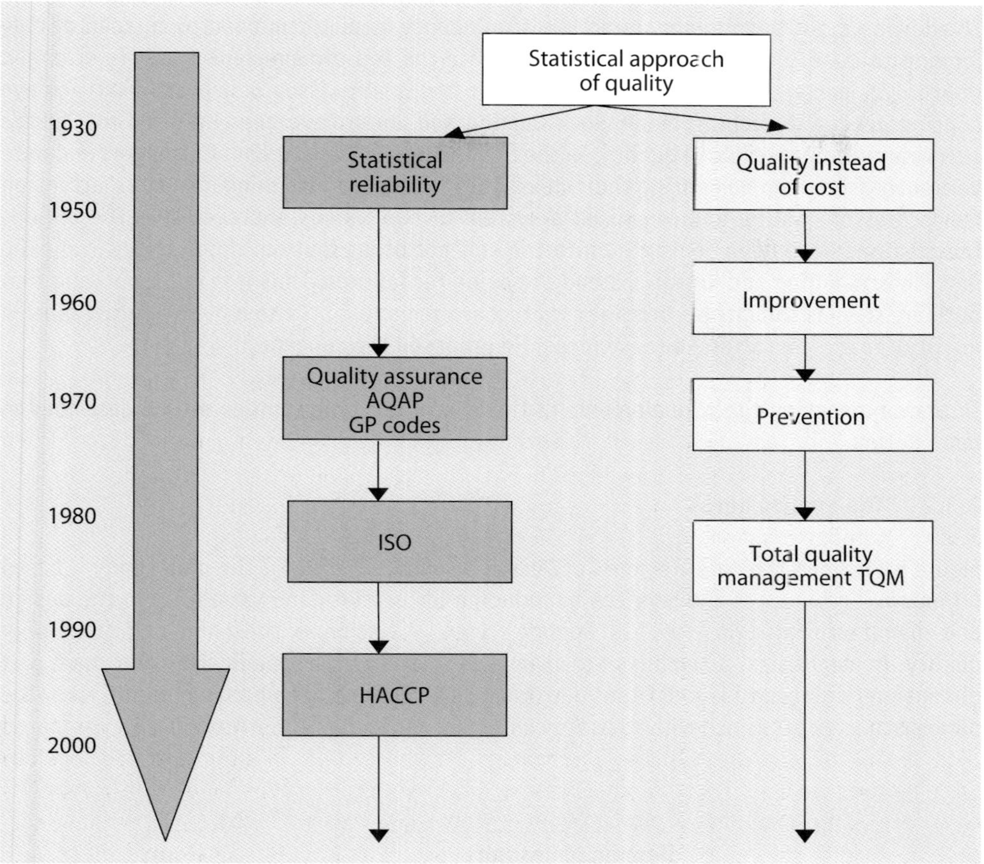

Figure 3.10. Historical overview of developments in quality management.

Quality Assurance Publications), a NATO system for assuring quality of incoming materials. AQAP is, more or less, the forerunner of the current ISO-system.

The International Standards Organisation (ISO) developed in the seventies the ISO 9000 series, providing a framework for quality assurance including norms for external certification. They were firstly published as a standard in 1987.

The HACCP system (Hazard Analysis Critical Control Points) originated from the need for safe food supply for manned space flights. It was not until the eighties that HACCP was seriously considered for broader application in the food industry. In the nineties the food industry in the EU was forced by law to apply the system (see Chapter 7).

On the right of the figure the development towards Total Quality Management is outlined. W. Edward Deming and Joseph M. Juran share Shewart's statistical viewpoint regarding the production process. Deming in particular lays significant stress on the statistical viewpoint.

In addition, both Deming and Juran focus on quality as an output factor, instead of only looking to quality as costs. They also emphasise the top management role very strongly. Only if top management commits itself fully to the quality issue is it possible to achieve continuous quality improvement. Both Deming and Juran have been credited for Japan's successes within the field of quality. In the fifties they advised Japanese managers and were very successful in implementing their ideas. They succeeded in turning managers' attention from costs to quality, and organised a system of continuous improvement. The lessons they learned have been translated into the concept of prevention: doing things right the first time, planning and design instead of paying for failures. Thus it is not surprising that both are seen as the fathers of total quality management. In the seventies another guru in quality, Crosby, also stressed these ideas. He proposed the concept of zero defects.

In the next section the core elements of the four gurus' philosophies will be described in more detail.

3.3.2 The quality gurus

Figure 3.11 shows the core elements of *Deming's philosophy* (1986). The philosophy is based on improving products and services by reducing uncertainty and variability in the design and manufacturing processes. In Deming's view, variation is the chief culprit of poor quality. In mechanical assemblies, for example, variation from specifications for spare part dimensions will lead to inconsistent performance and premature wear and failure. Likewise, inconsistencies in service will frustrate customers and so damage a firm's image.

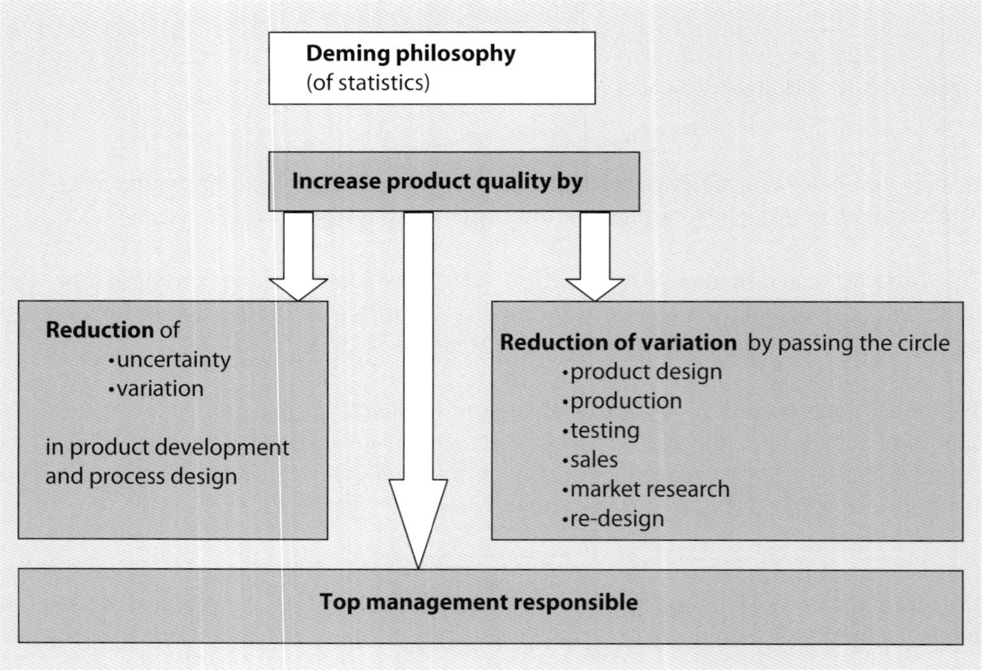

Figure 3.11. Deming's philosophy.

To achieve reduced variation he advocated a never-ending cycle of product design, manufacture, test and sales, followed by market surveys, then redesign. Deming stressed that higher quality will lead to higher productivity, which in turn will lead to long-term competitive strength. In his view, improvements in quality lead to lower costs because of less rework, fewer mistakes, fewer delays and better use of time and materials. With better quality and lower prices the firm can achieve a higher market share and thus stay in business, providing more and more jobs.

Moreover, Deming stated emphatically that top management has the overriding responsibility for quality improvement (Dean and Evans, 1994).

The core elements of *Juran's philosophy* are summarised in Figure 3.12. Juran (1951, 1964) emphasised the importance of customer perception of quality and paid much attention to management commitment to quality.

Juran suggested that employees at different levels of an organisation speak in different 'languages'. In contrast, Deming believed that statistics could be the common language.

According to Juran, top management speaks in the language of dollars, workers speak in the language of things, and middle management must be able to speak both languages and translate between dollars and things. Therefore, to attract the attention of top management,

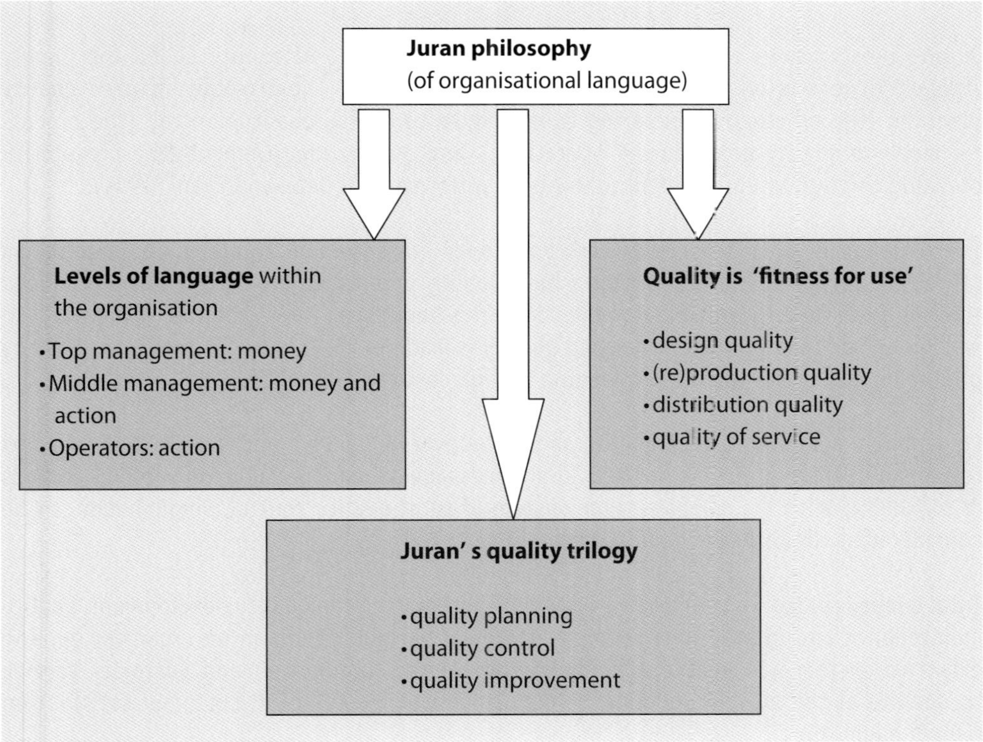

Figure 3.12. Juran's philosophy.

quality issues must be expressed in terms of money. Juran also advocated accounting and analysis of quality costs in order to focus attention on quality problems. At the operational level, Juran proposed increasing conformance to specifications through elimination of defects, supported extensively by statistical tools for analysis. His philosophy fitted well within existing management systems.

Juran defined quality as 'fitness for use'. This was divided into four categories: quality of design, quality of conformance, availability and field service. Quality of design related to market research, the product concept, and design specifications. Quality of conformance included technology, manpower, and management. Availability focused on reliability, maintainability and logistical support. Field service quality comprised promptness, competence and integrity.

Like Deming, Juran advocated a never-ending spiral of activities that included market research, product development, design, planning for manufacture, purchasing, production process control, inspection and testing, followed by customer feedback. Because of the interdependence of these functions, there is a great need for competent company-wide quality management. Senior management must play an active and enthusiastic leadership role in the quality management process.

Juran viewed the pursuit of quality on two levels: (1) the mission of the firm as a whole is to achieve high product quality, and (2) the mission of each individual department in the firm is to achieve high production quality.

Juran's prescriptions were focused on three major aspects of quality called the *Quality Trilogy*™ (a registered trademark of the Juran Institute). The triangle includes quality planning (the process for preparing to meet quality goals), quality control, (the process for meeting quality goals during operations), and quality improvement (the process for breaking through to unprecedented levels of performance) (Dean and Evans, 1994).

The major principles of *Feigenbaum's philosophy* are shown in Figure 3.13. Feigenbaum (1961) suggested in the fifties that a high-quality product is more likely to be produced through Total Quality Control (also the title of his book) than when manufacturing works in isolation. Feigenbaum stated that bad quality is made in the 'hidden factory' every day by producing, e.g. scrap and rework, leading to high costs and no profit.

Feigenbaum also stated that quality should be everybody's job and everyone should feel responsible. Thereby, he emphasised Juran's message on managerial responsibility (Bounds *et al.*, 1994). With his company-wide system of total quality control, Feigenbaum can be considered as the father of *quality assurance*.

Total quality control is an effective system for integrating the quality development, quality maintenance and quality improvement efforts of various groups in an organisation. The effective system will enable marketing, engineering, production and customer service to operate at the most economical levels, which will result in full customer satisfaction (Feigenbaum, 1961).

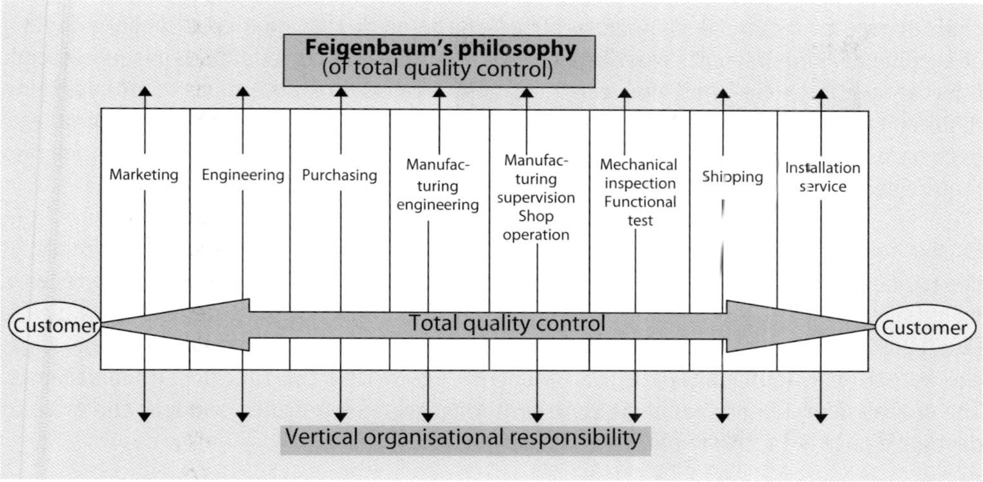

Figure 3.13. Feigenbaum's philosophy.

Feigenbaum proposed a systems approach to quality through the definition of a quality system. It was defined as 'a company and plant-wide operating work structure, documented in effective integrated technical and managerial procedures, for guiding the co-ordinated actions of work force, machines, information of the company and plant in the best and most practical ways to assure customer quality satisfaction and economical costs of quality'.

Feigenbaum expressed total quality control as a horizontal concept, stretching across the functional divisions of an organisation.

Crosby's ideas on quality management are represented in Figure 3.14. Crosby's (1979) quality philosophy is embodied in his 'Absolutes of Quality Management'. Quality means

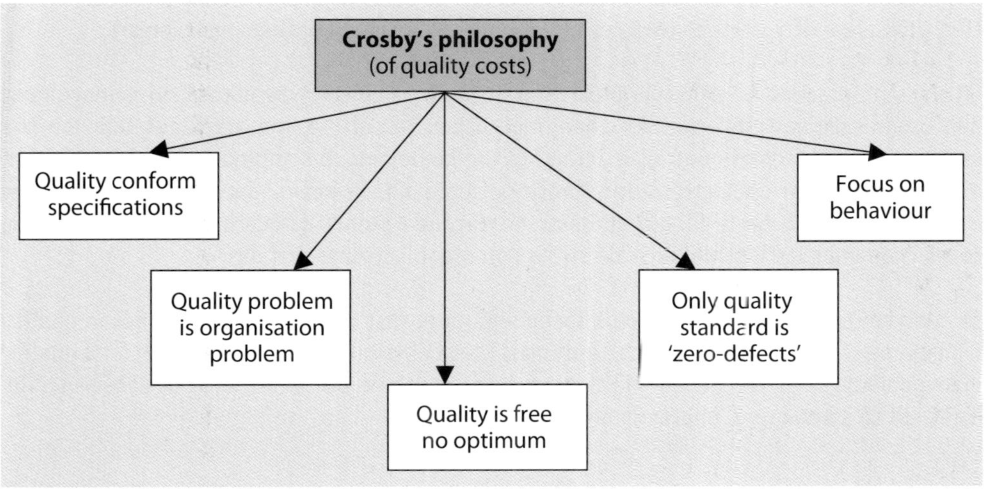

Figure 3.14. Crosby's philosophy.

conformance to requirements. Crosby dispelled the myth that quality is simply a feeling of excellence. Requirements must be clearly stated so that they cannot be misunderstood. Once a task is carried out, one can take measurements to determine conformance to requirements. The non-conformance detected is the absence of quality. Quality problems become non-conformance problems, because it consitutes a variation in output. Setting requirements is the responsibility of management.

Crosby suggested that there is no such thing as a quality problem. The individuals or departments that cause them must identify the problems. There are accounting problems, manufacturing problems, design problems, front-desk problems, but no quality problems. Quality is produced in functional departments, not in the quality department. Therefore, the burden of responsibility for such problems lies within the functional departments. The quality department should measure conformance, report results, and lead the drive to develop a positive attitude toward quality improvement.

Crosby also advocated that there is no such thing as the economics of quality: it is always cheaper to do the job right the first time. Crosby supported the premise that 'economics of quality' has no meaning. Quality is free. What costs money are all the actions that are involved in not doing jobs right the first time. The cost of quality is the expense of non-conformance. Crosby noted that most companies spend 15 to 20 percent of their sales money on quality costs. A company with a well-run quality management programme can achieve a cost of quality that is less than 2.5 percent of sales, primarily in the prevention and appraisal categories. Crosby's programme stimulated the measurement and publication of the cost of poor quality. Quality cost data are useful in attracting management's attention, selecting opportunities for corrective action, and tracking quality improvement over time. Such data will provide visible proof of improvement and recognition of achievement.

Crosby mentioned that the only performance standard is Zero Defects. The basic principle of zero defects is 'do it right the first time', which means concentrating on preventing defects rather than just finding and fixing them. People are conditioned to believe that error is inevitable, it is human to make errors. However, the Zero Defects principle does not accept this, just as we often refuse to accept in our personal life that others make errors.

Crosby's programme is primarily behavioural. More emphasis was placed on management and organisational processes for changing corporate culture and attitudes than on the use of statistical techniques. Like Juran but unlike Deming, his approach fitted well within existing organisational structures. Crosby's approach, however, provided relatively few details about how firms should address the detailed points of quality management. The focus is on managerial thinking rather than on organisational systems.

In addition to the four described philosophers (or gurus) other important names in quality management are Ishikawa, Taguchi and Imai. They contributed to specific aspects of quality management. Therefore, they will not be described here but dealt with in other specific topics of this book (e.g. improvement, policy).

3.3.3 Quality management activities

There are different definitions of quality management. There is a broad acceptance of the following definition that will be used in this book. In this broad description quality management is defined as the total of activities and decisions performed in an organisation to produce and maintain a product with the desired quality level against minimal costs. Quality management is that part of overall management activities that focuses on quality. When quality is defined as business performance, then quality management is a major part. When it is defined as technical product quality, quality management is a rather small part of total management activities.

Quality management always consists of different groups of activities, each with its own objective. Juran (1990) in particular focused on three major activities in quality management, the so-called Quality Trilogy (Figure 3.15).

Quality planning begins with identifying both external and internal customers, by determining their needs and developing product features that respond to these needs. Like Deming, Juran stressed that employees should know who uses their products, whether it is in the next department or in another organisation. Then quality goals are established, which meet the needs of customers and suppliers at a minimum cost. Subsequently, a process under operating conditions must be designed to deliver products that meet customer requirements.

Quality control involves determining what to control, establishing units of measurements so that data may be objectively evaluated, establishing standards of performance, measuring actual performance, interpreting the difference between actual performance and the standard, and taking action on the difference.

Managing for quality		
Quality planning	Quality control	Quality improvement
• Who are the customers? • Needs of customers? • Develop product features that fit consumer needs • Develop process able to realise product features • Transfer plans to operating forces	• Evaluate actual product performance • Compare performance to product goals • Act on difference	• Establish infrastructure • Identify improvement projects • Establish project teams • Provide teams with training, resources and motivation to: - diagnose cause - stimulate remedies - establish controls

Figure 3.15. Juran's Quality Trilogy.

Deming, Juran and Crosby placed the emphasis on identifying sources of variation and improving the work system. Juran specified a detailed programme for *quality improvement*. The quality improvement process involves the following steps:
- proving the need for improvement;
- identifying specific projects for improvement and organising guidance of the projects;
- diagnosing the causes and providing remedies for these causes;
- proving that the remedies are effective under operating conditions;
- providing control to maintain improvements.

Juran's assessment of most companies is that quality control has priority among the trilogy categories. Most companies feel that they are strong in this category. Quality planning and improvement are often a low priority and are significantly weak in most organisations. Juran proposed that more effort should be put into quality planning and even more into quality improvement.

Juran supported these conclusions with several case studies, in which Japanese firms used the same technology, materials and processes as American firms, but had much higher levels of quality and productivity. He explained these findings by the fact that, since the1950s, the Japanese have implemented quality improvement projects to a much larger extent than their Western counterparts. The result was that, in the 1970s, Japanese product quality exceeded Western quality and continued to improve more quickly.

Massive training programmes and top management leadership supported Japanese efforts on quality improvement. It encompassed training in managerial quality-oriented concepts, as well as training in the tools for quality improvement, cost reduction, data collection and analysis. The latter is one of the most important components of Juran's philosophy. Juran showed that the Japanese experience left little doubt as to the significance of quality training in competitive advantage, reduced failure costs, higher productivity, smaller inventories and better delivery performance (Dean and Evans, 1994).

In fact, improvement can be perceived as a learning process, where much more attention is paid to structural causes and solutions.

3.4 Food quality management

Food quality management is a special form of quality management. Activities and decisions deal with the special characteristics of food, food quality, and food production processes. These place special requirements on the quality management activities, which will be explained by discussing the food quality management functions.

3.4.1 Quality assurance

Due to the special characteristics of food products compared to normal products, quality assurance is an important addition to the regular activities of quality planning, control and improvement. Throughout the world, food manufacturing, distribution and retailing are becoming highly complex businesses. Raw materials are obtained from sources worldwide, an ever-increasing number of processing technologies are used, and a broad range of products is available to the consumer. Such complexity necessitates the development of comprehensive control procedures to ensure the production of safe and high-quality food. In addition, consumer expectations are changing with a desire for convenience, less processed and fresher foods, with more natural characteristics. Against this background, the total food chain has to ensure that the highest standards of quality and safety are maintained. All stages of the food chain, from acquisition of raw materials through manufacture, distribution and sales, whether it would be through retail or catering outlets, must consider quality issues associated with specific products, processes and methods of handling. There are a number of reasons why, especially in the agri-food business, the implementation of quality assurance systems is an issue of major importance:

- Agricultural products are often perishable and subject to rapid decay due to physiological processes and/or microbiological contamination.
- Most agricultural products are harvested seasonally.
- Products are often heterogeneous with respect to desirable quality parameters, such as the amount of important components (e.g. sugars), size and colour. This kind of variation is dependent on cultivar differences and seasonal variables, which cannot really be controlled.
- Primary production of agricultural products is performed by a large number of farms operating on a small scale.

Irrespective of size and complexity, all agri-food businesses must have an appropriate food quality assurance programme. Trading conditions and legislation require agri-food businesses to demonstrate their commitment to food quality and to establish an appropriate quality programme (Luning *et al.*, 2009).

3.4.2 Food quality management functions

From quality management literature five managerial functions (Figure 3.16) have been derived, that contribute to the realisation of food quality, i.e. design, control, improvement, assurance, and policy & strategy of quality (Luning and Marcelis, 2007).

Design, control and improvement are basically derived from Juran's Trilogy, where design is equal to the planning activities in Juran's model. Quality assurance is added due to the special requirements on the reliability of food products.

When looking at the management concerns (see section 3.2.1), the operations part is covered by quality control and the innovations part by quality design.The strategic part remains and includes decisions on quality, which can be labelled as Quality Policy and Strategy.

The food quality management functions can be summarised as:

1. *Quality policy and strategy*: this is the process of deciding on:
   - quality goals and objectives;
   - quality level of products;
   - quality level of resources;
   - the quality system.

   Quality policy and strategy is an important function because it relates the quality objectives to the goals and strategy of the company by setting requirements on product

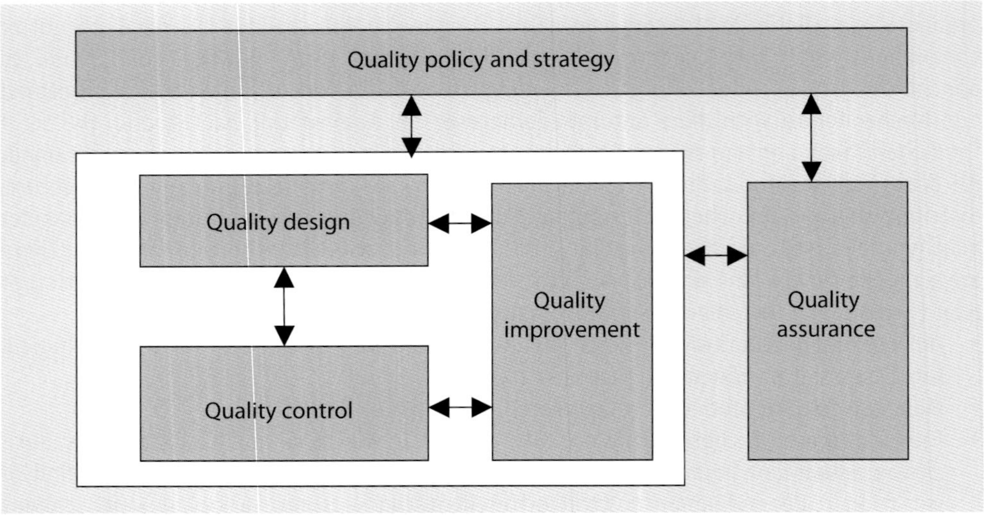

Figure 3.16. Food quality management functions.

quality levels and on the quality system. The importance of a quality policy and clearly establishing quality values in the company's strategy process is being increasingly acknowledged (Bergman and Klefsjö, 1994; Hellsten and Klefsjö, 2000). The objective of the function of quality policy and strategy is to determine long-term food quality goals and objectives, and how to achieve them by the quality system. It includes determining quality levels of products, and technological and human resources, and the type of quality system.

2. *Quality design* is the process of deciding on:
   - product and material specifications;
   - production process requirements;
   - quality of resources in terms of specifications.

   Quality design is important because high quality in complex products can only be achieved by starting at the source of the production cycle (Juran, 1990; Evans and Lindsay, 2005). Quality design aims at incorporating quality in food product and process design. It starts with specifying consumer and or customer demands and translating them into product and process specifications. Then requirements on resources are specified, e.g. suppliers, process and transport equipment, and people involved.
3. *Quality control* is the process of deciding on:
   - actual product and material quality;
   - actual production process quality;
   - actual quality of process environment (production rooms, storage facilities).

   Quality control is a basic activity of food quality management and the objective is to keep product properties, production processes, and human processes between certain acceptable tolerances. Quality control is the ongoing process of evaluating the performance of both technological and human processes and taking corrective action when necessary (Evans and Lindsay, 2005).
4. *Quality improvement* is the process of deciding on:
   - changes in product and material specification;

  - changes in production processes;
  - changes in quality of resources.

  Quality improvement is advocated by most quality philosophers as a cornerstone of quality management (Crosby, 1979; Ishikawa, 1985; Deming, 1986, 1993; Juran, 1990). Quality improvement involves a systematic approach to improving a system by paying attention to structural causes and solutions. It is aimed at changing people, processes, and resources in order to bring them to a higher level of quality performance and into 'new zones of control', which means working with lower tolerances (Juran, 1990).
5. *Quality assurance* is the process of deciding on:
  - requirements to the quality management system;
  - providing the quality management system related to organisation and technology;
  - actual performance of the quality management system;
  - necessary changes in the quality management system.

  Quality assurance is especially important for the food industry and agri-business since the typical characteristics of agri-food production (e.g. hidden safety risks, restricted shelf life, seasonal harvesting, heterogeneity of raw materials, complexity of supply chains) place great demands on assuring ultimate product quality, specifically on the intrinsic attribute of safety (Luning *et al.*, 2009). Quality assurance activities deal with setting requirements on the quality management system, evaluating its performance and organising the necessary changes. The objective is to control the quality management system, which includes all activities and decisions to realise quality, and to provide assurance to customers and consumers that quality requirements will be met (ISO, 2005). It should be noted that, in our view, each food quality management function has both a product-oriented and a resource-oriented part. In the following chapters on the different quality management function this distinction will be used frequently.

3.4.3 Food quality decisions

The food quality management (FQM) functions have different objectives and include different activities. However, all managerial functions are decision-making activities that activate the food production system (and thereby all technological functions) and the management system in a certain way. In this section, we have typified decisions on dynamics and conditioning of the food and human systems for each FQM function (Table 3.1).

For example, quality assurance deals with decisions on the design, implementation, and performance of the quality system (Luning and Marcelis, 2007):

- decisions on *food dynamics* concern identifying, measuring and evaluating which compounds or agents, and where in the process, are critical towards quality;
- decisions on *human dynamics* concern, for instance, requirements on the composition of quality system project team, and compliance behaviour towards rules and procedures related to critical quality points;
- decisions on *technological conditions* deal with assessing and validating the capability of equipment and processes related to critical points, selection and validation of process control parameters, and choice of monitoring systems; and
- decisions on *administrative conditions* concern the organisational position of the quality department, implementation of information systems for auditing purposes, and development of rules and procedures.

Table 3.1. Food quality decisions grid.

FQM-function	Decisions on			
	Food dynamics	**Technological conditions**	**Human dynamics**	**Administrative conditions**
Q-design	Quality standards and tolerances on product properties	Requirements on process parameters, equipment and buildings	Quality standards and tolerances on actions of people	Requirements on people's competencies, procedures, and information systems
Q-control	Out of tolerance of product properties and corrective action	Deviation from requirements above and corrective action	Out of tolerance of people actions and corrective action	Deviation from requirements above and corrective action
Q-improvement	Structural out of tolerance of properties and necessary modifications	Structural deviations from requirements above and necessary modifications	Structural out of tolerance of actions and necessary modifications	Structural deviations from requirements above and necessary modifications
Q-assurance	Critical properties, and critical points and parameters	Capability of processes, equipment and buildings at critical points, related monitoring systems	Critical decision and action points	Responsibilities and procedures in quality management system and auditing
Q-policy & strategy	Quality category of food products	Competence level of processes, equipment and buildings	Quality category of people's actions	Competence level of people, organisation, information system

The grid can be used to systematically analyse problems in the realisation of food quality from a broader perspective; it concerns more functions than just control and it explicitly acknowledges the complex food and human systems.

3.5 Factors influencing quality behaviour

Earlier we stated that the core activity in managerial processes is decision-making. So, all food quality management functions are dependent on the decision-making effectiveness of people involved. Decision-making processes are quite predictable on the one hand, e.g. programmed or routine decisions, but quite unpredictable on the other, e.g. non-programmed or innovative decisions. Administrative conditions reduce unpredictability by directing people's behaviour, and keeping it within certain limits (Luning and Marcelis, 2006). Nevertheless, different people react differently within the same administrative conditions, and individuals can react differently at different moments within the same conditions. Decision-making intentions depend on certain processes in people's minds, e.g. processes of perception and attitude forming. Perceptions and attitudes are not fixed, but they change slowly over time.

In short, human behaviour is dependent on human dynamics (i.e. variability in decisions due to factors such as variable perceptions, attitudes, or choice intentions), and administrative conditions (e.g. information systems and procedures) can influence these dynamics (see Figure 3.1). In this section they will be discussed in the context of food quality management.

3.5.1 Quality behaviour and leadership

Once people have made decisions about objectives and the ways to accomplish them, they turn them into action. People's behaviour is all these decisions and subsequent actions over the course of time. And when they restrict themselves to decisions and actions on quality, it is quality behaviour. In short, quality behaviour is how people act on quality related issues.

In his dissertation on quality behaviour, Gerats (1990) made an analysis based on the mental incongruent theory. This theory argues that in analysing behaviour, two conditions should be considered:
- *Disposition*, that is the employee's own disposition to behave in a certain direction.
- *Ability*, that is the objective opportunity to behave in a certain direction, i.e. the activity area.

Disposition refers to people's behavioural intentions. Ability refers specifically to physical conditions and personal skills, which enable people to really meet requirements. Disposition is the outcome of processes within people's minds, while ability is the result of organisational factors.

Gerats (1990) used this theory in his research into meat production and translated this into quality behaviour as shown in Figure 3.17.

The research was carried out in slaughterhouses and was focused on hygienic working behaviour. It was concluded that 60% of the workers did not comply with the conditions concerning disposition and ability. It was also concluded that the activity area (ability) for hygienic working behaviour was mainly limited by shortcomings in management with respect to hygiene, by low hygiene standards amongst workers, by low hygiene standards of first line supervisors, and by shortcomings in the hygiene facilities at the workplace. Disposition to hygienic working was mainly limited by the low knowledge level in the area of bacteriological contamination mechanisms, by the restricted social support from colleagues, by the lack of interest among supervisors in hygienic working practices, and by the limited opportunities for hygienic working. The research results indicated that in order to improve hygienic working behaviour, the activity arena should firstly be enlarged. Some important factors are briefly described below.

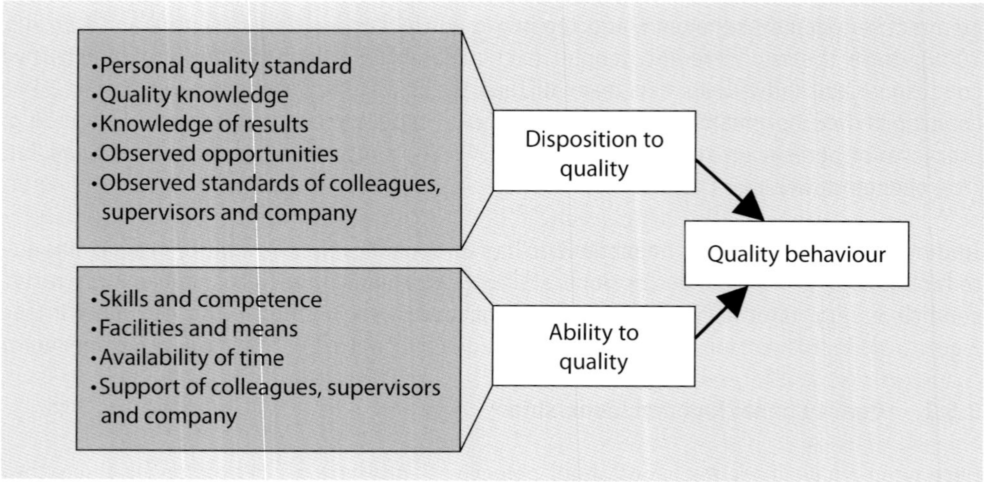

Figure 3.17. Conditions for quality behaviour (adapted from Gerats, 1990).

Knowledge, skills and competence

If is often said that quality begins with education and ends with education. All plant employees must first understand that quality equals conformance to requirements. Thus, the first requirement is to make certain that the supervisors and all the employees are trained to know what quality is, why quality has to be controlled, where to be concerned about quality, when to make changes and who is responsible (Gould and Gould, 2001). Training will be further elaborated on in Chapter 6 (quality improvement)

Availability of time and information

Clayton *et al.* (2002) identified barriers preventing food handlers from implementing food safety practices. Lack of time was the most quoted reason for failure to implement relevant practices, followed by lack of staff and a lack of resources. This could, in the food industry, be related to unreasonable demands made on employees by their employers. Some of the facilitators mentioned by respondents in order to carry out food safety actions can be divided into four groups: (a) personnel: more staff, recognition of problems by management, (b) time: less work, (c) design: more space, better design of work space, better location of sinks, (d) cost and resources: more cloths/towels, more/new equipment.

A second group of factors consists of providing the technical information. This and access to technical expertise make employees feel capable of solving any technical problems they encounter. In addition, they feel more competent to do their job and make decisions about how their job should be done.

Quality commitment in the company

According to Howard and Foster (1999), a necessary precondition for gaining employee commitment to quality might be leadership commitment to quality, or at least the perception of leadership commitment to quality. The previous statement is in agreement with Gerats' model, where the support of supervisors and the organisation will lead to a greater improvement in employees' quality behaviour. When employees perceive that management is committed to quality, they tend to be more committed to quality. When employees perceive strong organisational commitment to quality, employee work quality improves, and customer satisfaction increases. Therefore, gaining employee commitment to quality is critical, and that commitment might be gained by developing positive attitudes about management's commitment to quality.

A workplace climate which communicates respect for employee rights and needs will also have an impact on workers' commitment to the organisation and on quality behaviour. Showing respect for employee rights and needs and offering opportunities to fulfil employee needs through workplace arrangements are fundamental components of what has been referred to as 'quality of work life'.

Leadership

Behaviour is strongly influenced by leadership. To achieve goals, *leaders* work on aligning people, i.e. communicating the vision and developing the commitment to do it. Management in general achieves plans by controlling and problem solving, whereas leadership achieves its vision by motivating and inspiring (Schermerhorn, 2007; Dean and Evans, 1994).

Refering to the model of factors, influencing decision-making (Figure 3.4), it will be clear that the effectiveness of leadership depends not only on the personal characteristics of the leader, but also on the use of power, the communication process and the way of dealing with interests and conflicts.

Effective leadership creates clear values that respect the capabilities and requirements of employees and other company stakeholders. It sets high expectations for performance and performance improvement. It builds loyalty and teamwork based upon values and the pursuit of shared goals. It encourages and supports initiative and risk-taking of subordinates in the organisation. It avoids chains of command that require long decision paths.

Although quality management principles and practices may differ among firms and industries, there is unanimous agreement on the importance of leadership in implementing quality management. According to Juran, it cannot be delegated. Strong visionary leaders are an important element of a quality management approach (Ross, 1999).

Three different approaches to leadership, i.e. the trait, the behavioural and the contingency approaches, have tried to answer the question as to why some people perform very well as a leader, whereas others do not (Schermerhorn, 2007; Robbins and Coulter, 2007). The *trait approach* claims that effective leadership depends on the personal traits of the leader. However, researchers were unable to establish a defined profile of traits that consistently accounts for leadership success. The *behavioural approaches* tried to determine which leadership style worked best, i.e. the recurring pattern of behaviours exhibited by a leader. Blake and Mouton (1964) proposed a managerial grid based on the styles of 'concern for people' and 'concern for production' and concluded that it is best to combine both concerns.

The *contingency approaches* attempted to understand the conditions for leadership success in widely varying situations. A good example is the situational leadership theory of Hersey and Blanchard (Hersey and Blanchard, 1988; Schermerhorn, 2007), whose premise was that a leader's style should be contingent on subordinates' competence and commitment (Figure 3.18).

According to this theory the possible leadership styles that result from different combinations of task-oriented and relationship-oriented behaviours are as follows:

- Delegating: allowing the group to make and take responsibility for task decisions.
- Participating: emphasising shared ideas and participative decisions on task directions.
- Selling: explaining task directions in a supportive and persuasive way.
- Telling: giving specific task directions and closely supervising work.

Figure 3.18 shows follower characteristics from able to unable, and from willing to unwilling. Each of the leadership styles fits best with one of the four sets of these characteristics.

Another leadership model is the Transformational Leadership Theory. This is a model that fits in with modern quality management. According to this theory, leaders who wish to have a major impact on their organisation must take a long-term perspective. They have to stimulate their organisation intellectually, invest in training to develop individuals and groups, take some risks, promote a shared vision and values, and they should focus on customers and employees as individuals (Dean and Evans, 1994).

3.5.2 Human dynamics

Human dynamics can be summarised as variability in decision-making behaviour (Luning and Marcelis, 2006). Dynamic behaviour is caused by different factors.

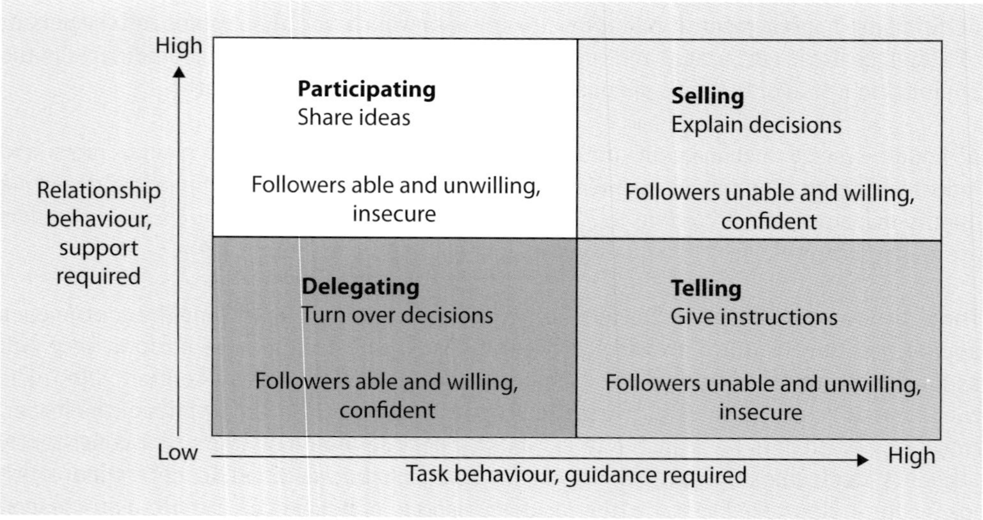

Figure 3.18. Hersey-Blanchard model of situational leadership.

Firstly, people have their own *characteristics*, and, when making decisions on quality, they have their own information, objectives, and interests. This is one reason why people show certain quality behaviour which is different in time and different from that of other people.

People are different because they have different *personalities*. An individual's personality is a unique combination of emotional, thought, and behavioural patterns that affect how a person reacts and interacts with others. Knowing someone´s personality helps us to understand why they behave the way they do.

Personality is most often described in terms of traits that a person exhibits (Robbins and Coulter, 2007). Research has shown that five basic personality dimensions encompass most of the significant variation in human personality (big five model):
1. Extraversion. The degree to which someone is sociable, talkative, and assertive.
2. Agreeableness. The degree to which someone is good-natured, cooperative, and trusting.
3. Conscientiousness. The degree to which someone is responsible, dependable, persistent, and achievement oriented.
4. Emotional stability. The degree to which someone is calm, enthusiastic, and secure (positive) or tense, nervous, depressed, and insecure (negative).
5. Openness to experience. The degree to which someone is imaginative, artistically sensitive, and intellectual.

Five other personality traits that have proven to be the most powerful in explaining individual behaviour in organisations are:
6. Locus of control. The degree to which people believe they control their own fate (internal locus of control, these people believe they can; external locus, these people believe they are controlled by outside forces).
7. Machiavellianism. The degree to which people are pragmatic, maintain emotional distance, and believe that ends justify means (e.g. use of power).
8. Self-esteem. An individual's degree of like or dislike for himself or herself.
9. Self-monitoring. An individual's ability to adjust his or her behaviour to external situational factors.
10. Risk taking. The degree to which someone is prepared to assume or avoid risk.

It is interesting to know that these dimensions may differ depending on culture. For instance, national cultures show differences in conscientiousness and locus of control (Robbins and Coulter, 2007).

A second reason for dynamic behaviour is formed by the *cognitive process of individual decision-making*. Figure 3.19 shows the different elements of this process, i.e. the processes of perception and attitude forming, and final decision-making, already mentioned in figure 3.1. They are always influencing decision-making by developing certain choice intentions, which show certain continuity, but always vary between people and over time. Additionally, learning processes and mental model development play a role. When people experience the effects of decision-making, learning processes arise. They include evaluation and analysis of the outcomes of decision-making, which results in modification of people's mental models. These, in turn, strongly influence perception and attitude forming. The different processes are shown in Figure 3.20, and are explained in more detail below.

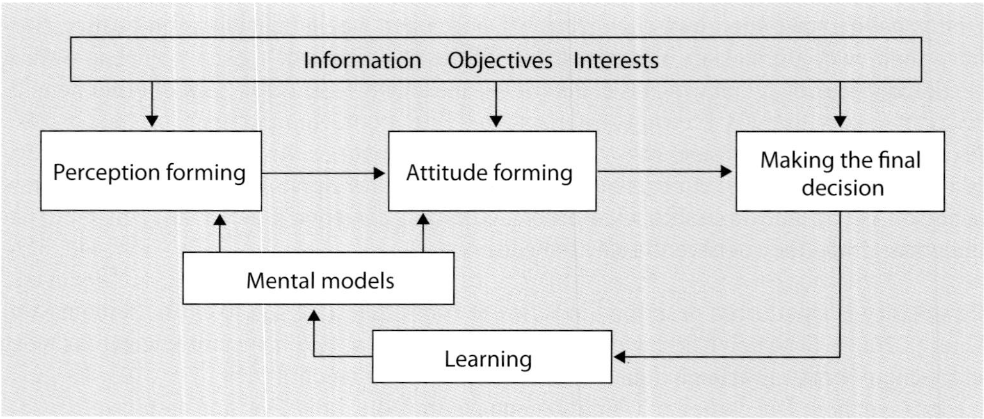

Figure 3.19. The cognitive process of individual decision-making.

Perception is a process by which individuals give meaning to their environment by organising and interpreting their sensory impressions. In Chapter 2 we discussed the sensory impressions, which are based on personal observation of the food product properties. In this section we also talk about observing other aspects of the environment, like production processes or other people's behaviour.

People use perceptual cues to make inferences and decisions about the state of their surrounding environment (Smith, Gilhooly and Walker, 2003). Cues are certain focus points, which serve as indicators about the circumstances of the environment. In the individual's mind mental representations of the situation are constructed, based on the information from these cues. So, the mind serves as a perceptual system, using selected information from eyes, ears and the other senses.

Research on perception consistently demonstrates that individuals may look at the same thing yet perceive it differently. It depends on the perceiver, the observed target, and the situation (Robbins and Coulter, 2007). Perceptions of people differ from perceptions of inanimate objects because we make inferences about the behaviour of people that we don't make about objects (Robbins and Coulter, 2007). Due to their dynamic character, food products have an intermediate position in this respect, because we perceive them as objects but they show certain behaviour.

During perception processes the quality or accuracy of a person's perception of the real world has a major impact on responses. Because people always make simplifications of the real world, perception distortions always occur. Examples of simple techniques that people use to simplify decision-making include:

- stereotyping: someone is judged on the basis of the perception of the group he or she belongs to;
- halo effects: only one attribute of a person is used to develop an overall impression; and
- assumed similarity: the perception of others is more influenced by the observer's own characteristics than by those of the person observed (Robbins and Coulter, 2007).

Reducing the negative effects of perception distortions is possible by adequate information supply and by changing individuals' mental models.

Attitudes are evaluative statements - either favourable or unfavourable - concerning objects, people, or events. They reflect how an individual feels about something. An attitude is made up of three components: a cognitive, an affective and a behavioural one. The cognitive component refers to the beliefs, opinions, knowledge, or information held by a person. The affective component of an attitude is the emotional or feeling part of an attitude. The behavioural component refers to an intention to behave in a certain way toward someone or something. Usually, the term attitude is only used to refer to the affective component.

Research has generally concluded that people seek consistency among their attitudes and between attitudes and behaviour (Robbins and Coulter, 2007). But unfortunately, we cannot exactly predict people's behaviour from their attitude. The reason is cognitive dissonance, which is any incompatibility or inconsistency between attitudes or between attitudes and behaviour. The theory about cognitive dissonance explains that behaviour can be different from the attitude when it is seen as important and thus acceptable, when people don't have enough influence on the situation, or when certain rewards let people accept the dissonance. As a conclusion, we can see that individuals change their behaviour depending on the situation in which they find themselves (Robbins and Coulter, 2007).

Attitudes result in behavioural intentions, which have an influence on *decision-making*. But individual behaviour also depends on the context of the choice environment. In the previous section about decision-making, it is stated that the availability of information and the existence of interests are major factors influencing the decision-making. Lack of information increases uncertainty, and existing interests result in constraints. Both uncertainty and constraints make some alternatives preferable, but allow others to be excluded.

In the ideal situation all the information needed is accessible so that people can take decisions with certainty. But it is nearly impossible in practice. Most organisations hold incomplete or ambiguous information.

In addition, during decision-making, interests can increasingly interfere with previous intentions. In this case interests are derived from both the organisation and the individuals. In many situations, the interests of organisation and individuals are conflicting. These conflicting interests may result in 'stressful situations for the decision-maker'.

Mental models are the unwritten assumptions and beliefs that people use when they think about a topic. Doyle and Ford (1998) came up with the following definition: 'a mental model of a dynamic system is a relatively enduring and accessible, but limited internal conceptual representation of an external system, whose structure maintains the perceived structure of that system'. These models determine what information people use and how it is interpreted. In order to understand the world and its phenomena, people construct mental models that are based on their prior knowledge, existing ideas or conceptions, and past experience. Mental models are useful or functional, because they allow people to make predictions or explain phenomena or events (Shepardson *et al.*, 2006).

People use more models and compare and select the most appropriate. But even when people select or construct mental models, they show satisfying behaviour, because they use as little information as possible in order to minimise the pressure on their working memory (Legrenzi *et al.*, 1993).

Learning is the process of acquiring and remembering ideas and concepts. People generate from their experience the concepts, rules, and principles that guide their behaviour in new situations, and they modify them to improve their effectiveness (Kolb, 1984).

There are different ways in which people can learn. According to Kolb (1984), four steps should be distinguished in the learning cycle: acting, valuing, thinking, and deciding. Acting includes executing tasks and acquiring experience. Valuing is aimed at identification and reflection of the effectiveness of the task execution. By thinking one tries to understand the routines and find alternatives and improvements. The last step is deciding and focuses on choosing an improved treatment for the next.

Both organisational and individual learning are linked to decision-making processes. Senge (1990) described a learning organisation as 'an organisation that by advantage of people, values and systems is able to continuously, change and improve its performance based upon the lesson of experience'. Individual learning is based on systematic evaluation and analysis of the outcome of decision-making processes, together with personal characteristics. Individuals face a very complex world and they do not possess all the knowledge to make the best decisions. Therefore people learn, because by learning they are better able to interpret their environment and know how to act accordingly (Denzau and North, 1994).

The third factor, causing dynamic behaviour, is formed by the administrative conditions. Despite their ability to reduce dynamics and create patterned behaviour, they do have different effects on people's decision-making, e.g. some do use available information, others neglect it. These conditions are explained in the next section.

3.5.3 Administrative conditions

Administrative conditions influence people regarding the decision-making processes in which they are involved by directing their behaviour. The administrative conditions are divided into three groups, based on the various ways that they influence decision-making (see Figure 3.1):
- people characteristics: commitment and competencies;
- information systems: information at the right moment and place, and systems to collect, organise and distribute data for decision support;
- organisation structure: division of tasks, responsibilities and authorities, and co-ordinating mechanisms like rules and procedures.

By changing these conditions it is possible to change the dynamic behaviour of people to a certain extent.

People characteristics

Commitment or motivation is the inner state that causes an individual to behave in a way that ensures the accomplishment of some goals. Since performance is a result of the behaviour of organisation members, motivating them is the key to reaching organisational goals. Over the course of time different motivation theories have been developed (Robbins and Coulter, 2007). Several theories explain motivation from contential factors, like:

- Maslow's Hierarchy of Needs, where higher needs like autonomy and recognition only become important when lower needs like food and safety are fulfilled.
- Herzberg's Two-Factor Theory, where hygiene factors eliminate job dissatisfaction, and motivators increase job satisfaction.
- McClelland's Three-Needs Theory, describing the need for achievement, the need for power and the need for affiliation.
- Other theories explain motivation from the processes people are involved in, such as:
- Goal-Setting Theory, proposing that specific goals increase performance and that difficult goals, when accepted, result in higher performance than do easy goals.
- Expectancy Theory, stating that an individual tends to act in a certain way based on the expectation that the act will be followed by a given outcome and on the attractiveness of that outcome to the individual.

The importance of commitment is certainly recognised in relation to quality. A firm asked: 'How do we enable workers to feel a natural sense of pride in their work and to be self-motivated?'. This approach to motivation is based on the assumption that employees inherently want to do a good job. In the quality-based management approach, the primary responsibility of management is to create and maintain a motivation system that supports each employee's natural ambition and pride in good workmanship (Ivancevich *et al.*, 1994; Deming, 1986; Crosby, 1979).

Competencies of people refer to individual knowledge, skills and abilities. They are acquired by education, experience and training. Education refers to basic education and school level education. Experience development can be influenced by mentoring and coaching, and by organisational measures like work selection, job rotation, co-worker and supervisor selection. Training is a planned process to modify knowledge (e.g. product knowledge, microbiology, technology), skills (e.g. supervisory skills, work skills, statistics), and abilities (e.g. presentation, team building, communication). Training is important both for workers and for the management staff of the organisation for three main reasons (Analoui, 1999):

- Training can increase the effectiveness by reducing learning time to achieve specified performance levels.
- Training can improve job performance levels and ultimately increases job satisfaction.
- Training ensures future succession of the needs of manpower.

Rapid technological advances in the workplace and the corporation's concern for profit in today's market place drive the emphasis on training employees. When properly used, training increases both the effectiveness and efficiency of employees (Swanson and Falkman, 1997). The final result of training is less formulated in terms of learning results but more increasingly in terms of work performance. When effectively executed and implemented, training also becomes a process of improving the decision-making behaviour related to quality issues.

Information systems

Information (see Figure 3.1) is also a critical enabler for quality management. Companies agree that information technology and information systems serve as keys to their quality success. Three categories of information are thought to be critical for quality management:

- *Operational information,* with the emphasis on process management, action plans and performance improvement.
- *Comparative information,* related to competitive position and best practices, both having an operational and strategic value.
- *Information that relates process management to business performance,* providing insight into cause/effect relationships (Ross, 1999).

Information technology has many effects, including improving performance and affecting organisational structure. It helps firms to retrieve information faster and more conveniently, which serves in decision-making. Moreover, it may help to improve organisational co-ordination. However, the rapid developments in information and communication technology (ICT) have resulted in an enormous amount of information sources and a tremendous volume of information, which are accessible through public (Internet) and private (Intranet) networks. As a consequence, the problem of technical access to information has been replaced by a problem of routing and selecting right and accurate information.

To gain access to accurate, timely, complete and relevant information, many organisations have turned to management information systems to manage the sometimes overwhelming amount of information. A management information system (MIS) organises past, present and projected data from both internal and external sources and processes them into usable information. This information is then available for people at all organisational levels. Because people have different needs, information systems must be able to organise data into usable and accessible formats (Schermerhorn, 2007).

A *decision support system* (DSS) uses special software to allow users to interact directly with a computer to help in making decisions for solving complex problems. With a decision support system (DSS), managers can anticipate the possible outcomes of alternative actions and so cope better with uncertainty. Decision support systems may incorporate artificial intelligence, which enables computers to 'think' as much as possible like the human mind. This has resulted in the development of expert systems, which mimic human decision-making processes by using a collection of thousands of 'if-then' rules to solve complex problems (Schermerhorn, 2007).

Organisational structure

The organisational structure is a formal system of relationships that both separates and integrates tasks. Separation of duties makes it clear who should do what. Integration of duties tells people how they should work together. Organisational structure includes four basic elements:

- *Specialisation*: identifying particular tasks and assigning responsibility for them to individuals or work groups who have been trained to do them.
- *Authority*: assignment of the right to decide and act to employees or work groups.
- *Standardisation*: developing procedures in such a way that employees perform their jobs in a uniform and consistent manner.
- *Co-ordination*: developing co-ordinating mechanism that integrate the activities performed by separate groups in an organisation, like rules and procedures, committees and 'liaison personnel' (Schermerhorn, 2007; Robbins and Coulter, 2007).

Specialisation and standardisation together are the basis for departmentalisation, i.e. subdividing work and assigning it to specialised groups within the organisation.

However, in order to achieve organisational objectives managers also need to co-ordinate people, projects and tasks. Besides using procedures, managers also should have a limited number of employees reporting directly to them (span of control) to ensure a clear and unbroken chain of command.

Authority is the right to act or make a decision and implies responsibility and accountability. Responsibility is an employee's obligation to perform assigned tasks; accountability is the expectation that each employee will accept credit or blame for results achieved in performing assigned tasks. Delegation of authority is the process by which managers assign the right to act and make decisions in certain areas to subordinates. They assign a task to a subordinate along with adequate authority to carry it out effectively (Hellriegel *et al.*, 2005).

It is worth noting in this context that behind every formal structure lies an *informal structure*. This is a 'shadow' organisation, which is the unofficial, but often critical, working relationship between organisational members. If the informal structure could be drawn, it would show who talks to and interacts regularly with whom, regardless of their formal titles and relationships. The lines of the informal structure would cut across levels and move from side to side. The informal organisation would show, for instance, people meeting for coffee, exercises in groups or friendly cliques. It is important to realise that no organisation can be fully understood without gaining insight into the informal structure as well as the formal one.

One should recognise that informal structures can be very helpful in accomplishing needed work. This is especially true during times of change when out-of-date formal structures may simply not provide the support people need to deal with new or unusual situations. Because it takes time to change or modify formal structures, this is a common situation. Through the emergent and spontaneous relationships of informal structures, people benefit by gaining access to interpersonal networks of emotional support and friendship that satisfy important social needs. They also benefit in task performance by being in personal contact with persons, who can help them get things done. In fact, what is known as informal learning is increasingly recognised as an important resource for organisational development. As a matter of fact, informal structures also have potential disadvantages. They exist outside the formal authority system, therefore the activities of the informal structures can sometimes work against the interests of the organisation as a whole. They can also be susceptible to rumour, carry inaccurate information, breed resistance to change and even divert work efforts from important objectives (Schermerhorn, 2007; Hellriegel *et al.*, 2005).

3.5.4 Organisational structure design

Designing an organisation implies that choices must be made in three different dimensions, i.e. complexity, centralisation and formalisation (Ivancevich, 1994; Schermerhorn, 2007).

The degree of *complexity* describes how much differentiation there is between structural units. Therefore, how specialised are the organisations' jobs and how geographically dispersed and large is the organisation. If the firm has many vertical levels, has a high degree of division of labour and is a multinational co-operation, then it is a complex organisation.

Lawrence and Lorsch (1967) stressed that increased differentiation among organisational subsystems creates the need for greater integration. However, integration becomes harder to achieve as differentiation increases (Schermerhorn, 2007).

Centralisation and *formalisation* are described in the administrative concepts (see section 3.2.5). Organisation concepts can be seen as the organisational answers to the requirements resulting from the administrative concepts (Figure 3.20). Concentrating decision-making authority in a few high-level positions is a reaction to centralisation. Decentralised responsibility and authority is the organisational answer to decentralisation. Rules and procedures, often accompanied by formalised information systems, are the organisational answer to formalisation as an administrative principle.

The more decision-making is decentralised, the more delegation is needed. Delegation is the process of distributing and entrusting work to other persons (Schermerhorn, 2007). There are three steps in delegation.

In step 1, the managers assign *responsibility* by carefully explaining the work or duties to someone else who is expected to do the work. This responsibility is an expectation that the other person will perform the assigned task.

In step 2 the manager grants *authority* to act. Along with the assigned task, the right to take necessary action (for example, to spend money or direct the work of others) is given to the other person. Authority is a right to act in ways needed to carry out the assigned tasks.

In step 3 the manager creates *accountability*. By accepting an assignment the person takes on a direct obligation to the manager to complete the job as agreed.

When formalisation is low (Robbins, 2000), employees have a great deal of freedom to exercise discretion in their work. The greater the standardisation, the less influence the employee has on how his or her work has to be done. Standardisation not only eliminates the possibility of employees engaging in alternative behaviours, it also removes the need for employees to even consider alternatives.

According to Figure 3.20, there are basic alternatives for the organisation structure. The *simple structure* is best characterised by what it is not, rather than what it is. It has a low degree of departmentalisation, wide spans of control and little formalisation. The simple structure is a flat organisation; it usually has only two or three vertical levels in

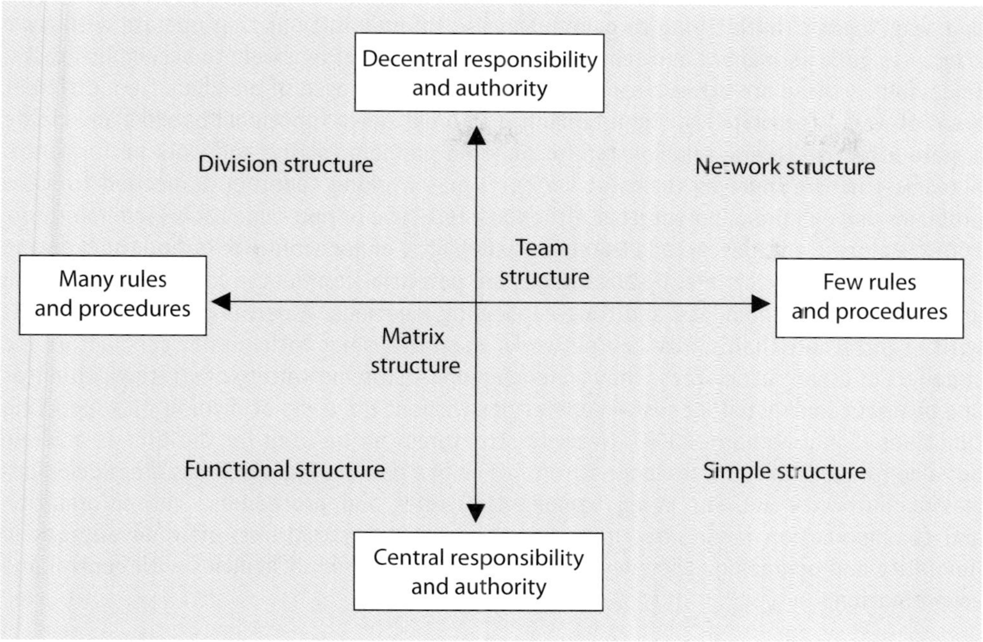

Figure 3.20. Organisation concepts.

hierarchy, a loose body of employees and one individual where decision-making authority is centralised. It is most widely practised in small businesses in which the manager and owner are one and the same. However, it is also the preferred structure in times of temporary crisis because it centralises control (Robbins and Coulter, 2007). In *functional structures* people with similar skills and performing similar tasks are formally grouped together, for instance in purchasing, production and marketing departments. Members of functional departments share technical expertise, interests and responsibilities. The key point is that members of each function work within their areas of expertise. Major advantages of the functional structure include efficient use of functional resources and a high-quality technical problem-solving capacity. The major disadvantages are lack of communication and co-ordination between different functions and loss of the total system perspective. The *divisional structure* groups people who work on the same product or process, serve similar customers, and/or are located in the same area or geographical region. Divisional structures are common in complex organisations that have multiple and differentiated products and services and/or operate in different competitive environments (Schermerhorn, 2007). The major advantages are product and customer focus and co-ordination, whilst the major disadvantage lies in increasing costs through the duplication of resources. The *matrix structure* is a combination of the functional and the divisional structure. In fact, it is an attempt to capitalise on the advantages of the two structures by using permanent cross-functional teams to integrate functional expertise with a divisional focus. Workers in a matrix structure belong to at least two formal groups at the same time - a functional group and a product or project team. They also report to two bosses - one within the function and the other within the team (Schermerhorn, 2007). The two-boss system is a major disadvantage as it is susceptible to power struggles between functional supervisors and team leaders, both trying to gain authority. In organisations that operate with *team structures* both permanent and temporary teams are used extensively to accomplish tasks. Importantly, these are often cross-functional teams composed of members from different areas of work responsibility. The intention is to break down functional barriers and create a more effective lateral relationship for ongoing problem solving and work performance. A team structure involves teams of various types working together as needed to solve problems and explore opportunities, either on a full-time or part-time basis (Schermerhorn, 2007). Major advantages include the speed and quality of decision-making and the breaking down of functional barriers. Disadvantages are potential conflicts in loyalty to both team and functional assignments, and time-consuming meetings. A *network structure* operates with a central core that is linked by 'networks' of relationships with outside contractors and suppliers of essential services. They are engaged in a shifting variety of strategic alliances and business contracts that sustain operations without the costs of 'owning' all supporting functions (Schermerhorn, 2007). Network structures can be used for the firm as a whole but also for parts of it, for instance a front office in a new market which has great decision-making authority without being hampered by rules and procedures. New information and communication technology, together with network structures, offer advantages in flexibility and operating efficiency. The disadvantages lie in difficulties with control and co-ordination.

In Chapter 6 on quality improvement, the team concept will be discussed, including self-managing teams as the most far-reaching concept. In Chapter 4 (quality design), the matrix structure and the team concept will be illustrated by discussing the role of project teams in new product development.

Quality department

In a typical organisation, *line authority* belongs to managers who have the formal authority to direct and control immediate subordinates who perform activities essential to achieving organisational objectives. Line authority thus flows through the primary chain of command. On the other hand, those with *staff authority* direct and control subordinates who support line activities through advice, recommendations, research and technical expertise. Staff managers and employees support line functions with specialised services and information. Line and staff conflicts can arise due to differences in roles and personal characteristics and differences in formal and informal authority.

With respect to quality control and assurance, most companies have their own quality department. Typical functional organisations organise quality knowledge in a quality control department, which is responsible for quality control of products and services. This type of quality control department will have all the disadvantages typical of the functional organisation structure (e.g. no connection with customers, no improvement opportunities, lack of responsibility, etc.).

In many companies, however, each individual within the firm is responsible for assessing and improving the quality of the process for which he or she is responsible. Because some managers may lack the technical expertise required for performing the necessary statistical tests or data analysis, technical specialists, usually in the quality department (Q-department), assist in the management in these tasks. Nevertheless, it must be emphasised that a firm's quality department cannot assure quality in the firm. Its proper role is to provide guidance and support for the firm's total effort toward this goal (Evans and Lindsay, 2005).

Tasks which can be conducted by the Q-department include:
- quality assurance and control by design, improvement and control of the quality control processes;
- collecting data and providing information (feedback) with regard to quality performance;
- assessing process analysis to establish effective and efficient quality control, such as HACCP;
- providing support to other departments in trouble-shooting, in setting complaints or recall procedures;
- keeping up the knowledge base with regard to quality management;
- providing support in improvement programmes, such as process and product improvement;
- providing support in defining and setting a quality policy.

Q-departments can be structured in different ways:
- *Product structure:* a structure which is based on the expertise of certain products. For each product there is a separate quality department. The advantage is the total approach with input of a variety of disciplines. A disadvantage is the splitting of expertise.
- *Expertise or functional structure:* a structure which is based on the similarity of expertise. Departments include microbiology, chemistry and process technology. The advantage is centralising similar expertise in one section. A disadvantage is that sections may not have any knowledge of other processes.
- *System structure:* a structure which is based on a process approach. For example, departments receive control, product control, and quality planning. The advantage is the bringing together of interrelated expertise that forms a whole unit. A disadvantage is that it may become less effective in problem-solving if feedback from other units is poor.

As discussed, the Q-department can be structured in several ways, but can also be positioned in several ways within the firm. Figure 3.21 shows several positions within the functional organisation structure.

The QA-department can be *staffed as an advisory department* for the management board. This position suggests that quality is of importance, that it has a major advisory function and also a great influence on the management board. However, it is not qualified to take major decisions.

The QA-department can also be *positioned equally* alongside other departments, such as production, procurement and marketing. This implies that the QA-department has the same influence on issues as other departments.

Finally, it can be positioned as *staff under the production department*. In this case it is an advisory department for production and its influence on issues is limited to this department. The advantages in this case lie in the acquirement of detailed knowledge of the products and processes and the commitment of staff employees to their 'own' production process.

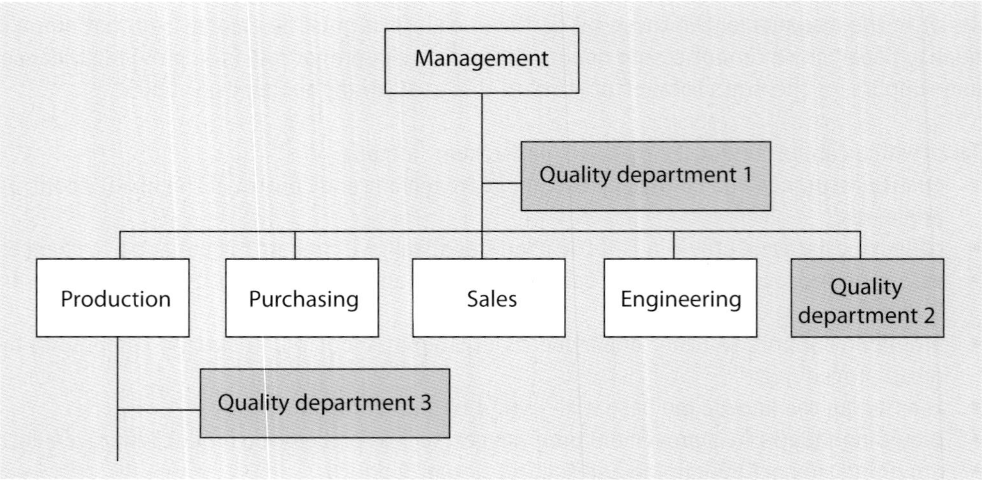

Figure 3.21. Position of the quality department within the organisation.

Organisation structure and quality management

From the viewpoint of quality management typical requirements are put on the organisational structure. When choosing the level of centralisation, the discussion is how much individual responsibility and authority is effective. When talking about formalisation, the discussion is about compliance with procedures, which makes procedures less effective than often expected.

Considering the organisation structures, Dean and Evans (1994) mentioned several inadequacies of the functional structure:

- It separates employees from customers because it tends to insulate employees from learning about customer expectations and their degree of satisfaction with the service or product the firm is providing.
- It inhibits process improvement because no organisational unit has control over a whole process, although most processes involve a large number of functions.
- It often has a separate function for quality; this may send a message to the rest of the organisation that there is a group dedicated to quality, so it is not their responsibility.

Dean and Evans suggest internal customer-supplier relations, team-based organisation, reduction of hierarchy and a high-level steering committee as good conditions for quality management. From that viewpoint, they consider decentralisation and low formalisation as favourable conditions for quality. Basically they talk about an organic design for the organisation, instead of a mechanistic design (Figure 3.22).

A *mechanistic design* is highly bureaucratic with centralised authority, many rules and procedures, a clear-cut division of labour, narrow spans of control and formal co-ordination. An *organic design* is decentralised with fewer rules and procedures, more open divisions of labour, wide spans of control and more personal co-ordination (Schermerhorn, 2007).

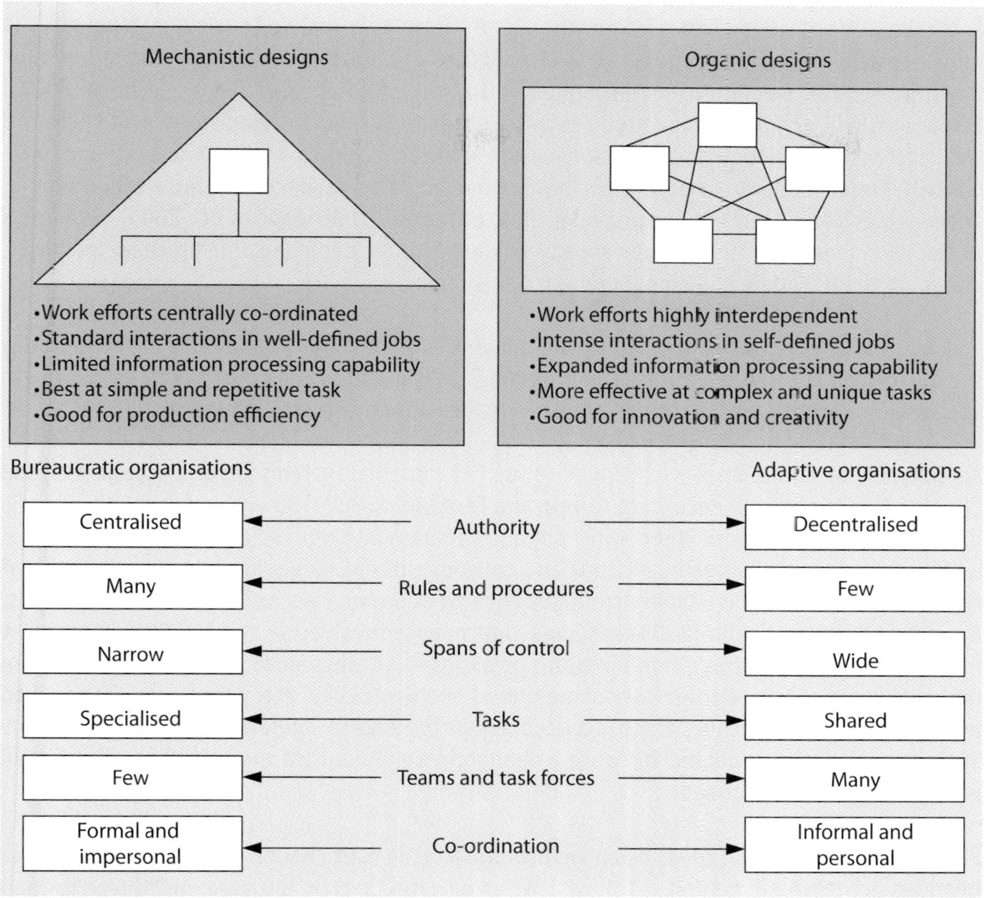

Figure 3.22. Mechanistic and organic organisation designs.

In Figure 3.20 the mechanistic design could be placed in the bottom left, and the organic design in the top right.

Organic designs work well for organisations facing dynamic environments that require flexibility to deal with changing conditions. Therefore, for quality management these designs are favourable, although high requirements in food safety and reliability at the same time ask for a more mechanistic design.

The concept of Total Quality Management (TQM, see Chapter 8) emphasises self-control, autonomy and creativity among employees and requires more active co-operation rather than just compliance (Ross, 1999). So, it favours organic designs which fit in with the principles of empowerment.

Empowerment is the process through which managers enable and help others to gain power and achieve influence. Effective leaders not only accept participation, but also empower others. They know that when people feel powerful, they are more willing to make decisions and take actions needed to perform their jobs. They also realise that to gain power it is not necessary for others to give it up. The success of an organisation may depend on how much power can be mobilised throughout all ranks of employees (Schermerhorn, 2007). Employees will acquire responsibilities and authority for making decisions, and therefore will act according to those decisions, thus gaining a sense of commitment and self-control. Empowerment satisfies basic human needs for achievement, a sense of belonging, and enables team members to increase their potential (Hellriegel *et al.*, 2005). Examples of decisions made in fully empowered teams are budget decisions, hiring team members, choosing team leaders and compensation decisions.

Procedures are often used in quality management, e.g. most quality systems (see Chapter 7) are basically a set of procedures. The problem is that people's *compliance with procedures* is variable. As a consequence, the effect of formalisation in this way is limited.

As an example, compliance with HACCP (one of the quality systems in food production, see Chapter 7) is explained. According to Vela and Fernandez (2003), an individual's behaviour is dependent on knowledge. When knowledge (awareness and understanding) is present, this will lead to a direct change in attitude and consequently a change in behaviour. So, each of these factors can cause barriers for adherence to behaviour according to the procedures. Azanza and Zamora-Luna (2005) assumed that people involved in the HACCP system must first become aware of the guidelines, and be exposed at least more than once to it, before understanding it. Once they fully comprehend the principles, the people are assumed to develop the right attitude which may subsequently lead to agreement and commitment. Such an attitude is expected to cause a change in behaviour through adoption and finally adherence to the guidelines.

Panisello and Quantick (2001) describe how difficult it is to change the attitude of people involved in the HACCP system as follows: 'When new HACCP principles are introduced to food processors, they do not see the need to change to a system that involves them in what they consider too complicated and beyond the capabilities of their staff. They take the view that there is no justification in changing their current procedures when those procedures have worked well for years and enabled them to produce "safe" food products'.

3.6 Quality management from a chain perspective

In the food production chain a far-reaching integration with customers and suppliers will go on. The traditional value-chain will in many situations be transformed into value-creating networks, where the strict borders between suppliers and customers will become unclear. Business will become knowledge intense. In this situation a great deal of trust between the interested parties is of the utmost importance (Bergman and Klefsjö, 1994, Simchi-Levi *et al.*, 2000).

In this section a restrictive overview is presented of quality-related chain topics, including supply chain management and customer-supplier relationships.

3.6.1 Supply chain management

Food and agribusiness firms, in general, are confronted with rapidly changing markets, new technologies and almost worldwide competition. A retailer purchasing internationally perishable foodstuffs needs reliable partners. And because of legislation on liability and food safety, only joint investments and co-operation can effectuate the development and introduction of new products. New distribution techniques enable large-scale production and frequent deliveries, but can increase the risk of being out of stock. In agriculture, technological developments have contributed to specialisation and large-scale production, but have also increased the need for and availability of capital. These developments have affected thinking about quality, which exceeds the individual organisation and requires a supply chain approach (Zuurbier *et al.*, 1996). Supply chain is defined then as all stages of the entire process from raw materials to final product.

The forces, which can be considered as the drivers for the supply chain management paradigm, can be divided into four groups:

1. Force of *consumer demand*, which concerns demands on processing methods (e.g. animal welfare, additives, radiation), food safety (free from chemical, microbiological and physical hazards), food quality (e.g. freshness, convenience, taste and variety) and environment.
2. Force of *productivity and technology*, which concerns the efficiency and effectiveness of processes (e.g. information technology, logistics technology, biotechnology, information and communication technology, measuring and monitoring technology).
3. Force of *government*, which concerns regulations and policies (e.g. shift of responsibility from government to business community, shift from domestic market protection to international market access, changes in traditional forms of subsidies, trend towards increased harmonisation).
4. Force of *resources*, which concerns demands on capital, resources and labour (e.g. economies of scale and size, resource and labour sustainability).

As a consequence, the world of farming, food manufacturing, distribution and retailing is becoming a dynamic and complex business, which requires management at the supply chain level. Specific market and production characteristics of food and agribusiness are additional motives for this approach. They include (Zuurbier *et al.*, 1996):

- restricted shelf life of produce;
- variability of quality and quantity of supplies of farm-based inputs;
- differences in lead time between successive stages;
- stabilisation of consumption of many food products;
- increased consumer awareness concerning both product and processing method;
- degrading intrinsic quality (intrinsic quality of raw material is highest attainable quality for fresh products); and
- availability of capital.

The restricted shelf life of many products puts great demands on the duration and conditions of storage, processing and transportation at all stages of the supply chain. Therefore, the existence of assured markets is very important to suppliers of perishable products. Moreover, capital intensive processing facilities put heavy demands on a continuous flow of supplied inputs to buyers. Differences in lead time between stages require efforts to match these stages. For example, pigs need time to grow, to reach optimal productivity and they cannot be stored alive. Meat production is a process that will inherently lead to a wide range of final products. In addition, some products cannot be made in isolation. For example, hams cannot be produced without pork chops. As a consequence, it is not always possible to produce exactly what is demanded because of side-products inherent to the required product. This complicates the supply process.

3.6.2 Customer-supplier relationships

From the quality management perspective, every company is a part of many long chains of suppliers and customers. One implication of this concept is that a company must focus on both their immediate and subsequent customers in the chain. Food companies, for instance, work hard to satisfy the needs of both the people who use their products, and the retailers that sell them, labelling the former 'consumers' and the latter 'customers'. Companies should also try to establish similar productive relationships with their suppliers to that with their customers.

By developing partnerships, customers and suppliers can build relationships that will help them to satisfy their shared customers further along the customer-supplier chain. The idea of creating mutually beneficial relationships with both customers and suppliers is a major difference compared to the traditional approach to customer and supplier relationships. In the traditional approach, self-interested competitors negotiate against each other, to maximise their slice of the pie at the expense of others. In modern quality management approaches, the focus is on expanding the pie rather than arguing over its division, thus developing a win-win relationship based on trust (Dean and Evans, 1994).

Very often the decision to co-operate is inspired by the existence of mutual interdependencies (Figure 3.23). Several interdependencies are at issue, e.g. technical or technological knowledge and social interdependencies (Kamann, 1989; Zuurbier *et al.*, 1996).

Activities of actors are linked within a partnership. They are continued in the scope of efficiency, in such a way that interdependencies will be developed and/or will be strengthened. The structure and organisation of a partnership will act as a control mechanism, which makes some changes easier and others more difficult. Besides the mutual interdependencies, the success of a partnership is also affected by:

1. *Bounded rationality of actors*. Firms are restricted in obtaining and processing all available information.
2. *Opportunist behaviour of actors*. Firms may take advantage of their position and impose on others by providing incomplete or incorrect information deliberately, which is also affected by the uncertainty and complexity of the business environment and degree of market concentration.
3. *Rigidity of organisational structures*. Within partnerships routine structures may be developed, which impede organisational learning. Firms may withhold or manipulate information to retain acquired positions.
4. *Balance of power between actors*. Relations can vary from symmetric to asymmetric. A relation is asymmetric if control is unbalanced. These types of relations can be found in

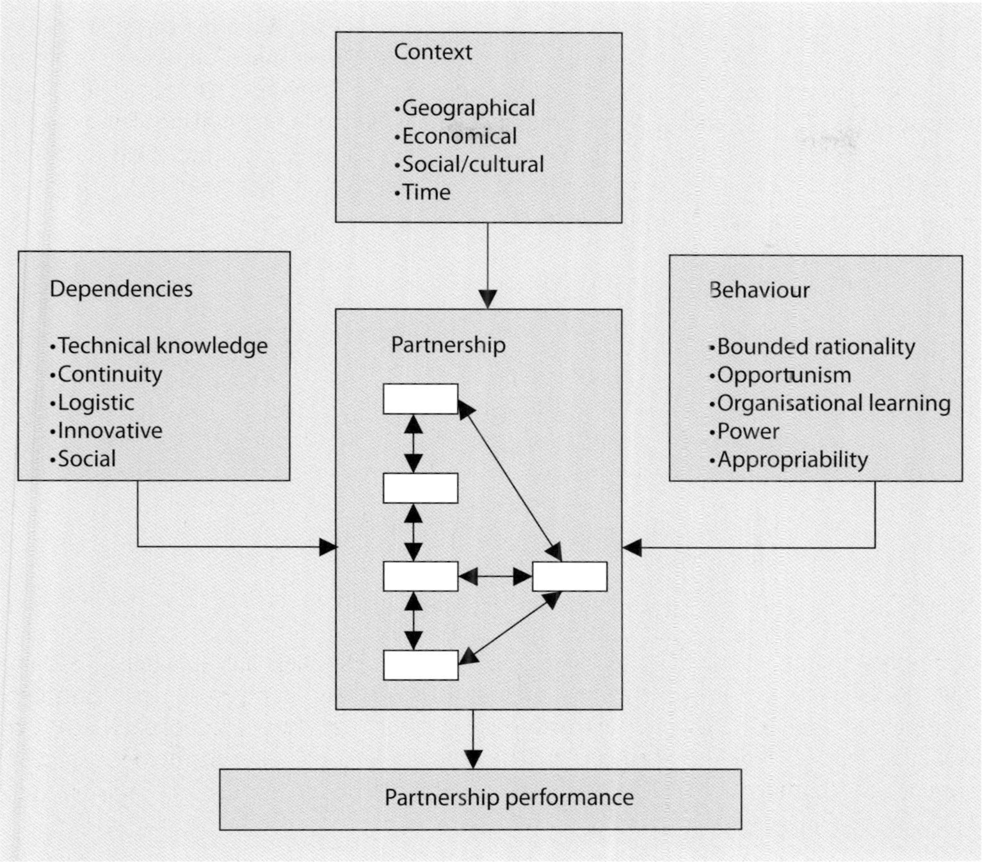

Figure 3.23. Factors influencing partnership performance.

a dual production system, where a dominant firm ties down a group of dependent firms. As a consequence, new partnerships may be developed to create countervailing power.

5. *Appropriation of resources*. The inability to account for costs and benefits to each of the actors may hamper the development or duration of partnerships.

Partnerships will only last if they are able to meet the key factors of successful partnerships. Changes often occur during economic recession in order to maintain continuity and control of acquired positions. Functional, strategic and political issues should also be involved. If common objectives are split up and co-ordinating actors lose their influence, then the partnership might deteriorate.

Chapter 4
Quality design

This chapter focuses on relevant technological and managerial aspects that should be considered to achieve quality in food products and process design. Firstly, the principle and general steps of the design process will be outlined (sections 4.1 and 4.2). Subsequently, typical variables that influence quality in food products and process design will be explained (section 4.3). The importance of customer orientation is described in the next section (4.4). Then, tools and methods that can be used to support the design process are briefly described (section 4.5). Typical managerial topics with respect to quality design include cross-functional design and project management. These topics are discussed in detail (section 4.6). In the last section the typical situation in the food industry with respect to quality design is briefly described and considered from a techno-managerial perspective (section 4.7).

4.1 Introduction

Nowadays, companies in agribusiness and food industry are increasingly focused on product design and updating design at very short intervals to stay competitive. Almost all firms increase the speed with which they design products as well as increase the quality of those products. Nevertheless, in order to stay competitive, costs of production should be low and principles of mass production are still important.

According to Juran (1992) and Evans and Lindsay (2005), high quality can only be achieved by starting at the source of the production cycle. In other words, quality should be built in at the design stage of products and corresponding manufacturing processes. Quality is generally summarised as 'meeting or exceeding customer expectations'. Therefore, it is essential to incorporate consumer requirements and expectations, also called 'voice of the customer', into the design. Moreover, minimal costs, optimal productivity, minimal environmental impact of product and production, and compliance with legal requirements should also be guaranteed.

In Figure 4.1 the quality design function is described in its main elements. The design process includes the activities wherein quality design objectives are translated into product and process requirements, and into requirements for resources that will be used for production of the product. This translation is realised by designing quality into the product and process specifications, and by setting requirements on resources, like suppliers, process and transport equipment, and people to be involved.

In the design process (Figure 4.1) product variables have to be taken into account, e.g. safety aspects, decay processes in the product, and packaging-product interactions. Process variables also play a role, e.g. temperature conditions, process parameters and their tolerances, and organisation of the production (degree of automation, production series, etc.). Typical tools and methods support the design of food products. Quality Function Deployment (QFD) is a method to translate customer requirements into technical product specifications. Other tools and methods, e.g. product tests, predictive models and

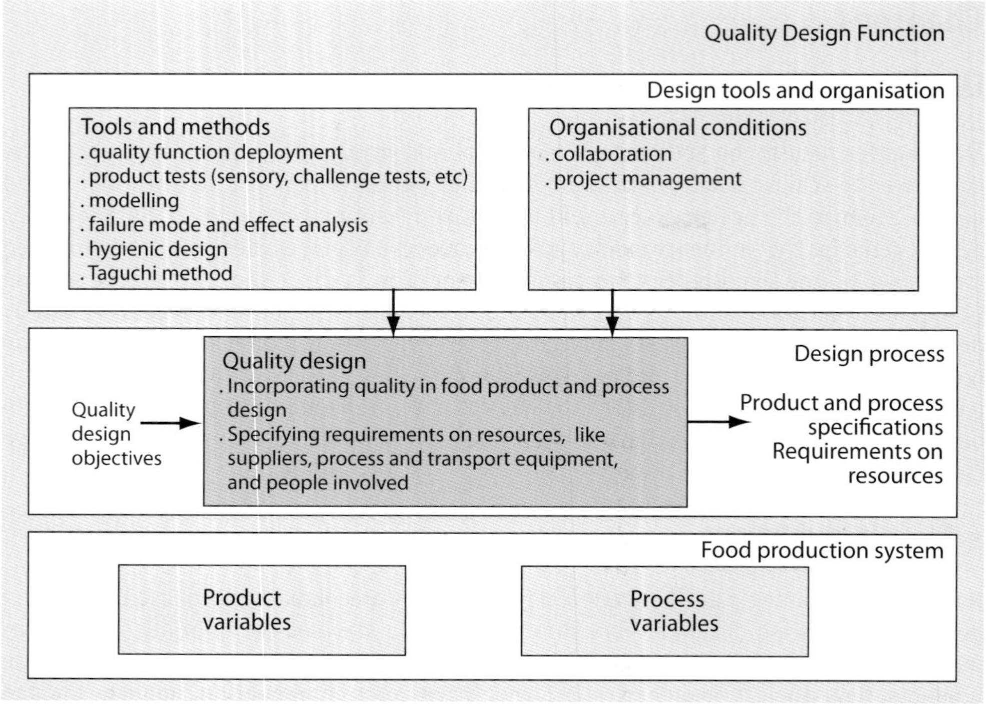

Figure 4.1 Quality design function.

Taguchi methods, also contribute to effective product and process development. Failure Mode and Effect Analysis and Hygienic Design principles are especially meant to support and or improve process design. Typical organisational conditions for design processes are collaboration, due to the high number of different departments involved in the design process, and project management, due to the project character of the quality process. In the next sections these factors will be explained and discussed.

4.2 The quality design process

Quality design deals with building quality into products and processes.

> *The core mechanism of quality design is building in quality into the product and production process by setting specifications on product properties, process parameters, and technological and administrative conditions in a way that final quality will meet or exceed customer/consumer needs and legal requirements.*

In this section the activities in the quality design process, as well as the decisions involved, are described.

4.2.1 Quality design activities

Quality design starts with specifying consumer and or customer demands and translating them into product and process specifications. Then requirements on resources are specified, e.g. suppliers, process and transport equipment, and people to be involved.

Design activities

It may be clear that quality design is closely related to and broadly built into the activities of different departments. The marketing/sales department receives customer and consumer information and translates it into product requirements; the research and development department develops new knowledge, product concepts and prototypes; the engineering department develops new production technologies; human resource management develops training programmes; and the production department improves production processes by analysis and modification. So, quality design needs well-organised decision-making processes, and is typically dependent for its success on communication and collaboration within the company. Different product development procedures are being used in practice and various ideas exist about how it should be conducted. Some consider product development as a procedure of sequential steps, e.g. where screening is depicted as a second step after idea generation. Others suggest a more iterative process, where screening occurs at each step of product development (e.g. Urban and Hauser, 1993; Fuller 2004; Beckley *et al.*, 2007). The product is becoming more and more refined using feedback information at each step, from technologists, marketing, finances, production, quality control, etc. In traditional product development much attention was paid to testing of concepts and prototypes, while neglecting previous steps of product development, such as collecting information on consumer demands. However, this approach does not integrate marketing science, quality management and product design concurrently. Additionally, many companies make a fatal mistake by starting with preconceived ideas of what they *think* consumers will want (Fuller, 2004).

Product development can be described as all activities that are carried out to translate the voice of the customer into a product that can be effectively manufactured. Process design involves not only design of process equipment, but also planning of physical facilities and development of information and control systems required for manufacturing of the product (Evans and Lindsay, 2005). Ideally, the design process should be an interactive process of product development and process design. Products-in-development set requirements on the process design, whereas process requirements limit or facilitate product development opportunities.

Design process

Although there is no univocal procedure for product development and process design, typical steps that are commonly included (e.g. Fuller 2004) are depicted in Figure 4.2. The different phases are described below:

- The design procedure commonly starts with an *inventory* of consumer needs and demands and defining the target group (i.e. for whom the product will be designed). Boundary restrictions, such as the company's strategy and objectives, legislative regulations, and

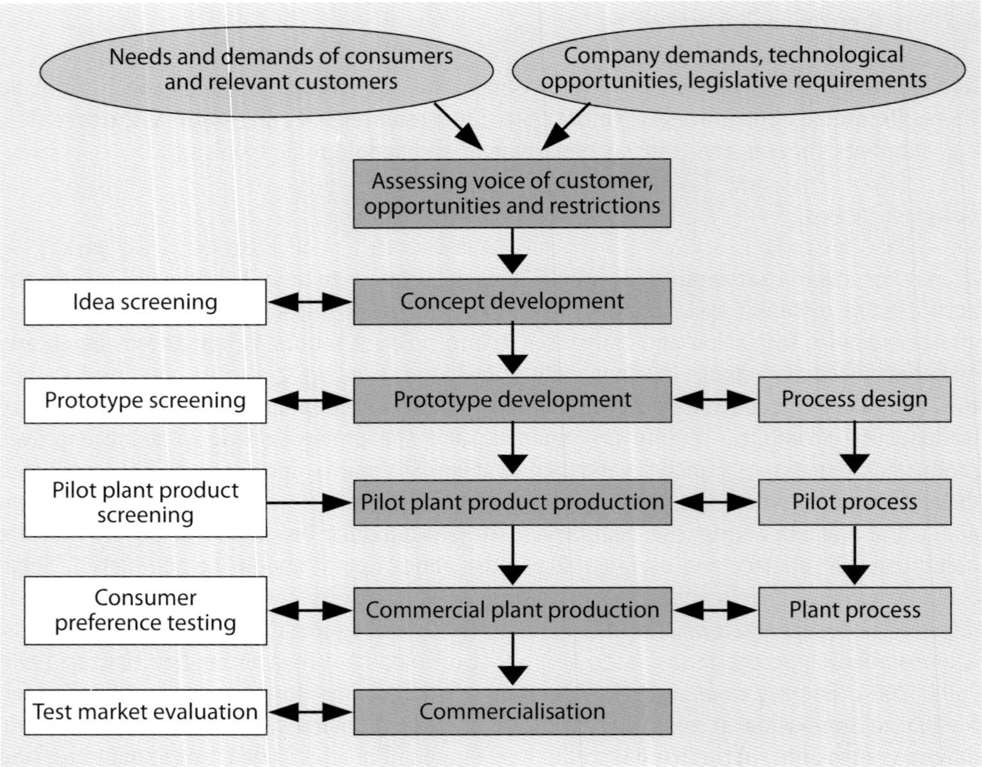

Figure 4.2. Quality design process and activities.

technological opportunities should also be considered. At this stage the 'voice of the consumer' and the 'voice of the company' must be clarified.

- In the *concept phase* product ideas are generated and screened by the product development team based on the inventory of consumer needs and demands and the boundary restrictions. Then product ideas are embodied in a product statement or concept, which includes all attributes as desired by the defined 'target group' of consumers (i.e. the expected end users), but also by related customers such as distributors (who have to transport the products) and retailers (who have to sell the products).
- In the *prototype phase*, (food) technologists and (increasingly) also culinary experts usually start with a basic recipe. These recipes can be obtained from various sources such as (international) cook books, traditional recipes, ingredient suppliers and or analytical data of competitor's products. Subsequently, the basic recipes are modified with ingredients to obtain and maintain desired product attributes. After the preparation of several prototypes, the quality attributes must be evaluated and screened by objective testing. Objective testing includes quantitative measuring of the quality attributes, e.g. how sweet, how fruity, how creamy it tastes, or how smooth or crispy is the texture, etc. Prototype development provides input for process engineers to design the process, for instance which handling (e.g. cutting, mixing), preservation techniques (e.g. heating, drying, packaging), and process conditions are required. On the other hand process engineers must give feedback to the product developers on design restrictions.

- In the *pilot production phase* the product is made in a real process line. Commonly, the packing process is included at this stage. Typical aspects that should be considered in the pilot production phase include:
  - Determination of product shelf life with respect to safety and sensory features.
  - Subjective evaluation of quality attributes of pilot products by preferences testing, where consumer panels must state which product they prefer, which taste they find most acceptable, and which product they would buy, etc.
  - Finding reliable sources for raw materials, spices, food additives, and packaging materials at acceptable costs and the whole year round.
  - Establishing other resources such as required equipment and tools.
  - Assessing potential hazards in the production process, with respect to food safety, and how to measure and control them (i.e. quality control).

  The pilot production phase provides input for in-process specifications. All process treatments that can influence quality attributes must have specified limits and tolerances. Typical process parameters in the agri-food industry that require specifications are:
  - Time-temperature conditions of the product at each step in the production process. For example, the type of heat treatment, flow rates, and product viscosity affect the required time-temperature conditions.
  - Shearing action on the product (e.g. during mixing action, due to flow rate, depending on pipe diameters, etc.).
  - Pressure changes on the product (e.g. during mixing action, due to flow rate, depending on pipe diameters, etc.).
- In the *commercial plant production*, the product is made under the actual process conditions and production circumstances (i.e. with all equipment and people involved in production). In this phase the following aspects should be included: labelling, actual packaging (primary and secondary), transportation conditions, organisation of operators and staff involved, quality control systems and information support systems, plant maintenance and plant sanitation programmes.

  Furthermore, recipes must be modified to make them suitable for commercial production. Problems due to up-scaling from a pilot to a commercial plant should be kept to a minimum. The actual quality attributes of the manufactured product must be evaluated by consumer preference tests. Moreover, real shelf life and product safety must be confirmed by shelf life tests. Finally, the definitive product and process specifications and product price should be established.
- The *commercialisation phase* commonly starts with a test market. Test markets are introductions of the new product(s) into carefully selected regions for a variety of geographical, marketing, and company reasons. Basic questions should be where (is the product peculiar to the test market?), when (e.g. which season) and how (e.g. advertising, promotions) to introduce. Test market results are often misinterpreted because it is difficult to fully understand consumer behaviour and the role of competition. After evaluation of the test market, the product is launched in the real market place and sales are monitored.

Critical aspects of design process

It is important to realise that product development and process design are not once-in-a-lifetime events but should be a major and continuous process in the organisation. As a matter of fact, product innovations in food production have become an important competitive tool to maintain or increase market share in a saturated market. Major motives for innovation of food products include (Fuller, 2004; Jongen and Meulenberg, 2005; Beckley *et al.*, 2007):

- *Decrease in product life cycles*. Each product has a life cycle, which consists of the following stages: (1) product introduction, (2) increasing of sales by new consumers and repeated buyers, (3) declining growth because of market saturation, (4) stable sales period and, finally, (5) decline due to competitive products. The time between introduction and product decline has been reduced in the last decades, therefore continuous innovation is required.
- *Market conditions*, which force companies to change their product concepts, in order to anticipate the constantly changing requirements. For example, several forms of retail outlet have recently been developed, such as internet shopping, shop-around-the-clock, and convenience food shops at fuel stations, to anticipate changing market demands.
- *Company's policy and strategy*, which may adopt an aggressive growth programme to comply with long-term goals. The strategy may therefore stimulate product development.
- *Technological developments*, which may offer opportunities for new product concepts, which were not possible before. For examples, current developments in new mild preservation techniques (e.g. High Intensity Light or HIL, and Ultra High Pressure or UHP) might offer new opportunities for producing healthy products with a fresh character.
- *Various external sources*, such as legislative changes, health programmes, and/or agricultural policy, which may necessitate product innovation. For example, new packaging regulations stimulated the food and packaging industry to redesign their packaging concepts.

In addition, design processes have not always been (and are still not) very successful. For years and years, product development and process design have been technology-driven. This meant that technologists invented new products, which were subsequently launched on the market supported by advertising. After a test period, consumer perception of the new product was evaluated and, if necessary, the product concept was modified or removed. This common approach to product development appeared to be expensive and time-consuming, and it has often resulted in a high rate of product failures in the last few decades. Up to 75-80% of all product ideas never reached the market stage and 30-40% of the finally launched products did not survive the test market (Cooper, 2001), so that fewer than 1% of the original ideas ever resulted in a successful product in the market.

Several studies have attempted to assess the causes of failure and success in product development. Van Trijp and Steenkamp (2005) summarised determinants for success and failure of new products, which were analysed in different studies. They distinguished determinants related to consumers, organisational aspects, competition, and marketing:

- Key factors to success related to consumers are proper defining of product concepts and adding a higher value to a product for target consumers.
- Proper structuring of the design process, appropriate embedding in the organisation and commitment from top management are typical organisational key factors for success in product development.
- Competition-related factors include competitive activity and turbulence, size and attractiveness of the market.

- Accurate assessment of market size and desired positioning are typical marketing-related determinants.

The dynamic environment in which companies operate require flexibility on the one hand, because one should anticipate the changing environment by putting great effort into product and process design to survive in the competitive market. On the other hand the high demands on quality, especially food safety (see also Chapter 7), require efforts in the area of strict control and assurance so that products comply with safety and quality requirements.

4.2.2 Quality design decisions

The food quality decisions grid (Figure 3.17) shows which decisions can be seen as the result of quality design activities, i.e. decisions on:
- Quality standards and tolerances on product properties (influencing food dynamics), e.g. final product specifications, and specifications for incoming materials.
- Requirements on process parameters, equipment and buildings (influencing technological conditions), e.g. required storage temperature, required pasteurisation time and temperature, and hygiene level of equipment.
- Quality standards and tolerances on actions of people (influencing human dynamics), e.g. required employee behaviour for certain production activities, and required compliance with procedures.
- Requirements on people's competencies, procedures, and information systems (influencing administrative conditions), e.g. required procedures and instructions in the production, required training and education, and required information for product tracing.

These different decisions show the various ways in which food quality can be influenced in the design phase. This is especially important because design decisions build up the starting points and thus are boundaries for control and improvement decisions.

4.3 Variables in product and process design

In order to develop products that meet or exceed customer expectations it is necessary to know which quality attributes and dimensions determine the perception and judgement of overall quality by target consumers and/or customers, and which technological conditions and or manual operations can influence these attributes during food production (see also Chapter 2). As demonstrated by the extended quality triangle (Chapter 1), product quality consists of various dimensions such as (physical) product quality, quantity/availability, and costs. Service, flexibility and reliability are additional factors that are important for the overall quality of production as realised by the firm. Product and process design variables mainly affect physical product quality, whereby a distinction between intrinsic and extrinsic attributes should be made. Typical intrinsic attributes are safety, health, and sensory features, whereas the character of the production system and the environmental impact of products are typical extrinsic attributes. Product costs and availability, service, flexibility, and reliability of the firm can be greatly influenced by management activities.

In this section some important product and process variables, relevant to product development and process design of food products. are described. We start, however, with a short overview of the different types of new product.

4.3.1 New food products

The continuous process of developing new products is of major importance for the food industry. Whereas years ago products used to stay in the market for years, nowadays the life cycle of products has been considerably reduced due to rapidly changing consumer demands and other environmental factors, e.g. legislation, competitive markets, etc. (Jongen and Meulenberg, 2005). New products can range from a new packaging concept for an existing product to a completely new product. New products can be divided into different categories; each puts special demands on the design process.

Fuller (2004) distinguished various categories of products and described their major characteristics with respect to the design process:

- *Line extensions* are new variants of an established (food) product. Typical examples are new flavours for existing products (e.g. flavoured potato chips) or new tastes in a family of products (e.g. new taste in series of dried soups). The design process of these products can be characterised by relatively little effort and development time, small changes in the manufacturing process, little change in marketing strategy and minor impact on storage and/or handling techniques.
- *Repositioned existing products* are current products that are again promoted in order to reposition the product. For example, with the increased interest in health products, a margarine brand was repositioned because of its naturally high levels of tocopherol (vitamin E). The development time for repositioned products can be minimal and only the marketing department need make an effort to capitalise on the niche market.
- *New form of existing products* refers to existing products, which are then transformed into another form (e.g. soluble, granulated, concentrated, spreadable, dried or frozen). An illustration of this is the transformation of frozen ready-to-eat pizzas into pizzas that can be stored in the refrigerator; the latter have a much shorter shelf life. These types of products may require an extensive development time because the physical properties of the product change drastically. In the case of the pizza it also affected storage and distribution conditions. A frequent cause of failure is that the consumer does not always appreciate the change. The change must have an added value.
- *Reformulation of existing products* involves existing products with a new formula. There can be several reasons for reformulation like reducing the cost of ingredients, limited availability of raw materials, or new ingredients with improved characteristics. Examples are products with better colour, improved flavour, more fibre, less fat, etc. The design process for reformulation is usually inexpensive and needs a relatively short development time. However, for food products minor changes in composition might have major consequences, e.g. for their chemical or microbial shelf life. Therefore, a profound knowledge of food technology is required to estimate and anticipate possible changes.
- *New packaging of existing products* involves existing products with a new packaging concept. For example, the technique of modified atmosphere packaging created many opportunities for extension of product shelf life. With respect to the design process,

new packaging concepts may require expensive packaging equipment, but products may also be reformulated for the new application (e.g. microwave packaging).
- *Innovative product* is defined as a product resulting from changes made to an existing product other than described above. The changes must have an added value. The design process is generally longer and more expensive, when more product changes are required (higher innovation degree). Marketing may also be costly because consumers may have to be educated about the novelty. However, in some cases time and cost of innovation were relatively little, e.g. assembling of frozen vegetables and a frozen pastry on a tray resulted in a very successful innovative ready-to-cook product.
- *Creative product* is described as a newly introduced product, i.e. the never-seen-before product. Typical examples are novel protein foods (or meat substitutes) which are based on vegetable proteins. Creative products commonly require extensive product development, tend to be costly (major marketing efforts, new equipment), and have a high risk of failure. However, once a product appears to be successful, imitators will rapidly flood the market.

Before starting the development process, it is thus necessary to assess the character of the new product because it can have different consequences for various development activities, e.g. technological investments, marketing efforts, and or packaging technology, which may be required.

Furthermore, one should recognise that nowadays companies commonly develop an assortment of new products, e.g. breakfast fruit drinks with various aromas.

4.3.2 Product design variables

Food products are complex, perishable, living systems and are therefore distinctively different from non-perishable products. Product properties are even different under the same conditions and thus variable and not easily predictable. When designing new food products and ingredients one should take into account the typical characteristics of food products that require special attention, here called *product design variables*. Some major product design variables are explained below and summarised in Figure 4.3.

Instability quality attributes

A typical aspect of food materials is the phenomenon of the decay process, which starts immediately after harvesting (i.e. fresh materials) and processing (i.e. manufactured products). Several decay processes, e.g. chemical, microbial, physiological, enzymatic and physical reactions, can cause a decline in intrinsic quality attributes (e.g. changes in colour, odour, taste, texture, and appearance) (see Chapter 2). These dynamic processes are a major source of variation in quality attributes of final products for consumption.

In product development, therefore, it must be established which and how quality attributes can be stabilised, in order to retain the desirable intrinsic attributes for the required shelf life. Stabilisation can be obtained by modifying the composition (e.g. regulating water activity (a_w), acidity (pH), and using additives), optimising process circumstances (e.g. appropriate time-temperature conditions), and selecting suitable packaging concepts (see

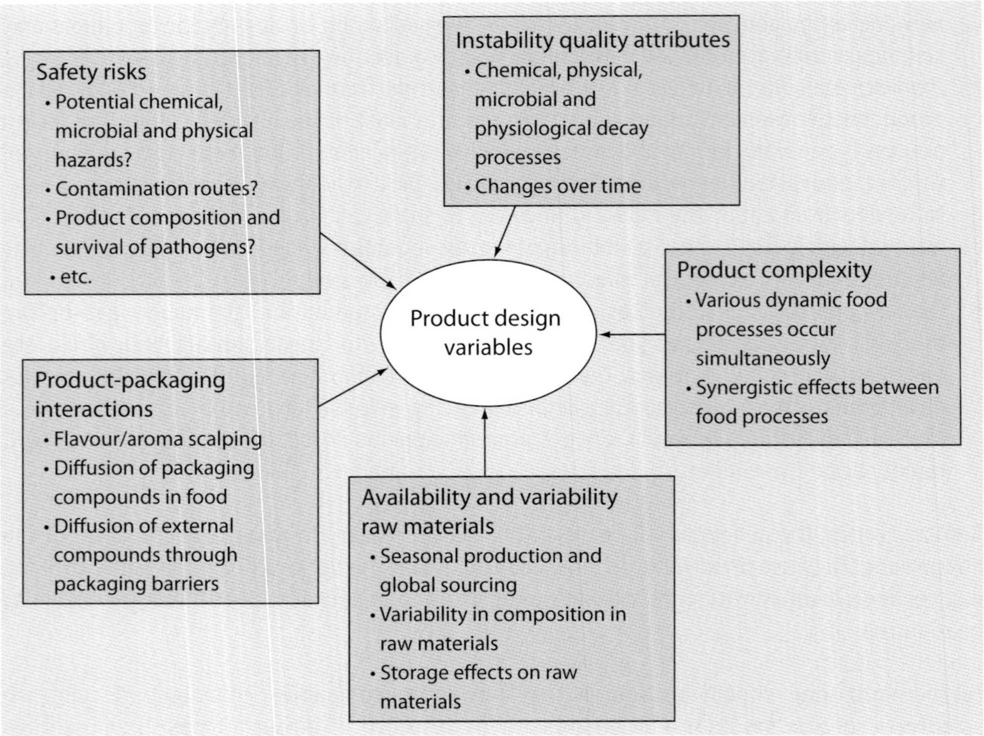

Figure 4.3. Product design variables.

Chapter 2). Examination of the stability of intrinsic attributes should start as early in the development process as possible.

The possibilities and limitations for stabilising quality attributes can have a great impact on further processes in the food chain, such as choice of distribution channel. For example, fresh export beef is commonly transported by air but the development of innovative packaging concepts has enabled a much cheaper alternative, i.e. transport by ship. Distribution conditions can also be driven by microbial stability properties, e.g. chilled food must be transported through a cooled chain to maintain the safe microbial state of the product. Therefore, distribution requirements should also be considered at an early stage of product development because they can be determined by the restricted stability properties of the product.

Safety risks

Safety risk refers to the probability, during any step of food production, of a hazard contaminating food, which if ingested by a consumer can cause a health disorder. Product safety can be affected by the growth of pathogen micro-organisms, by toxic compounds (e.g. microbial toxins, environmental contaminants or pesticide residues), by foreign objects and allergenic compounds (see Chapter 2). These agents can contaminate (new) products via various routes. Safety risk must therefore be assessed and critically evaluated along each step of product development from concept to final product. All factors that can significantly influence the safety of the new product should be considered, such as possible initial (bacterial and or chemical) contamination of raw materials, compositional factors (e.g. below pH 4.5 there is seldom growth of pathogens), appropriateness of process parameters (e.g. correct core temperature of product), hygienic operation, and hygienic process design. The latter three factors need to be considered during process design as well.

Moreover, a safe product, produced by a company, can be turned into an unsafe product due to e.g. inaccurate distribution or storage conditions or misuse by consumers (e.g. too high a temperature for a longer period can favour the growth of pathogen bacterial spores). Therefore, during product development there should be an examination of the risk of potential hazards that might occur due to production and/or due to external factors (e.g. distribution, consumers) (see also Chapter 7).

Product complexity

Raw materials, fresh produce, and manufactured products in particular are usually very complex, due to the wide range of components and food processes occurring in the composite food matrix. Various chemical, enzymatic, physical, and physiological reactions may occur simultaneously and may even interfere with one another. As a consequence, slight modifications in the recipe (e.g. reformulation of existing product composition) may result in surprising changes. For example, the water activity (a_w) of dried powders was increased to prevent lipid oxidation. However, the product turned brown due to the Maillard reaction, which was favoured by the increased water activity (see Chapter 2). In such situations the optimum conditions need to be found. For product development, one therefore requires profound knowledge about reactions and synergistic effects in food matrices, which need to be tested and evaluated as early as possible in the design process.

Product-packaging interactions

Another aspect typical for food products is product-packaging interaction, which may negatively influence intrinsic quality attributes. Interactions can be divided into three major mechanisms: (1) diffusion of food compounds into packaging material, also called flavour/aroma scalping; (2) diffusion of foreign compounds through the packaging materials into the product; and (3) diffusion of packaging compounds into the product. A typical example of diffusion of food compounds is the diffusion of flavours into polymeric packaging materials, which can modify product flavour and in some cases even changes packaging properties. Diffusion of foreign compounds through the material depends on the packaging characteristics. Each material has its specific diffusion properties; some are permeable for water vapour whereas others are more permeable for oxygen. Improper use of packaging materials may result in rapid loss of intrinsic attributes (e.g. softening of cookies due to water diffusion or desiccation of soft products). In addition, packaging compounds, such as colour additives, plasticisers and heavy metals, may diffuse into the food product and so affect flavour or even product safety. It should be noted that migration limits for packaging compounds are enacted in the European Directive (90/128/EEC) for food contact materials.

In product development it is of essential importance that the choice of packaging concept should hinge on product properties and legislative restrictions as early in the development procedure as possible, and not at the end of the design process.

Availability and variability of raw materials

In contrast to most other industries, the availability of raw materials in the food industry is usually seasonal. As a consequence, raw materials must be obtained from other countries or stored under proper conditions to close the gap between harvest and use. For example, some fruits and vegetables can be stored frozen or dried until processing, whereas others can be stored under controlled conditions to extend their shelf life. However, storage conditions and different origins can affect the composition of raw materials, and thereby influence the required intrinsic quality attributes of the final product.

In product development, it is important to critically evaluate and/or analyse the effects of changes in raw materials on final quality attributes. A typical approach to this problem is the development of blends with raw materials from different origins. This principle is commonly applied in the coffee and tea industry.

Furthermore, variability in the composition of raw materials (e.g. due to weather and seasonal conditions) may put special demands on the recipe. Variability of raw materials should be considered when establishing product specifications and tolerances. The contribution of the variation in composition of raw materials to the variation in the end product may differ, depending on recipe and process conditions. This variation must be considered during product development as well.

Last but not least, *extrinsic attributes*, like characteristics of production methods to be used and potential environmental impact of selected food production circumstances, contribute to quality perception of food products as well. Therefore, the possible effects of production system characteristics (e.g. use of genetically modified food, origin of raw materials, animal welfare, environmentally sound) on consumer perception should also be evaluated and/or considered during product development, e.g. by acceptance studies. For food, the environmental impact is commonly reflected in the selected packaging concepts (e.g. minimal packaging, recycleable, reusable), which must be considered along with the functional aspects of packaging (e.g. diffusion properties).

More recently, the *cradle-to-cradle* concept has attracted attention. The current approach to sustainable product development is focused on reducing negative side effects. An attempt is made to reduce the environmental burden by selecting less demanding raw materials, more efficient production systems and more effective recycling processes. The basic philosophy behind cradle-to-cradle is that all used materials in one product can be used again after their 'life' in another product without loss of quality. This vision was developed and described by William McDonough and Michael Braungart in their book 'Cradle to cradle: remaking the way we make things' (McDonough and Braungart, 2002). Currently, the cradle-to-cradle concept has not yet been applied in food supply and production systems.

4.3.3 Process design variables

As with product design, when designing (or modifying) the food plant production with its specific equipment and tools necessary to make the new products, it is essential to recognise that the specific food characteristics have an impact on process design as well, here called the *process design variables*. It is important to know that process design is not only restricted to process engineering, but also includes design of control and information systems, organisation of production, and organising material and stock control (Evans and Lindsay, 2007). For each of these aspects we describe typical aspects of food which should be considered when designing a process. An overview of the process variables is given in Figure 4.4.

An important activity in process design is *process engineering*, which includes the design (or modification) of equipment and tools together with the parameter and tolerance design, in order to produce the new (modified) product (assortment). Typical process engineering questions include: which sequence of processing steps and associated equipment are necessary, which process parameters affect the product (e.g. temperature, shear pressure), what are the optimal conditions for process parameters, what are acceptable tolerances, is process equipment capable of producing within these parameters, and are there specific hygienic design requirements?

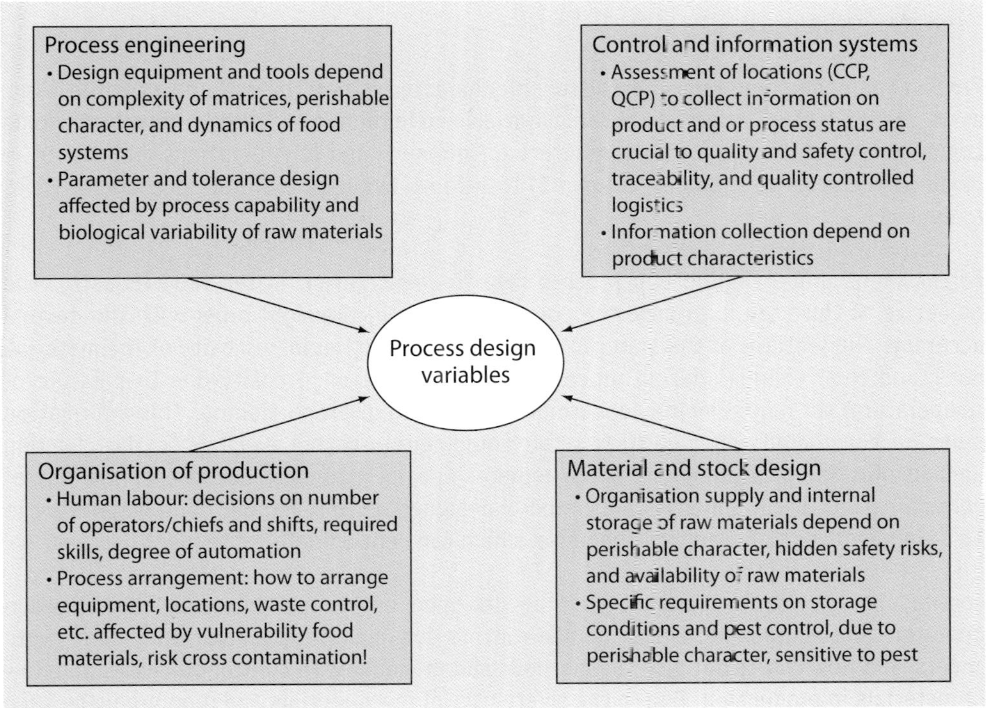

Figure 4.4. Process design variables.

When designing (or modifying) equipment for the food industry process, the following food characteristics should be considered on the basis of their consequences for the process design:

- *Complexity of matrixes:* raw materials and ingredients used for food production often contain many components; they have complex matrices and there may be (many) interactions between components. The materials cannot therefore usually be considered as a homogenous mass, which has consequences for the design of processing equipment.
- *Perishable character:* raw materials and ingredients are commonly perishable (due, for example, to spoilage bacteria) but may also contain pathogens (hidden safety risks), which put special demands on accurate processing, and hygienic design of process equipment (see also section 4.5.3). For example, adequate cleaning is of major importance for the safety of the products made.
- *Dynamic systems:* raw materials are dynamic systems, where various biochemical (enzymatic), physical (e.g. migration), and physiological food processes continue to take place. Depending on the type of end product that has to be made from the same raw material, different processing conditions might be necessary due to these dynamic food processes. For example, the heat load during pasteurisation of milk for the production of low-pasteurised milk is much more critical than for milk powder production; for low-pasteurised milk some enzyme activity must remain after the heating process. Moreover, the properties of the raw materials may change due to the process conditions applied. For example, fluctuation in pressure and/or temperature may have considerable consequences for the viscosity of the product.
- *Biological variance (variability)*: agri-food materials are variable in nature and variations of 10% are common for such materials, a fact which needs to be considered when setting system tolerances.

Process engineers should consider the specific characteristics of the end products to be made, as well as the typical properties of raw materials and ingredients. Table 4.1 represents some common process steps in food manufacturing (column 1) and considerations that should be made by the process engineer (column 3) to achieve the required process characteristics (column 2).

For example, proportioning is a process step frequently used in the food industry. The objective of this step is to dose the correct amount, in the right order with the desired accuracy. The features of the material (e.g. a powder or a liquid, viscosity of the material, etc.), and their changes during processing (e.g. change of viscosity due to pressure or temperature) set requirements concerning the method of proportioning. This information must be considered when designing the equipment. Another example is the cleaning and disinfecting of equipment, which requires special attention because of the risk of contamination. If equipment is not properly designed, then remaining food materials can be a source of spoilage and toxic bacteria, which can contaminate subsequent batches.

Control and information systems must be designed as a communication tool between operators and equipment. These include routine documents like operation sheets, route sheets, and flow charts but also the methods and procedures to control processes and flow of materials in production. Due to the *safety risks* of raw materials and food products, such control and information systems should be designed and/or adapted based on profound risk assessment, and identification of *critical points to control safety* (CCP) and *critical quality points* (CQP, i.e. locations where loss of control results in unacceptable variation in desired quality attributes). Depending on *the nature of the food products* one has to search for information that can be easily monitored and is representative of the actual product status. For example, T-t conditions can be monitored to ensure that pathogens and spoilage bacteria are adequately inactivated, whereas product samples and subsequent microbial analyses are commonly only executed to verify whether the set process parameters are properly functioning.

Collecting appropriate information about product and process status at these critical points is essential to control quality but also to support traceability systems. Moreover, a more recent concept, *Quality Controlled Logistics* (QCL), requires the collection of information on physical chemical properties in combination with models to predict quality and collection of information about availability/quantity and demands in the support market. The combined information helps better quality and logistic decisions to be made, thereby improving the quality of the products at consumption and attaining better chain revenues (e.g. lower costs, less waste).

Organisation of production involves decisions about the use of *labour*, such as how many operators and production chiefs, how many shifts, the degree of automation, required skills for operation and/or use of automated systems. It also involves decisions on the *arrangement of processes*, e.g. process-based (similar processes in the same area) or product-based (entire production lines are dedicated to specific products). Consideration of the specific food product and process characteristics places specific requirements on the organisation of production. Typical requirements are that raw and processed materials may not come into contact with each other, that *vulnerable products* require differentiated and cooled production zones, that operators must wear special clothing and comply with strict hygiene procedures to prevent *cross-contamination* (e.g. in case of assembly of ready-to-eat meals, or food for special target groups). Moreover, the *perishable character* of food materials put special requirements on waste control as well.

Material and stock design includes the organisation of the supply and internal storage of raw materials, ingredients (e.g. spices, additives, etc.) and other supporting materials (e.g. packaging). Here, it is extremely important to consider the *perishable character* and *hidden safety* risks of many raw materials for food production. These characteristics put demands on the design and control of the storage facilities. Moreover, the character of raw materials and food products (energy rich, proteins, carbohydrates) make them *sensitive to pests*, which puts demands on pest control. The *variable supply* (due to seasonal harvesting, global sourcing) also put demands on material and stock control. In this respect it is worth mentioning the concept of JIT. Just-in-time (JIT) is a Japanese philosophy on material management, which aims at keeping the inventory as low as possible, because perishable products and internal transports cost a lot of money (Evans and Lindsay, 2007). There is also a link here with the concept of quality controlled logistics (QCL), which affects the way material and stock can be managed, and thereby impacts the design of storage houses and facilities.

Table 4.1. Common process steps in agri-food manufacturing and related points of attention for process engineering.

Process step	Process characteristics
Proportioning of ingredients/ materials	To dose the *correct amount* (weight, volume, density, etc.) In the *right order* With desired *accuracy*
Composing ingredient mix, e.g. mixing, homogenisation, etc.	To obtain *equal (homogenous) distribution* of ingredients To obtain *desired structure* and *rheological properties*
Transportation of (intermediate) product(s)	To get the product at the *right place* At the *right time* And in a *proper condition*
Separation processes	*Separation* of *desired streams* of (raw) materials
Chemical conversions	To *change* the *molecular composition* e.g. backing and cooking, hydrogenation of lipids, or hydrolyses of proteins
Heating and cooling processes	To obtain the product at *accurate temperature* For the *right amount of time* To obtain the *suitable heat transfer process* (i.e. correct energy load per time unit)
Filling and packaging process	To *accurately* dose the product To obtain an *integral* (no leaks), *safely* (e.g. no glass pieces) and *correctly* packaged product
Cleaning and disinfection	Equipment must be made *microbiologically, physically* and *chemically clean*

Engineering aspects

Product properties (e.g. type of product, homogeneity) and required accuracy determine the possibilities for proportioning e.g.:
- liquids can be dosed by volume or time
- powders and granular materials can be dosed by volume, weight or time

Effects of shear tension on bulk properties put demands on the design of the stirring (or homogenisation, mixing) equipment, like:
- shear velocity of e.g. the stirrer or screw
- shape and size of stirring blades or screw

Product characteristics determine the choice of transportation facilities:
- transport via pipes e.g. for fluid products
- transport via conveyor belt e.g. for solid products

And the transport conditions like:
- shear tension during transport, which are affected by e.g. pump capacity, pipe dimensions
- temperature changes during transport
- residence time and residence time distribution (RTD)

Material properties and their structure determine the level of separation and the separation mechanism, i.e.:
- macro level, often mechanical separation by e.g. sieving, air circulation
- meso level, depending on particle size separation can be performed by e.g. filtration, hydro cyclones
- molecular level, depending on material is the separation based on e.g. thermodynamic properties (boiling point), molecular weight, etc.

Depending on the product characteristics and the required conversion the following should be considered
- process conditions i.e. temperature, residence time and residence time distribution, which are affected by type of heat transfer, equipment design (e.g. length, diameter) and pump capacity
- type of (bio)catalyst required

Product characteristics (e.g. product tolerance for local over-heating) during manufacturing constrains process design. Important engineering aspects include:
- type of heat transfer i.e. direct/indirect and medium used (steam, air, cool water, etc.)
- residence-time and residence time distribution, which are affected by equipment design (e.g. length, diameter) and pump capacity

Product characteristics require certain packaging properties and both put demands on the filling/packaging equipment, such as:
- velocity of filling and packing process
- accuracy and unit (weight, volume) of dose measuring device
- seal conditions

Details in section hygienic design

Designing processes for food applications requires knowledge of - or input from - food technologists, food microbiology, and food process engineering to be able to take appropriate decisions. This section is not intended to give an in-depth description of food engineering but we have attempted to outline some major points of attention for process engineers with respect to quality engineering of food processes.

For details on food plant and process engineering the reader is referred to Saravacos and Kostaropoulos, (2001) Maroulis and Saravacos (2003), Lopez Gomez and Barbosa Canovas (2004).

4.4 Customer-oriented design

Although design processes in the food industry deal with complex technological problems, customer-orientation is attracting more attention. Firstly in this section customer information is introduced as a vital condition for customer-orientation. Secondly, the process of customer-oriented design is described.

4.4.1 Customer-oriented design processes

Steenkamp (1989, 1990) is one of the first authors who attempted to get more of a grip on food quality from a consumer perspective linked to technological characteristics and marketing. In his *quality guidance model* (Figure 4.5) he shows that quality perception is formed at two different moments: (1) when the product is purchased consumers form a quality expectation and (2) when the product is consumed the actual quality is experienced.

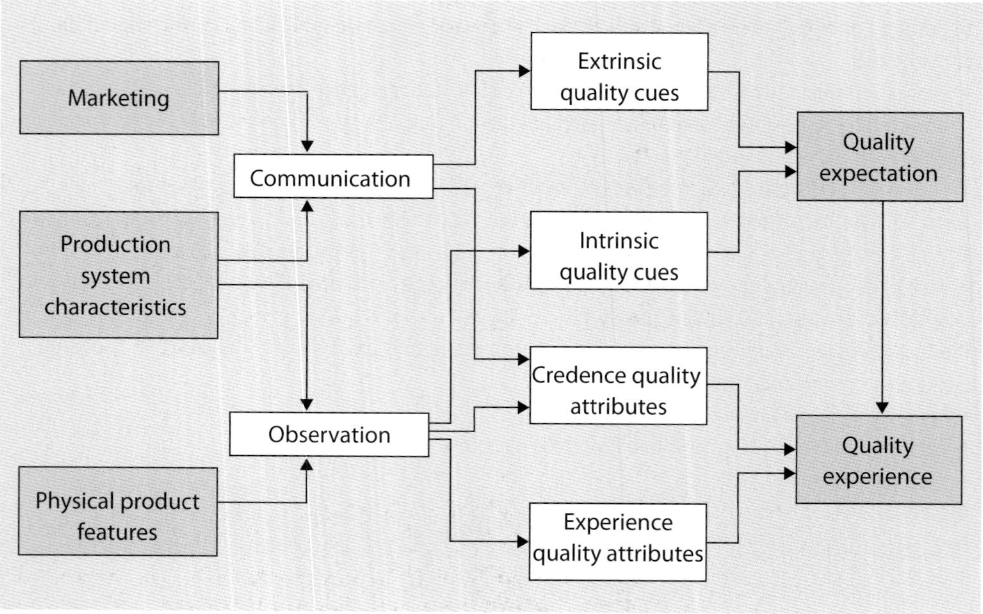

Figure 4.5. Quality guidance model (from Steenkamp, 1989).

The model shows that judgement of expected quality can be affected by quality cues. *Quality cues* are informational stimuli that can be verified by consumers through the senses prior to consumption. Intrinsic cues are described as part of the physical product like colour and appearance. Extrinsic cues are related to the product but are not part of it like price, brand or product information. These cues have a predictive value for the consumer and so influence the quality expectation. Steenkamp (1989) found that intrinsic cues are usually weighted more heavily than extrinsic cues. The actual quality experience is supposed to be based on the integration of perceptions of quality attributes, which can be divided into credence and experience quality attributes (Steenkamp, 1990). Experience quality attributes are determined by the sensory stimuli that can be checked by consumers based on actual consumption (e.g. taste, flavour, texture, etc.). *Credence attributes* cannot be verified through personal experience but do affect quality experience, such as the absence of additives, healthiness or environmental friendliness. The model shows that the extrinsic quality cues (price, brand) and credence quality attributes (e.g. safety, organically produced) are affected through communication, i.e. informational stimuli (marketing). The physical properties that can be observed as sensory stimuli (taste, odour, appearance, etc.) affect intrinsic quality cues and experience quality attributes. According to Van Trijp and Steenkamp (1998, 2005), these two types of attributes have a different impact on the overall judgement of quality performance.

This model gives insight into how the overall judgement of food quality is affected by different types of cues and attributes through communication by marketing and observation of physical product features.

George and Weimerskirch (1994) discussed how a company that is focused on customers is *customer driven*, which means that customer requirements drive organisational activities and customer satisfaction is the goal of these activities. Customer requirements and customer satisfaction dictate a company's direction and actions. In quality management,

Table 4.2. Customer focus quality management practices (adapted from Sousa, 2003).

Customer relationships	Establishing strong relationships with customers by emphasising partnership arrangements, direct customer contacts (face-to-face meeting, plant visits) and integration of the plant's operations with customers (logistics co-operation, single sourcing arrangements, mutual technical assistance, organisation of the plant's activities around customers)
Customer involvement in new product design/ introduction	The involvement of customers in the new product design/introduction process
Collection of information on customer needs	The collection of information on customer needs via frequent and close interaction with customers, including forward looking information (for example, about new requirements, services or technologies needed by existing and potential customer), information on the importance placed by existing customers on several requirements (e.g. price, delivery)

Sousa (2003) explained that customer focus practices involve the establishment of links between customer needs and satisfaction and internal processes. As described in Table 4.2, a customer-focused company should always include the customer in its management practice by creating a relationship with customers to understand their needs, involving the customers in product design by collecting information on customers' needs and responding to them in order to deliver satisfaction for customers.

The *customer-oriented design approach* is not new, but is attracting more and more attention in the food industry. Juran (1988) was the first one to present the design process starting with customer needs. In his 'Quality planning roadmap' the first step included identifying customers and discovering customer needs. Juran also paid much attention to translating customer needs into units of measure, thus providing the basis for the concept of Quality Function Deployment.

Urban and Hauser (1993) developed a design process model and stated that the new product's superiority could be summarised in the 'Core Benefit Proposition'. This core benefit proposition can be described as the unique benefits of the product that comply with consumer needs, and those benefits that are required to meet and surpass competition. Van Trijp and Steenkamp (1998) supported this product design process model and adapted it to the model presented in Figure 4.6.

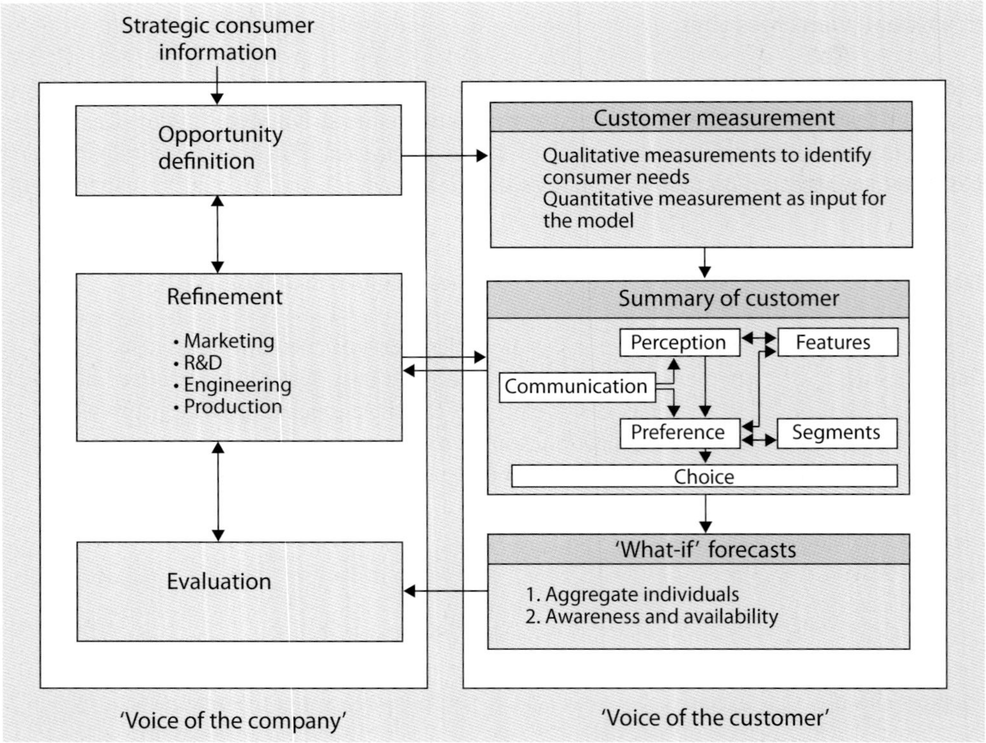

Figure 4.6. Product design process model of Urban and Hauser (1993) as modified by Van Trijp and Steenkamp (1998, 2005).

In the model, a distinction is made between the managerial and consumer behaviour research component, which was called 'voice of the company' and 'voice of the consumer', respectively by Van Trijp and Steenkamp (1998, 2005).

The *company part* concerns decisions with respect to opportunity definition (the market), refinement (benefit delivery and optimisation through the integration of marketing, R&D, engineering and production) and evaluation, to assess business opportunity. Van Trijp and Steenkamp (1998, 2005) added pro-active strategic consumer research to the model to make explicit that opportunity definition should find a strong basis in a continuous and deep-rooted understanding of the consumer in the relevant product categories.

The *consumer part* concerns customer measurement (to make an assessment of market potential before substantial funds are invested in product development), modelling consumer behaviour (to judge and verify product or service alternatives) and 'what-if' forecasts to make an assessment of strengths and weaknesses of the new product concept/ prototype. In the modified 'product design process model' it is assumed that consumer-oriented behaviour finds its basis in consumers' evaluative reactions to the product (product preference). Product preferences find their basis in product attributes as perceived by the consumer (perception). These attribute perceptions, in turn, may by generated through manipulation of the physical product features or through communication (marketing). It is assumed that consumers attach different weights to their attribute perceptions in preference formation. Therefore, recognition of the differential weighing of benefits may provide a relevant segmentation basis. Finally, it is important to recognise that choice does not automatically follows from preference. Factors such as availability and point-of-sale promotion may interfere in this relationship and should be taken into account early in the product development process (Van Trijp and Steenkamp, 1998, 2005).

4.4.2 Customer information

A major key determinant for success in product development is the degree of fit between the new product and consumer needs. As a consequence, customer-oriented design takes consumer needs as the starting point for product and process design. In other words, the new product and its production technology are not a goal as such, but rather an instrumental means to the fulfilment of customer and/or consumer wishes and needs (Van Trijp and Steenkamp, 2005). Van Kleef and co-authors (2004) indicated that, although incorporating the needs of the consumer at the first stages of the design process has been acknowledged as a critical success factor for new product development, proper consumer research is often neglected, particularly in the early stages of the process. Consumer research can be performed during all stages of the design process: opportunity (market) identification, product development, testing and launch; it is most widely applied during the last three stages and happens rarely in the opportunity identification stage.

There are different consumer research methods and their appropriateness depends on the stage of the process. Each consumer research method generates different information and the type of response received differs accordingly. Van Kleef and co-authors (2004) evaluated different methods to obtain information on consumer perception, preference and value satisfaction. The choice for one particular method depends on the purpose, e.g. marketing

Table 4.3. Sources of information for generation of product ideas.

Direct information		Indirect information	
Surveys and interviews Focus groups Field intelligence Direct customer contact		Complaint analysis Analysis competitor's product Purchase data analysis to detect gaps Collective memory of company	
Supporting information			
Societal developments • Governmental publications • Research publications	Technological developments • Conferences, symposia • Trade literature • Scientific literature	Changes in agri-food chain • Trade literature • Sales representatives	Changes in legal requirements • Governmental publications • Workshops • Trade literature

and R&D will use different consumer research methods (MacFie, 2007). The results of such consumer studies (customer needs) have to be translated into the language of the R&D and marketing departments to enable them to undertake appropriate actions.

A broad range of methods and sources, which provide input for product ideas, can be consulted as shown in Table 4.3.

Some methods give *direct information* about consumer needs, like surveys, focus groups or field intelligence (Evans and Lindsay, 2007). A *survey* is aimed at measuring consumer satisfaction but often includes questions to indicate the importance of particular quality attributes. A *focus group* is a panel of individuals who answer questions about a company's product as well as those of competitors. This approach aims at comparing experiences with expectations. Typical questions are: What do you like or dislike about the product? If you had the opportunity, how would you change the product? *Field intelligence* means that all employees, ranging from sales to receptionists, who have contact with customers (or consumers) obtain useful information just by listening.

Indirect information about consumer requirements and needs can be obtained by using collective memory, analyses of complaints, evaluation of competitors' products and of analysis of purchase data. *Analysis of purchase data* may reveal gaps in product assortment offering opportunities for product development. *Collective memory* includes all knowledge and experience present in the organisation. This collective memory should be organised into accessible catalogued information, which can be consulted for product development.

Furthermore, supporting information can be obtained, which can help in directing product development (Table 4.3). This information is described as research guidance (Henning, 1988), and includes topics, such as:

- *Societal changes,* which might have consequences for future consumer demands. For example, the reduction in family size, the increasing number of dual income families and economic processes are pushing the development of ready-to-eat meals.
- *Technological developments,* which may offer new opportunities. For example, new preservative techniques, like high-pressure treatment, electric field pulses, modified atmosphere and active packaging may offer new opportunities to satisfy consumer demands for fresher, more tasty and less processed products.
- *Changes in the food supply chain* may have consequences for *customer* demands in the chain and *consumer* needs in the market place. For example, a trend in retail to open shops at fuel stations might have consequences for packaging concepts, e.g. packaging must be easy to handle in the car, and barrier properties may be required against fuel components in the air!
- *Legal developments* at the national and international level may have consequences for the boundary restrictions for product development. For example, an EU directive on packaging forced the packaging and food industry to develop new packaging concepts with less material in order to reduce packaging waste.

All these sources give input and restricting conditions (e.g. legal limits) to the product development team.

4.4.3 Quality function deployment

One method that has attracted attention in the food industry in its support of methodology in product development is quality function deployment (QFD). QFD was developed in Japan in 1972 as a planning tool for the development of electronics, computer software, defence, and health care. Nowadays, the method is also used in design processes in the food industry but has its limitations (e.g. Costa *et al.*, 2000; Benner *et al.*, 2003; Van Boekel, 2005).

Quality function deployment is a concept that provides the means to *translate customer (or consumer) requirements into appropriate technical (or technological) requirements.* In fact, it is a planning process that ensures that quality is incorporated into a product and its process at the design stage. In the QFD process a matrix is constructed as illustrated in Figure 4.7. The matrix depicts interrelationships between the 'whats' (e.g. what do consumers want) and the 'hows' (e.g. how can these consumer demands be translated to physical measurable units). Moreover, it shows the priorities of the 'whats' and the 'hows'.

The matrix is called the 'House of quality' due to its typical form. The house of quality consists of six sections as described below.

- Section 1: The 'WHAT' in the house of quality is a qualitative part and contains a structured list of consumer (or customer) wishes and needs. Consumers often provide different types of needs, ranging from sensory properties and health aspects to product convenience. These needs must be sorted, classified and structured. The needs can be sorted by asking questions like, what, when, where, why and how. For example, what does convenience mean for you? Easy-to-open, a short preparation time or other? Needs are then classified, for example, by function (e.g. packaging properties), benefit (e.g. healthy aspects) and targets (e.g. maximum amount of fat is 5%). The final result of this section is a detailed, clearly defined and structured list of consumer requirements.

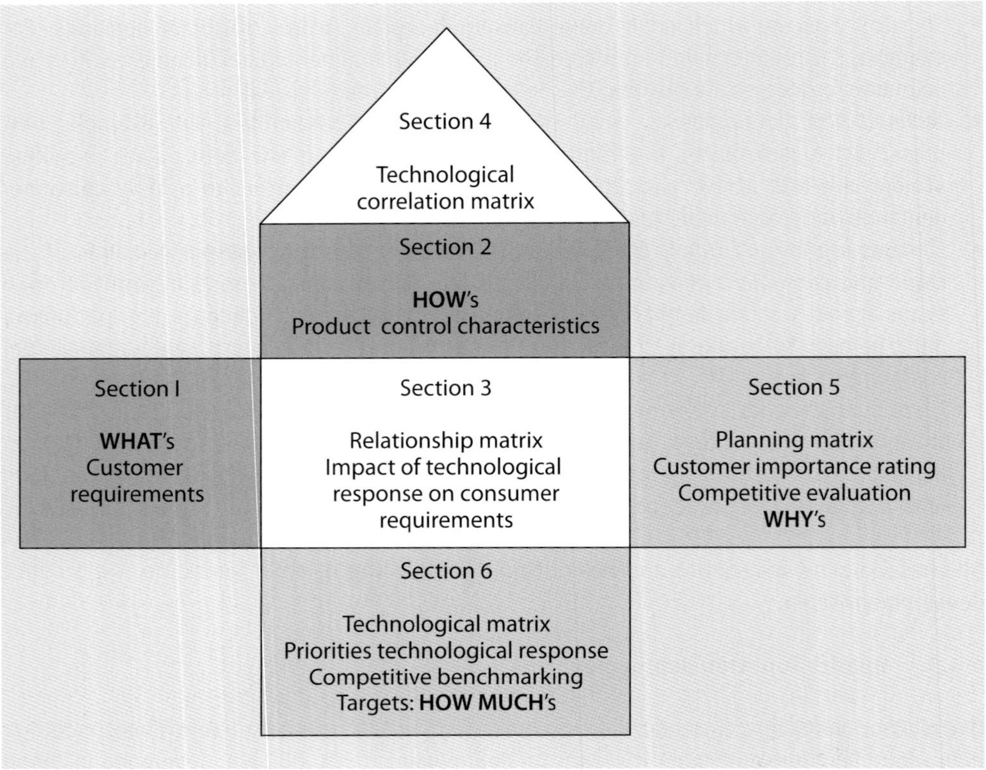

Figure 4.7. House of quality (HOQ) matrix.

- Section 2: The 'HOW' are the measurable characteristics described in the words of the (food) technologist. So the consumer language (i.e. the wishes and needs) must be translated into the technological language, i.e. the product specifications requirements. In this section, the following aspects are considered: unit of the specification (minutes, kilos?), direction of the specification (the more the better, or a specific target) and how it can be measured.
- Section 3: The *relationship matrix* shows the strength of the relationship between the consumer needs (WHAT) and technological product specifications (HOW). The relationship can be strong, medium or weak. Thus, the relationship matrix indicates to what extent the product specifications fulfil specific needs or wishes of consumers. For example, the percentage of fat in the product (HOW) is strongly negatively related to the need for a healthy (low fat!) product but is often positively related to desirable sensory properties (WHAT).
- Section 4: The *correlation matrix* is the roof of the house and shows interrelationships between different product requirements; changing of one can affect the other requirement(s). For example, sugar content is strongly related to sensory perceived sweetness. Decreasing the sugar content will, therefore, weaken the perceived sweetness. If consumers prefer a strong sweetness and a low sugar content than special technological modifications are required because the product requirements are negatively related. The correlation matrix can provide insight into areas that require

more technological research to fulfil consumer wishes and needs, which are not yet technologically feasible.

- Section 5: The *planning matrix* (WHY) forms the *quantitative* part of the 'WHAT'. It contains quantitative marketing data indicating the relative importance of the different needs to the consumer. In this part of the matrix consumer wishes and needs are ranked from high to low priority. Furthermore, the matrix contains information about how far the company's product and the competitor's product fulfils specific needs and requirements. It gives insight into which product scores best on which needs and requirements.
- Section 6: The *technological matrix* with target values (HOW MUCH) forms the quantitative part of the 'HOW'. It contains three types of information, i.e. technological priorities, comparative information and performance targets.

The technological priorities are based on the relative importance of attributes (i.e. needs and requirements) as set by consumers (section 5), and the relationship between these attributes and the technological response (section 4). Thus, a technological measurement having a strong relationship with an important consumer need (high priority score) receives a high technological priority score. Comparative information includes analysis of the competitor's product and comparison with own product performance. Performance target values (HOW MUCH) are established based on the technological priorities and comparative analysis data.

Similarly, additional matrices can be constructed for each stage of product development, including marketing strategies, planning, product design and engineering, prototype evaluation, production and sales (e.g. Cohen, 1995).

The house of quality matrix can support product development by systematically incorporating consumer needs in the design process. It is a structured method, which helps translate consumer needs into the different technological/technical languages (e.g. of the food technologist, process engineer, production planning manager, etc.). It requires that people with different disciplinary knowledge sit together and discuss this complex translation process. Moreover, the HOQ matrix forces prioritisation, e.g. which attributes are rated highest by consumers, how does the company's product score as compared to the competitor's, and what might be unique selling points for the company. Furthermore, it provides insight into technological barriers (in the correlation matrix), which should be overcome to fulfil consumer requirements in the future.

However, a general comment on the QFD methodology is that it is focused on an initial diagnosis of what consumers want, whereas during the subsequent development process there is no feedback to the consumer. QFD designs for and not with the consumer. The methodology is also less suitable for innovative products because it requires the comparison with competitors' products.

More specifically, from a food perspective, the methodology has its limitations because the complex interactions between components and the dynamic food processes in the food matrix complicate a straightforward assessment of relationships between desirable quality attributes and the assignment of technological specifications (Costa *et al.*, 2000, Benner *et al.*, 2003).

4.5 Tools to support quality design

After consumer (customer) needs and demands have been systematically analysed and translated into initial product concepts, the design of the physical products and associated processes starts. During this process many decisions need to be taken in the development, selection and evaluation of, amongst others, prototypes, pilot products, and market test products, but also in the design and testing of the process facilities and equipment. In the product development procedure, the (food) technologist and quality manager can use several methods and techniques that can help assess e.g. sensory attributes, shelf life characteristics, and or potential hazards during product and process development. Some major methods and techniques focused on product assessment are described in section 4.5.1. In section 4.5.2, we briefly explain the use of modelling techniques to support product development and process design. As with product development, there are also supportive 'design tools' for process engineers, e.g. the technique of Failure Mode and Effect Analysis (FMEA) (section 4.5.3), the concept of hygienic design (section 4.5.4), and the Taguchi method for product and process design (section 4.5.5). Whereas, FMEA and the Taguchi method are not specific for food but generally used in process engineering, the hygienic design concept is specifically for process engineering of vulnerable raw materials and products in agribusiness and the food industry. It is not a specific method or technique, but it provides general criteria and detailed guidelines to support the design process.

4.5.1 Methods and techniques to assess product characteristics

The methods and techniques described in this section can help in making decisions with respect to the choice of prototypes, products from pilot scale and full scale, and test market products, from a technological point of view. One should recognise, however, that other constraints like consumer demands, legislative regulations, finances, as well as the actions of competitors can have a considerable influence on the selection procedure during product development (see also Chapter 4.6).

Sensory techniques

Sensory techniques are primarily aimed at evaluating the sensory properties of (food) products (like taste, colour, appearance, odour and texture). They are commonly used in product development but also for quality control purposes. A wide range of techniques are available to support decisions on sensory attributes of products. However, depending on the sensory test, different information is obtained and different types of conclusions can be drawn. Therefore, it is important to select an appropriate technique for the intended purpose.

Sensory techniques can be categorised in various ways. One classification is based on the type of judgement that should result from the research. This can be an *objective* or *subjective* judgement. Objective refers here to a judgment about the intensity of certain sensory attributes or a description of the characteristics of sensory attributes, but without judgement about liking and acceptance. This type of sensory research is called *analytical sensory research*. The liking and acceptance questions are typically subjective research questions, which is called *hedonic sensory research*. Other terms used, instead of hedonic, are respectively acceptance, affective, and preference research. This distinction between hedonic and analytical is very important because it determines the types of sensory tests and panels to be used. For example, for analytical research one needs a panel, with people that are sensitive to the attributes under study, and they must be able to properly express their observations. This requires usually intensive training. For hedonic research, one needs people with little knowledge about sensory research, but just give their opinion about the test products under study.

Another distinction that is commonly made is between *difference (discriminative) tests* and *descriptive tests*. The objective of difference tests is to judge whether differences in certain sensory attributes between products can be observed. This is usually to compare own products with products of competitors, or to compare products from the line, with standard products (quality control). Descriptive tests aim at judging the character and/or intensity of sensory attributes. Difference tests are used both in analytical and hedonic research, whereas descriptive tests are more common in analytical research.

The people who participate in the sensory panels play a crucial role in sensory research. Different types of panels can be distinguished, i.e. consumer panels, outsider panels, trained panels, and expert panels:

- *Consumer panels* consist of normal consumers, and they are only used for hedonic sensory research. A consumer panel should consist of at least 100 people, but in practice lower numbers are often used. If consumers have to test products at home (in-home use tests), then the researcher has less insight into how they use and judge the product as compared to consumer panels tests executed in a sensory research studio.
- *Outsider panels* consist of people with interests in nutrition, food, cooking, etc., and have good sensory observation properties, without specific education or training. Such outsider panels commonly exist of 50 people, and they can be used for both analytical and hedonic sensory research. Such panels are often used to obtain a quick indication about possible consumer judgements on products in development. However, when a systematic analysis or consumer opinions are needed in a critical phase of the product development process, then a trained panel(for analytical research) or a large consumer panel (for hedonic research) is required.
- *Trained panels* consist of people that are specifically trained to observe certain sensory attributes or certain products, and to do certain tasks. These people have a normal potential for sensory observations but are intensively trained. Such panels are exclusively used for analytical sensory research. Trained panels commonly exist of 5-20 people, these people are often recruited from the company itself, but it can also be an external trained panel. During training and subsequent tests they use specific consensus-based terminology to systematically judge the products.
- *Expert(s) (panel)* consists of experts in a certain area, e.g. experts for judging coffee aroma, or meat quality, or wine. Such experts judge the products daily, they have a unique capacity to observe the sensory attributes (low threshold detection, high differentiation capacity, etc.) of that specific product, and they commonly use their own jargon. Such expert(s) (panels) are only used for analytical sensory research and are of great value to the company.

The selection of a certain sensory technique and panel depends greatly on what information is required and where in the quality design or control process. It requires specialist knowledge to design appropriate sensory tests, but also to process and interpret data to support design and control processes. For details about sensory techniques and sensory research (in general) the reader is referred to Moskowitz and co-authors (2006), Brinkman and Hogendoorn, (2006), and MacFie (2007).

Shelf life tests

Shelf life tests are storage experiments to estimate the storage life of a new product. The shelf life is often assessed by the just noticeably difference (JND). This is defined as 'the earliest time when a difference between the quality of test and control samples can be detected by a trained sensory panel'. This criterion can be applied to specific quality attributes or to the overall perceived product quality. The degree of difference between the test product and the control can be assessed at any predetermined confidence level, using statistical analysis (Singh, 2000).

In order to design appropriate storage experiments one should:
- determine *which food processes* (e.g. microbial growth, physiological processes like ripening, decolouration due to enzymatic reaction, drying due to diffusion of water, etc.) restrict the shelf life;
- determine the *technological conditions* (e.g. temperature, relative humidity) and packaging circumstances that affect the shelf life, and which are representative of the real situation in practice;
- select the *testing criteria and acceptable loss*. In other words, what parameters need to be measured and or analysed (e.g. colour, texture, vitamin concentration, spoilage bacteria, migration of packaging compounds, etc.), and when they are unacceptable (using statistics).

There are two typical shelf life tests. *Direct monitoring*, where the shelf life is determined and monitored of the products stored under the conditions that will be applied in practice. And *accelerated tests*, where the shelf life is determined of those products subjected to extreme conditions to speed up the ageing process and so reduce experimental storage time (Ellis and Man, 2000; Kilcast and Subramaniam, 2000).

In *direct shelf life monitoring*, samples should be taken at significant stages in the development or modification of a product. These samples should be examined under controlled environmental conditions imitating the actual situation in practice. For example, products that should perform under tropical conditions must be tested at appropriate temperature and humidity conditions. The application of temperature fluctuations should imitate actual distribution and storage conditions. Testing of various packaging materials is important when the packaging concept is an integral part of maintaining product shelf life.

Product samples can be taken at various stages of product development. Shelf life tests of pilot products give insight into the effect of recipes, product formulations, processing and/or packaging changes, without incurring the costs of factory time and material quantities. Products from full-scale factory batches will provide data for setting the actual shelf life and specification standards. Shelf life evaluation of products from first continuous production trials provide data that should confirm earlier samplings. Products from early runs of standard production not supervised by development personnel form a major part of the development review.

Accelerated shelf life tests are more appropriate for shelf life evaluation of long-life products, because results are required in a much shorter time to meet product launch schedules.

The accelerated test conditions must be selected very carefully because normal ageing or spoilage processes can be changed by the extreme conditions. To accelerate the shelf life test temperature can be raised because it accelerates many ageing processes. Use of high oxygen atmospheres is commonly used to accelerate decay processes that are due to oxidation (e.g. aroma and colour changes). These tests are often used to compare the shelf life of prototypes with existing products. However, shelf life testing under the actual long-term storage conditions should be used to confirm accelerated shelf life test results.

As part of the shelf life surveillance system, samples should be taken at suitable intervals for storage trials. The sampling interval is commonly about 20% of the shelf life providing 6 different stages from fresh (just produced) to full shelf life. For long-life products more frequent intervals may be necessary to examine any modifications or changes during product storage (Ellis and Man, 2000). Shelf life is commonly assessed by the subjective assessment of appearance, texture, aroma, taste, and colour. The problem with such assessments is that it is difficult to compare new products with a suitable control sample. Although there is no univocal method to preserve reference or standards samples, freezing is often used, but such standards are often freshly prepared each time. Sensory analysis plays an important role in the subjective examination of shelf life by the assessment of the aroma, texture and taste attributes using trained panels or experts. Where possible, quantitative measures should be used to underpin the subjective assessments, e.g. analysis of colour compounds, measurement of firmness or viscosity (as an indication of texture attributes).

Additional tests may be necessary in shelf life analysis, for instance the analysis of spoilage bacteria, content of vitamins (and other health compounds), concentration of migrated packaging compounds, etc.

Shelf life schemes generate a substantial body of information, and the organisation and documentation of this information is very important in supporting (new) design processes. Shelf life test schemes can also be part of a quality control programme. In that case, shelf life tests are done to check (verify) if the current recipes and applied technological conditions can still comply with shelf life requirements.

Challenge testing

Challenge testing is a technique which provides information about the microbiological status of a product during its normal or expected life before consumption. It gives information about product safety (related to growth of pathogens and or production of toxins) and stability (related to growth of spoilage bacteria and production of undesirable compounds). In practice, challenge testing comprises the direct inoculation of a food with a micro-organism. For food safety testing, one should select a suitable pathogen that is likely to growth and survive the typical product and process characteristics. For stability testing (linked to shelf life), those micro-organisms that are known to cause the problems in the specific (new) product should be selected. The choice of micro-organisms for inoculation, their culture and method of inoculation are therefore important considerations for challenge testing. Challenge testing is rather time-consuming and requires sophisticated laboratory skills, which may not always be available in food companies (Jones, 2000). To support laboratories in food companies, guidelines are being and have been developed for specific challenge tests of foods, e.g. for *Listeria monocytogenes* (Scott *et al.*, 2005). Such tests are also used to validate the effectiveness of intervention processes aimed at inactivating pathogens to acceptable levels.

4.5.2 Modelling to support product and process design

This section is not aimed at providing a detailed overview of modelling of food quality but gives a brief overview of different types of modelling that can be used in the food industry to support the quality design process. For details, the reader is referred to specific literature about modelling in food by Tijskens and co-authors (2001) and Van Boekel (2008).

Time is often a restricting factor in quality design processes due to, amongst other things, market competition, pressure from customers, new legislative requirements, food safety analysis, or (new) environmental concerns. However, shelf life experiments, challenge tests, and optimisation of process parameters are typical time-consuming activities, which are necessary for taking decisions on product and process conditions during the design process. Therefore, food companies are increasingly on the look-out for tools that can shorten their design processes.

The use of mathematical models can help:
1. to obtain indications about the behaviour of new products and processes in an earlier phase of the design process;
2. in understanding which parameters are more or less critical (e.g. by doing simulations), which give indications about critical control locations, and which parameters should be monitored;
3. in focussing experimental design for the testing of new product and process conditions. Fewer experiments are needed because it is not based on trial and error but on predictions obtained from initial modelling studies.

Table 4.4 gives a brief overview of typical modelling areas that can support product development and process design in the food industry. The first column shows typical areas of modelling, e.g. modelling of kinetics of chemical reactions (kinetics is the study of reaction rates), microbial growth and inactivation (relevant in terms of safety, spoilage, and fermentation processes), enzyme kinetics (enzymes play a crucial role in various processes in food), respiration of fresh produce (gives an indication about their shelf life as a function of transport and storage conditions), but also modelling the impact of packaging properties. The second column describes the major objectives, and the last column gives some typical models used in this area. This information is in no way comprehensive, therefore readers are referred to the specialists' books and articles (see some sources of Table 4.4).

Table 4.4. Brief overview of modelling areas of interest to support product development and process design.

Modelling	Objective	Models
Chemical reaction kinetics	To describe and/or predict the time course/ rate of reaction of chemical processes occurring in food • during processing, e.g. rate of formation of compounds, conversion rate of reactants, dependent on product composition and (technological) process conditions • during shelf life, e.g. rate of degradation (deterioration) of products, dependent on initial product properties and applied storage conditions	Empirical models, i.e. fitting zero, first, second order models to data Mechanistic models are based on understanding of the reaction mechanism Multi-response modelling confirms that more reactions occur at the same time (network mechanism) Expert systems, including combinations of empirical, mechanistic models, and expert knowledge (or knowledge rules)
Microbial changes, i.e. predictive microbiology	To describe and or predict the extent and rate of growth of an undesirable micro-organism (kinetic models) To estimate probability of some event (e.g. growth of pathogens, formation toxins) occurring within a certain period of time (probability-based models) To estimate fermentation processes	Growth and inactivation models Models to describe impact of product (e.g. pH, aw)and or process conditions (temperature) Integrated models (decision support systems), including above types of models and expert knowledge Simulation models (like Monte Carlo simulation)
Enzyme kinetics	To describe and/or predict the activation (e.g. protease in raw milk) or inactivation (e.g. lipase in milk) of enzymes that occur in food which affect its quality	Michaelis-Menten kinetics to model reaction rate, formation or degradation of compounds Heat denaturation kinetics to model denaturation of enzymes
Respiration rates in fresh produce	To describe and/or predict the respiration behaviour of fresh produce under various storage conditions	Respiration models based on Michaelis Menten kinetics Respiration models based on physiology approach
Diffusion of packaging compounds	To describe and/or predict migration of packaging components into food To describe and/or predict barrier properties of packaging concepts (water and oxygen barriers)	Diffusion models describing diffusion of compounds through packaging materials, and from packaging materials into the food matrix
Thermal processes in food	To describe and/or predict heat and mass transfer processes in food products during thermal processes, e.g. heating, cooling and freezing.	Models describing heat and mass transfer processes

Sources: Tijskens *et al.* (2001), Lee *et al.* (2006), Devlieghere *et al.* (2006) and Van Boekel (2008).

When using models in quality design processes, one should be aware of the fact that models are always a simplification of the 'real' situation. They are useful but intrinsically wrong (Van Boekel, 2008). This means that one should have a critical attitude towards the development of models and interpretation of the data. There is often a limited understanding of how processes occur exactly, which means that models cannot be built on insight into underlying reaction mechanisms (i.e. mechanistic models). In such situations, empirical models (black box) are frequently used, which describe certain behaviour in time, and then the only criteria is how well data fit with the models (i.e. fitting capacities).

Modelling can be a useful tool in quality design processes but expert knowledge is required to design and interpret data properly.

4.5.3 Failure mode and effect analysis (FMEA) in process design

Besides specific methods to assess product characteristics, there are also tools to support process design, like failure mode and effect analysis (FMEA). FMEA is a systematic analytical technique to detect potential failures at the design stage of products, services or processes in order to minimise loss due to failures (McDermott *et al.*, 2008). There are two forms of FMEA:
- *Design FMEA*, which involves the analysis of potential failures of a new product or service.
- *Process FMEA*, which includes failure analysis of the manufacturing (or service) process.

This section focuses on the principle of process FMEA. The procedure for design FMEA is very similar but will not be discussed here; details are available in the literature (McDermott *et al.*, 2008). Process FMEA must preferably be carried out by a team. This team should include employees from relevant areas, such as process engineering, manufacturing, maintenance, sales and marketing.

Process FMEA consists of eight steps as shown in Figure 4.8, and is briefly described below:
1. The first step includes the *identification* and *establishment* of the different components (i.e. steps, elements or chains) of the process by the FMEA team.
2. In the second step, the various *potential failures* must be *identified* for each process component. For example, potential failures of pasteurisation could be due to a too high or too low temperature combined with a too short or too long heating time. Too low temperatures combined with too short heating times may result in inappropriate inactivation of pathogenic bacteria, whereas a too high temperature and too long heating time can have negative effects on sensory properties or nutritional value.
3. In the third step, *possible causes* of each potential failure must be identified, e.g. inappropriate heating may be due to malfunctioning of the steam valves, which regulate the temperature.
4. In step four, *the impact of each failure* for each process component on the internal and/or external customer should be identified. E.g. inappropriate heating may result in reduced product safety and a short shelf life, because of the risk that pathogenic and/or spoilage bacteria may survive and grow during storage and distribution.
5. Step five includes *identification of existing or proposed controls*, which should be applied to monitor or prevent the mode of failure. Regular inspection of the steam valves may prevent improper heating.

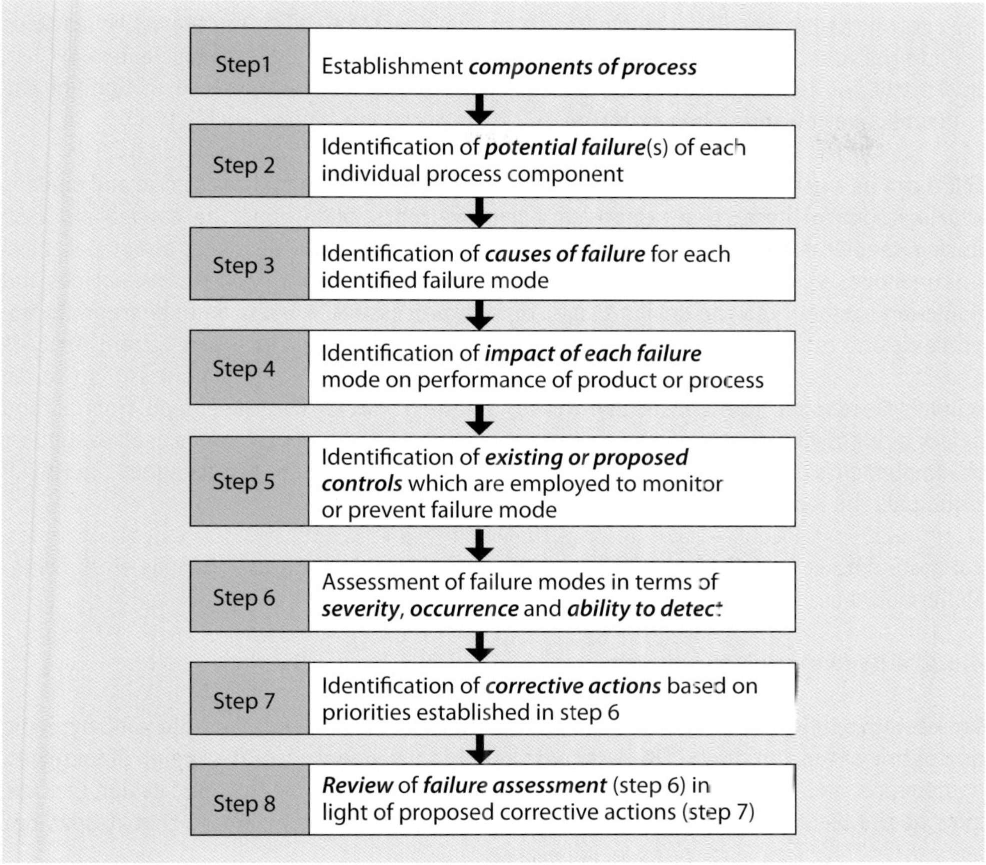

Figure 4.8. Steps in failure mode effect analysis (FMEA).

6. In step six, *failure modes must be assessed* in terms of severity [S] and occurrence [O] of the failure and the ability to detect [D] the failure. Severity, occurrence and ability to detect can be graded on a scale of 1-10. For example, an occurrence [O] classified at scale 1 means that the risk of a defect occurring is 'very rare'. The product of severity [S], occurrence [O] and ability to detect [D] is called the Risk Priority Number (RPN):
   - RPN = S × O × D
   - The ratings for severity, occurrence and ability to detect are scored on a scale from 0-10 according to descriptive and quantitative criteria.
   - If RPN >90 then a corrective action should achieve a high priority.
   - The RPN only gives an indication of the importance of each process step with respect to the consequences of potential failures. FMEA is not a quantitative method and RPN just underpins estimations, knowledge and ideas of the FMEA team (Banens *et al.*, 1994).
7. In step seven, *corrective actions* are *identified* to improve significant failures based upon the relative priorities (i.e. failures with an RPN >90) as set in step 6. If the RPN exceeds 90, then a corrective action requires a high priority, e.g. by implementation of charts to

control the process. By implementing a control chart the ability to detect [D] increases and the value for [D] decreases resulting in a lower RPN as illustrated in Table 4.4.
8. Finally, in step eight, the *failure assessments* of step 6 are reviewed in light of the corrective actions proposed in step 7.

FMEA can be a useful tool in evaluating potential failures in the process design and enables a prioritising of corrective actions and improvements that should be carried out first during specific steps of the process. Process FMEA provides information about potential failure modes, which are ranked in order of priority, so as to identify corrective actions and improvements early in the design stage. In fact, process FMEA helps in understanding the process. One technique, which is based on the process FMEA principle, is Hazard Analysis and Critical Control Points. Whereas FMEA was developed for economic reasons (i.e. to minimise economic loss caused by process failures), HACCP was developed from a food safety point of view (i.e. to minimise health risks). The FMEA analysis can be used to support the implementation of HACCP principles (Scipioni *et al.*, 2002). For details about the HACCP principles the reader is referred to Chapter 7.

For more details on FMEA principles one is referred to the literature (Banens *et al.*, 1994; McDermot *et al.*, 2008).

4.5.4 Hygienic design concept

Increasing public awareness about food safety has encouraged the safe and hygienic processing of food products. This has considerable consequences for the design of facilities, equipments, and tools in agri-business and the food industry. The hygienic design concept aims at the design, construction, and installation of such equipment so that it does not adversely affect food quality, especially safety.

Nowadays the legislative requirements on hygienic design have been enacted at different levels, i.e. international, European, and national level.

- At the international level, the World Trade Organisation (WTO) is responsible for regulations with respect to food hygiene. The SPE agreement (Sanitary and Phytosanitary measures) of WTO contains a broad range of measures focused on protecting human and animals against food-borne diseases (see also Chapter 2).
- At the European level, since January 1995 all new equipment has had to comply with the Machinery Directive 89/392/EC. The Machinery Directive includes safety requirements for equipment with respect to the user and a hygiene section for agri-food machines. The latter includes requirements for the production of safe and hygienic products
- In EC member states, the EC Directive for hygienic process design has to be implemented in their national regulation.

Several international and national standard organisations produced standards for, amongst others, the hygienic design of food equipment, e.g. ISO (International Organisation for Standardisation), CEN (European Committee for Standardisation), NIN (Nederlands Normalisatie Institute) and DIN (Deutsches Institut für Normierung).

The formulation of the hygiene requirements in the Machinery Directive is rather general. Therefore, the European Hygienic Equipment Design Group (EHEDG) has developed guidelines to provide support as to how to design, construct, and install equipment in such a way that it prevents contamination. The guidelines apply to durable equipment used for batch and continuous, open and closed manufacturing operations. EHEDG is an independent consortium set up to develop guidelines and test methods for safe and hygienic processing of foods. The group includes representatives from research institutes, the food industry, equipment manufacturers, and government organisations in Europe. EHEDG reports contain detailed instructions for design engineers of agri-food machines.

Hygienic design requirements as recommended by EHEDG are described in the functional requirements and state that equipment:

- Must be easy to clean and disinfect, preferably cleaning in place (CIP).
- Should protect the product from microbial and chemical contamination.
- Must prevent the ingress of micro-organisms in case of aseptic design.
- Can be monitored and controlled for equipment's functions that are critical to microbiological safety.

To fulfil these functional requirements a document is drawn up, describing detailed criteria for the hygienic design of equipment intended for the processing of foods. Its fundamental objective is the prevention of the microbial contamination of food products. Such contamination may, of course, originate from the raw materials, but the product may also be contaminated with micro-organisms during processing and packaging. If equipment is of poor hygienic design, it will be difficult to clean. Residues (soil) may be left behind in crevices and dead areas, allowing the micro-organisms which they harbour to survive and multiply. These may then cross-contaminate subsequent batches of product.

The three major topics for which criteria are described in detail by EHEDG (http://www.ehedg.org/guidelines/doc8.htm) include:

- *Criteria on materials of construction*. Materials used in the construction of food machinery must fulfil certain specific requirements. Product-contact materials must be inert both to the product and to detergents and disinfectants under the conditions of intended use. They must also be corrosion resistant, non-toxic, mechanically stable, and their surface finish must not be adversely affected under the conditions of intended use. Non-product-contact materials shall be mechanically stable, smoothly finished and easily cleanable. It is worthwhile maintaining an awareness of new developments in materials and products for the food industry and seeking the advice of materials suppliers where appropriate.
- *Functional requirements*. Hygienic food processing equipment should be easy to maintain to ensure it will perform as expected to prevent microbiological problems. Therefore, the equipment must be easy to clean and protect the products from contamination. In the case of aseptic equipment, the equipment must be pasteurisable or sterilisable (depending on the application) and must prevent the ingress of micro-organisms (i.e. it must be bacteria tight). It must be possible to monitor and control all of its functions which are critical from a microbiological safety point of view.
- *Criteria on hygienic design and construction*. In the design, fabrication and installation of equipment the basic criteria must be taken into consideration as listed in Table 4.5.

Table 4.5. Compilation of hygienic design requirements as developed by EHEDG (http://www.ehedg.org/guidelines/doc8.htm).

Criteria on materials of construction	
Non-toxicity	The designer has to take care that only non-toxic materials of construction are used in direct contact with the product. It is imperative to check legislative aspects.
Stainless steel	The designer has to consider good resistance to general corrosion. Stainless steels offer excellent corrosion protection. Selection of most appropriate grade depend on corrosive properties (in terms not only of the chemical ions involved but also the pH and the temperature) of the process and of the cleaning and antimicrobial chemicals.
Polymeric materials	When choosing polymeric materials a number of criteria should be considered, e.g. compliance with regulatory requirements, compatibility with food stuffs and ingredients, chemical resistance, temperature resistance in use, etc.
Elastomers	Same parameters as listed in the 'polymeric materials' section above will apply for the selection of an elastomer.
Adhesives	Adhesives used should always comply with the FDA regulations and with the recommendations of the supplier of the equipment for which those gaskets are used.
Lubricants	Equipment should be designed such that lubricants do not come into contact with products.
Thermal insulation materials	Thermal insulation of equipment must be carried out in such a way that the insulation material cannot be wetted by ingress of water from the outside environment (e.g. hosing down, condensation on cold surfaces).
Signal transfer liquids	Liquids used for signal transfer may come into contact with the process fluids if the barrier between them fails. Therefore these liquids must be food grade.
Functional requirements	
Cleanability and decontamination	Equipment which is difficult to clean will need procedures which are more severe, require more aggressive chemicals and longer cleaning and decontamination cycles.
Prevention of ingress of micro-organisms	Ingress of micro-organisms into products must be avoided in general, to meet requirements of public health and required shelf life.
Prevention of growth of micro-organisms	Under favourable conditions micro-organisms grow very rapidly. Consequently any areas, e.g. dead areas, gaps and crevices, where micro-organisms can harbour must be avoided.
Compatibility with other requirements	A design with excellent hygienic characteristics but lacking the ability to perform its functional duties is of no use; hence a designer may have to compromise. Such action, however, will have to be compensated by more intensive cleaning and decontamination procedures and these must be documented so that the users are aware of the nature of the compromise. The cleanability of the equipment, including the CIP where appropriate, must be demonstrated.

Table 4.5. Continued.

Functional requirements (Continued)	
Validation of the hygienic design of equipment	Irrespective of the amount of know-how and experience with hygienic design which is applied when designing and fabricating, practice has shown that inspection, testing and validation of the resulting design to check if the requirements are met is very important. In critical cases it may be necessary to check the hygiene level as part of the maintenance procedures. The designer has to make sure that relevant areas are accessible for inspection and/or validation.
Hygienic design and construction	
Surfaces and geometry	Surfaces must be cleanable and must not present a toxicological hazard by leaching of components into the food. All product contact surfaces must be resistant to the product, and to all detergents and disinfectants under the full range of operating conditions (the intended conditions of use).
Surface finish/ surface roughness	Product contact surfaces should have a finish of an acceptable Ra value and be free from imperfections such as pits, folds and crevices. A roughness of Ra >0.8 µm is acceptable if test results have shown that the required cleanability is achieved because of other design features, or procedures such as a high flow rate of the cleaning agent.
Drainability and lay-out	The exterior and interior of all equipment and pipework must be self-draining or drainable and easily cleanable.
Installation	The risk of condensation on equipment, pipe work and the internal surfaces of the building should be avoided wherever possible. If unavoidable, the design should be such that condensate is diverted away from the product. Equipment and support structures must be sealed to the supporting surface (floor, walls, columns, ceiling) in such a way that no pockets or gaps exist.
Welding	Permanent metal to metal product contact joints must be continuously welded and free of imperfections.
Supports	Supports for piping or equipment must be fabricated and installed such that no water or soils can remain on the surface or within the supports.
Insulation	Insulation materials should be clad with stainless steel, which must be fully welded, so that no ingress of air or moisture is possible, as this may encourage microbial growth and hence increase the risk of microbial contamination or corrosion of the cladding if the insulation materials release chlorides.
Testing the hygienic characteristics of equipment	A series of EHEDG test methods for assessing the hygienic characteristics of equipment has been published: • a method for assessing the in-place cleanability of food processing equipment; • a method for the assessment of in-line pasteurisation of food processing equipment; • a method for the assessment of in-line sterilisability of food processing equipment; • a method for the assessment of bacteria tightness of food processing equipment; • a method for the assessment of in-place cleanability of moderately-sized food processing equipment.

The manufacturer may mark equipment with the special CE (Conformité Européenne) symbol, if the equipment complies with all requirements of the EC Machinery Directive. In fact, without the CE mark and evidence that its use is justified, equipment may be neither sold nor put into use (Lelieveld *et al.,* 2003).

Another important aspect is the maintenance of food-processing equipment. Product contact surfaces must be controlled on a regular basis for signs of wear, such as corrosion and crevices. In addition, ageing of materials (such as compression capacity of elastomers) must be regularly checked. Furthermore, the hygienic status of equipment can be tested by several assessment methods, as developed by EHEDG. The maintenance of equipment takes time and production often has to be put on hold. However, the costs of equipment defects, because of poor maintenance, can be much higher then the costs of regular maintenance.

Hygienic design is becoming increasingly important because of increased public awareness on food safety. Adequate hygienic design of equipment and facilities support effective food safety management systems, because it contributes to the prevention of (cross) contamination which may result in unacceptable safety problems for human health. The long-term benefits of hygienic design are not only product safety but also the potential to increase the life expectancy of equipment, reduce maintenance and consequently lower operating costs.

4.5.5 Taguchi method for product and process design

In order to estimate the true cost of quality and to improve quality cost-effectively the Taguchi method provides designers with a tool to improve the product as well as the process by which it is made (Evans and Lindsay, 2005).

Taguchi asserted that the quality of a *product* is a function of key product characteristics referred to as performance characteristics. The ideal value (or state) of a performance characteristic is its target value. A high-quality product performs consistently near target values throughout its life span and under all operating conditions. Taguchi's quality loss function estimates the loss to society from the failure of a product to meet its target for a particular performance characteristic (Figure 4.9). This loss can be incurred by the consumer (e.g. short product life, increased maintenance and repair costs), by the firm (e.g. increased scrap, rework and warranty costs, damage to reputation, lost market share) or by society in general (e.g. pollution, safety). Either exceeding or not reaching the target will result in a loss even when the product falls within specifications. The closer each product is to its target value, the better it fits with adjacent products, the better it looks and the better it performs its intended function. The better fit with adjacent products is especially important for assembled goods (e.g. cars, computers, etc.). The quality loss function enables a firm to quantify cost savings arising from product and process improvement. Another issue that might arise is the reduction of a product's sensitivity to environmental fluctuations. Taguchi described a three-step approach:

1. *System design*. Firstly, an initial product design is developed, which includes preliminary settings for parameters that affect the value of performance characteristics
2. *Parameter design*. Secondly, the factors that contribute most to variation in the end product must be identified. A set of experiments must be carried out to determine the

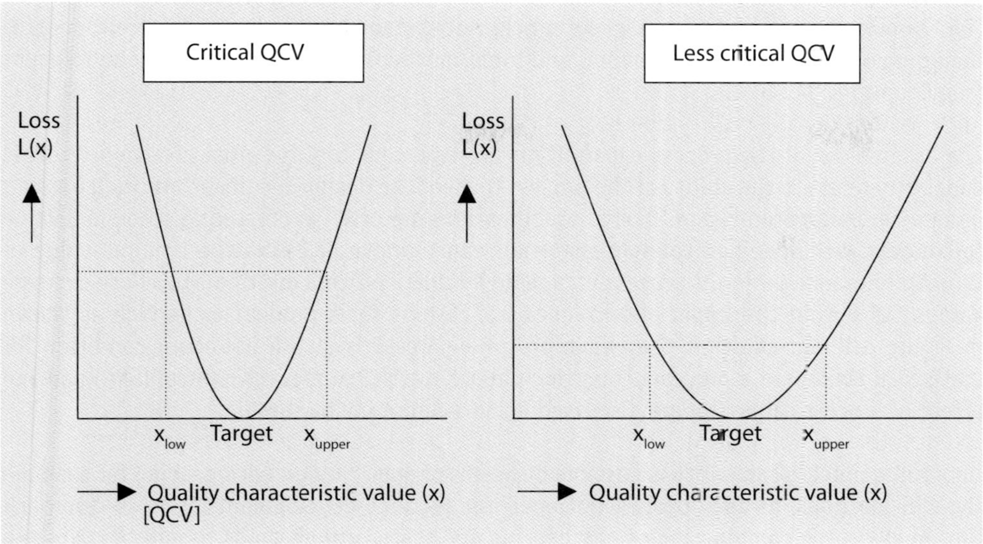

Figure 4.9. Schematic illustration of the principle of the Taguchi's quality loss function.

settings that minimise the sensitivity of engineering designs to the sources of variation. And thus to minimise the expected loss to society. Taguchi also used these statistically designed experiments to identify design parameter settings that reduced costs without hurting quality.

3. *Tolerance design.* Acceptable tolerances are determined around the values of the target design parameter as set in step two. Not all tolerances need to be tight, only those around design parameters whose variation will result in significant variation in the product's performance characteristics should be tight (Noori and Radford, 1995).

The resultant product will have acceptably low levels of product variation despite fluctuations in materials, manufacturing (equipment, operators) and environmental conditions. Since performance variation is minimised by reducing the influence of the sources of variation rather than by controlling them, Taguchi methods are cost effective.

As previously mentioned, designing the *process* not only includes system design (i.e. design of equipment and tools) but also parameter and tolerance design and both are important aspects that must be considered. *Parameter design* involves the translation from the designer's concept into concrete manufacturing specifications, which have nominal dimensions (e.g. gram, millilitre, °C, minutes, etc.). A typical aspect of food products and food packaging materials is that the nominal dimensions (i.e. parameter design) are often set by legislative regulations, whereas for mechanical products the designer usually sets the parameters.

Tolerance design includes the determination of tolerances around the nominal settings as identified by parameter design. Tolerance is the permissible variation in the dimension. Tolerances are necessary because it is not possible to produce food products or components, all of which comply exactly with specifications, due to common causes. Natural variations (i.e. common causes) in a production process or system can be due to different sources including people, materials, machines and tools, methods, measurements, and environment (see Chapter 5).

The assessment of the process capability is an important tool for process design. Process capability refers to the ability of the process to meet the design specifications for a product or service; parameter design. Design specifications are often expressed as a nominal value (or target) with specified tolerances (or allowance above or below the nominal value or target): tolerance design. A process is capable if it has a process distribution whose extreme values fall within the upper and lower specifications for a product or service as shown in Figure 4.10. In addition, the capability of a process is also a boundary condition for statistical control of a process. The principles of statistical process control (SPC) and the effect of process capability are discussed in more detail in Chapter 5.

Improving process capability is possible by means of quality engineering, using for example Taguchi methods, as previously mentioned for product development. It is an approach that involves the combination of engineering and statistical methods to reduce costs and improve quality (Krajewski and Ritzman, 2007). In practice, the consequences of tolerance setting on product functionality, manufacturability and the economic effects are often not considered. Taguchi related tolerance design to economic implications and advocated designing narrow tolerances preferably at critical (processing) steps. In other words, narrow tolerances must be set where exceeding the upper and/or lower tolerance limit has major economic consequences. This approach is called the Taguchi's quality loss function as previously illustrated in Figure 4.9. The loss increases as a quadratic function, when values do not meet the target value. Therefore, according to Taguchi the only meaningful specification is being on-target.

However, for food products narrow tolerances are often not feasible, because of large natural variations in raw materials due to, amongst other things, seasonal effects, weather conditions, genetic predisposition or transport conditions (see Chapter 2). Nevertheless, the principle of Taguchi's quality loss function is also applicable to food products. For

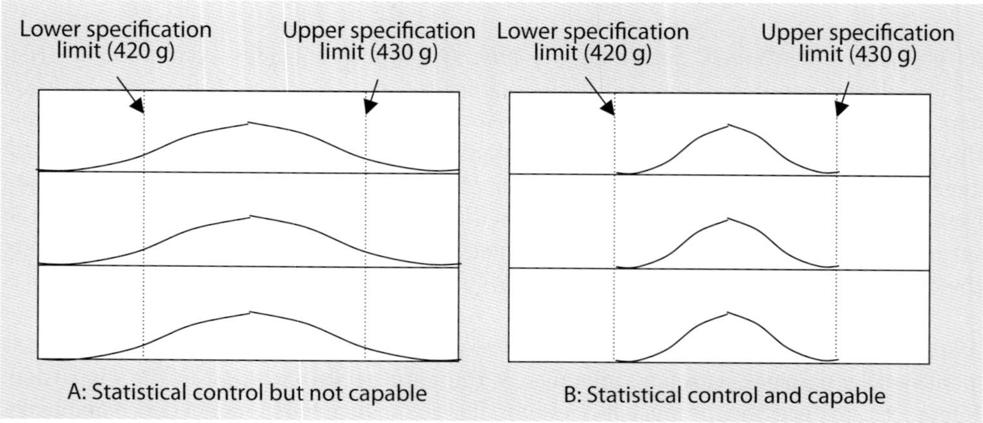

Figure 4.10. Illustration process capability concept.

example, if an expensive aroma ingredient is used then exceeding the upper limit will cost a lot of money, whereas too low levels will have less economic consequences until it is noticed in sensory experiments or by consumers. So when designing the critical limits (i.e. tolerance design) the costs of exceeding these critical limits should be considered.

4.6 Managing the design process

In previous sections of this chapter the steps in the design process have been described, including different tools to support the design performance. Nevertheless, design processes will not succeed if they are not properly managed. In this section three important management activities will be discussed. The *product development strategy* provides the basis for embedding product development in the organisation and in the long-term strategy of the company. *Collaboration* is described in the concept of concurrent engineering. Finally, *project management* provides techniques and organisation principles for effectively and efficiently realising design goals and objectives.

4.6.1 Product development strategy

The design process has to be managed actively in order to reach design goals both in qualitative terms and in delivery time. In Figure 4.11 design management is described as product development strategy (Wheelwright and Clark, 1992).

Wheelwright and Clark (1992) invoked the image of a funnel, illustrated by the dotted lines, to depict activities that support a fast-cycle product development activity. At the front end of the funnel, two ongoing business processes must initiate the process. Technology assessment and forecasting predict technological developments that a business should

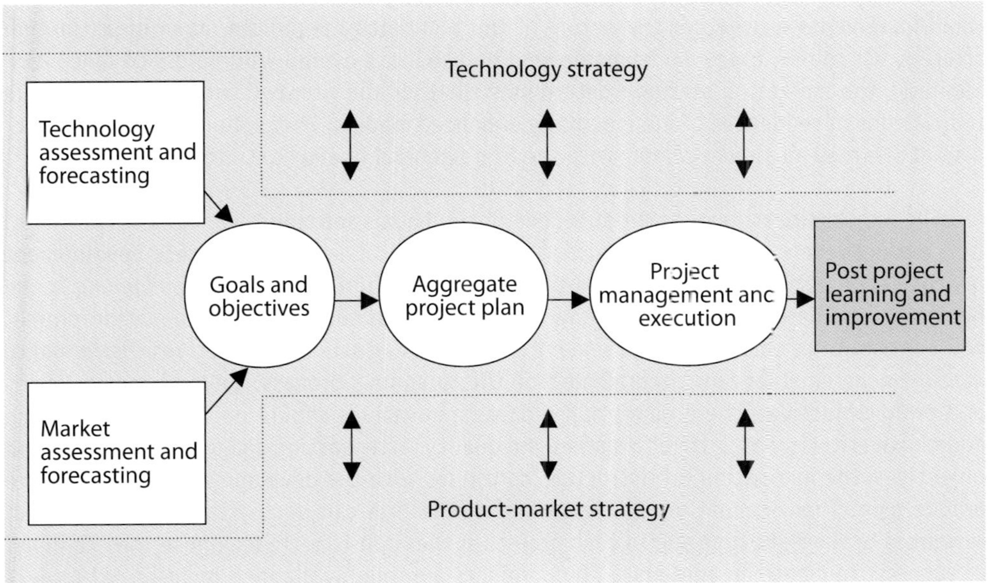

Figure 4.11. Product development strategy.

anticipate. These functions evaluate new process technologies that will emerge from research and development actions, in order to find practical applications in the market place. This technology assessment then continues by weighing the benefits of acquiring essential capabilities from outside suppliers, for example by strategic alliance, by simple purchase, or by developing them internally (Melnyk and Denzler, 1996).

In 1991, Roussel and co-authors already emphasised the importance of creating strategic technology plans. In a competitive environment there should be an ongoing effort to estimate the strengths of competitors in technologies, which are important to the business and which are different in respect to the successful conduct of R&D. They described three types of R&D:

a. *Incremental R&D*, i.e. clever exploitation of existing scientific and engineering knowledge in new ways; characterised by low risk and modest reward.
b. *Radical R&D*, i.e. creation of knowledge new to the company, and possible new to the world, for a specified business objective; characterised by higher risk and high reward.
c. *Fundamental R&D*, i.e. creation of knowledge new to the company, and probably new to the world, to broaden and deepen understanding of a scientific or engineering arena by the company; characterised by high risk and uncertain applicability to business needs.

A product development strategy must continuously consider the contribution of R&D projects to the creation of unique knowledge.

In addition to information about technology, the product development strategy needs an assessment of the firm's current markets and those that it would like to enter. First, the firm must complete these assessment processes, because their outputs become major inputs to the firm's technology strategy and product/market strategy. Both of these component strategies specify directions for introductions or innovations into the marketplace. The technology strategy determines whether the company will introduce new technology in small increments or whether the nature of the technology permits a major breakthrough strategy. Of course, these decisions depend on the firm's overall willingness to take risks. Similarly, the marketing strategy determines whether the firm will introduce large-scale innovations or sequences of incrementally enhanced models. The technology and product/market strategy must also comply with the firm's general business strategy.

From general business strategy, product development goals and objectives must be translated into project levels. *Statement of goals* and objectives also specify process development because each activity competes with the firm's innovation-related funds. *Planning* is the next step and here it is estimated how much and when each product-innovation project will contribute to the firm's achievement of its overall goals and objectives. *Performance measurement* must be an integral part of this planning process. Beyond routine sales and profit objectives, the product and process innovations should be evaluated on three additional criteria: costs, time to market and quality. After setting and qualifying goals and objectives, the innovation infrastructure continues with the development of an *aggregate project plan*. Then it must be checked whether the firm can deploy the right amount of resources to the right type and to each project at the right time, to maximise the likelihood of success. To create an aggregate plan, projects are roughly planned by means of project planning techniques. However these project plans are subject to considerable uncertainty.

Quantifying, at least approximately, is necessary to avoid the risk of the projects running out of control (Melnyk and Denzler, 1996).

4.6.2 Collaboration in design processes

Current design management includes collaboration between different disciplines inside and outside the company. In this section collaboration is discussed with the concept of *concurrent engineering*. Separately *cross-functional teams* are described as being a basic condition for effective exchange of knowledge.

It has already been mentioned by Griffin and Hauser (1996) that different departments have different perceptions due to, amongst other things, different personalities, cultures, languages, organisational responsibilities and physical barriers, which was confirmed by Shaw and Shaw (1998). Shaw and co-authors (2003) stated that the main reason for the existence of any conflict is education and training. Other factors such as goals and priorities are also mentioned as reasons for different perceptions.

The culture in marketing and research and development (R&D) departments differs, due to their different educational background and training. Marketing professionals are usually recruited from business schools, whereas R&D professionals are acquired from process engineering and technological science schools. Education and training in business schools focuses on general problem solving, combining data and intuition to make decisions that lead to profitable corporate performance. On the other hand, the education and training at technological science and engineering schools are focused on the scientific method of hypothesis generation and testing and solving technical problems (Griffin & Hauser, 1996). For production professionals, it is assumed that they have more or less the same background as the R&D professionals since they probably come from science and engineering schools too. Thus, they will also focus principally on the technical and technological problems. Shaw and co-authors (2003) who conducted research on German engineers confirmed that, production people focus on technical and technological issues and consider that marketing people lack knowledge in this area. This lack of familiarity will lead to a different way of setting priorities.

As an illustration of how interests differ in various departments, Table 4.6, as compiled from Griffin and Hauser (1996) and Shaw and Shaw (1998), explains the different interests and goals of marketing, R&D, and production. This table illustrates the assumption that differences in interests, due to different perception formations occur between the three departments.

In most companies, there is a certain specialisation in function for each department: marketing is expected to be the key source of information regarding user need (identification of the customer needs) and competition. R&D is the key source concerning technology, and manufacturing (production) is the key source of the production process (Song *et al.*, 1998; Schroeder, 2000).

Song and co-authors (1998) stated the importance of the *information flow* between different departments:

Table 4.6. Differences in goals and interests between Marketing, R&D and Production compiled from Griffin and Hauser (1996) and Shaw and Shaw (1998).

Department	Goals	Interests
Marketing	• knowledge as a source of value to mankind • research for research's sake • peer evaluation and recognition	• prefer short time horizon of incremental project • focus on the market • accept a high degree of ambiguity and bureaucracy • loyal to the firm
R&D	• organisational survival and growth • all activities relevant to firm's objective • organisational recognition	• prefer the long time horizon of advanced project • focus on scientific development • low tolerances for ambiguity and bureaucracy • loyal to scientific profession
Production	• to develop the perfect product regardless of whether it is what the customer wants	• prefer the long production run • focus on technical detail

- R&D provides information to marketing regarding product characteristics and to manufacturing regarding product and process specifications.
- Marketing informs R&D about consumer needs and manufacturing about customer delivery requirements.
- Manufacturing provides information to R&D concerning production limitations and to marketing concerning production capacity and lead times.

In the product design process, the information system is very important to ensure the cooperation between multiple functions to reduce uncertainty and improve the quality of decisions. Song and co-authors (1998) indicated that frequent communication and exposure to a variety of perspectives will increase the possibility of generating new valuable knowledge, which will in turn increase the likelihood of producing successful new products.

Sequential versus concurrent engineering

Transforming an idea for a new product or service into an actual product is usually time-consuming and expensive. The costs associated with making major changes to a product concept or even rejecting a concept increase dramatically as the product development process progresses. In the recent past, product life cycles were so long that mass production and a long product life cycle easily gave a return on the costs of development. However, trends like fragmentation of markets, mass customisation, rapid technological advancements, shortening of product life cycles, and intense global competition have changed the mass-production approach. In fact, these challenges call for a drastic change in the way in which products are designed and in the role of the production department and experts in the development process. In many companies design processes were performed according to a sequential approach. However, this concept is not suitable in a dynamic market where time and costs of design processes should be kept as low as possible. For this purpose the concept of concurrent engineering offers more opportunities. Both concepts are explained below.

In a *sequential approach* the information flow is predominantly downstream (Figure 4.12). The product concept is transformed into functional product and process specifications and turned over to production and purchasing. At each stage decisions are made but the available options are usually dependent on earlier decisions. As a consequence, by the time production is involved in product and process design, it might be too late to make changes without initiating seemingly endless engineering changes. This can lead to the looping effect, which rapidly lengthens the product development process. The design process becomes less flexible and costs of incorporating engineering changes can increase by a factor of 10 as the product moves from concept stage to market launch. Studies have shown that the design of a product accounts for only about 5% of total costs. However, the cost impact of additional quality changes, manufacturing and equipment modification and

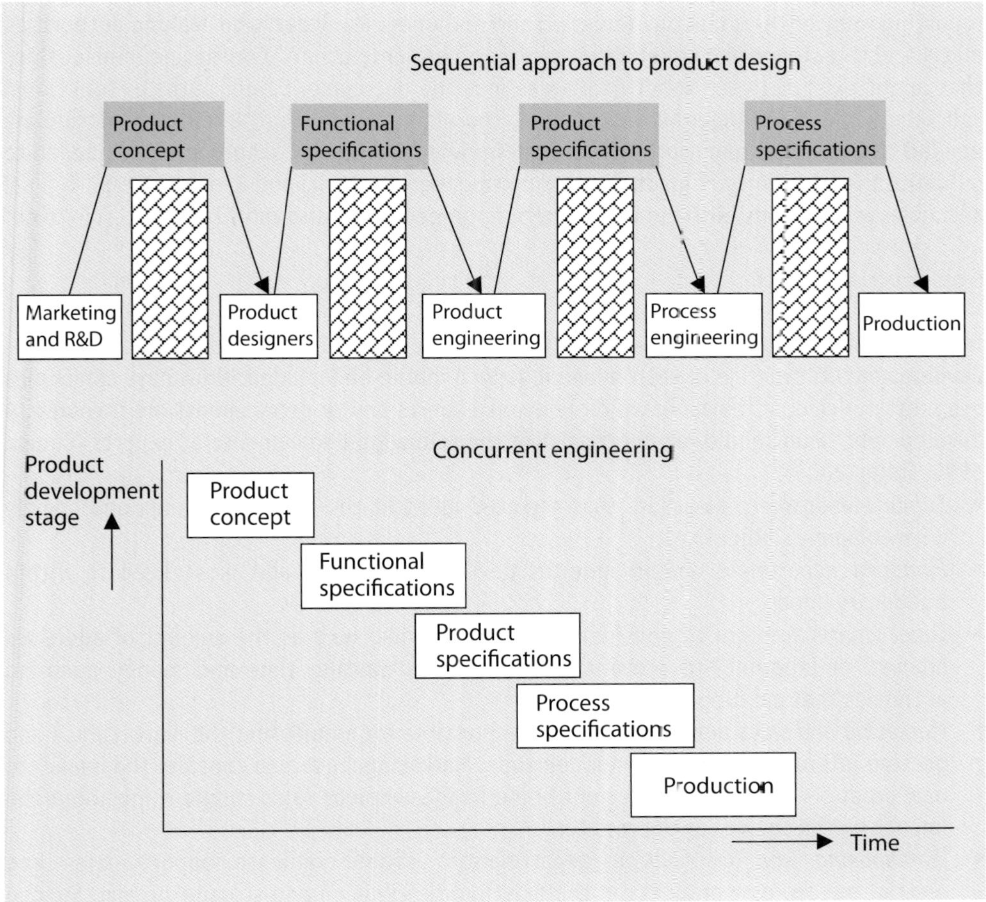

Figure 4.12. Sequential versus concurrent engineering.

technology acquisition can be at least 70% of the total product. The sequential approach is also known as the *over-the-wall approach* (Noori and Radford, 1995).

An alternative to this approach is called *concurrent or simultaneous engineering* (Figure 4.12, Table 4.7). It is an organisational tool that can facilitate integration by creating cross-functional teams. Functional product and process specifications are developed concurrently. Ideally, concurrent engineering involves the firm's customers and suppliers along with a broad range of representatives of different functional areas. Inclusion of customer representatives in the design team might help to ensure that a product really meets their needs. Suppliers may have a major influence on product costs and quality, therefore, their involvement from the beginning can reduce costs and improve quality. Collaboration with suppliers is even more crucial when a firm transfers partial or even full design responsibilities for specific components to its suppliers (Noori and Radford, 1995).

In order to make concurrent engineering successful, it is necessary to change the manner of operation and communication in the functional areas. Functional managers must delegate decision-making authority to their representative in the team so that the team can review, modify and approve designs quickly. As a consequence, current lines of authority and reporting relationships must be reviewed and realigned. Many decision-making authorities might feel threatened and be reluctant to pass on power to others. Conflicting relationships that might exist between functional areas must be overcome. Career paths of functional specialists might no longer be well defined. Therefore, changes in the reward structure are needed to encourage teamwork, rather than individual efforts. Changes are also needed to minimise the possibility of inconsistent functional goals (Noori and Radford, 1995). Typical activities and responsibilities in concurrent engineering are shown in Table 4.7.

Cross-functional teams

In order to make proper decisions, in the multidisciplinary process of product and process development, knowledge of each relevant aspect should be included. Nowadays companies use cross-functional teams, in which relevant experts and or departments are involved. A development team should ideally consist of the following departments or experts (Cohen, 1995; Fuller, 2004):

- *Senior management* to assure that selected ideas fit the corporate image and comply with company's objectives.
- *Financial department* should monitor development costs and must keep it within budgetary limits.
- *Legal adviser* in order to consider legal implications, such as the amount of additives allowed or labelling, to prevent the team from wasting time and money pursuing activities that are illegal.
- *Marketing and sales* department(s) are responsible for monitoring the marketplace and getting information on consumer demands. Marketing must also consider the impact of new products on company's current brand image, whereas sales should think about the impact on the shelves at the retailer.
- *Warehousing and distribution* department(s) should indicate opportunities and restrictions for new products in the existing distribution and storage system. Special

Table 4.7. Activities and responsibilities in concurrent engineering.

Conceptual design	
Marketing	Proposes and investigates product concepts
Engineering	Proposes new technologies and stimulates performances
Operations	Proposes and investigates manufacturing and delivery processes
Product design	
Marketing	Defines markets and specifies objectives
Engineering	Chooses components and key suppliers
Operations	Defines process architecture and estimates costs
Product and process engineering	
Marketing	Conducts customer tests on prototypes
Engineering	Builds full-scale prototypes for evaluation and refinement
Operations	Builds system to manufacture prototype, plans full-scale system, tests tooling and new procedures
Pilot development and testing	
Marketing	Prepares for market roll-out, trains sales force
Engineering	Evaluates and tests pilot unit
Operations	Builds pilot units in commercial processes, refines the system, trains personnel, checks suppliers
Volume production and launch	
Marketing	Fills distribution channel, sells and promotes, gains feedback from target customers
Engineering	Evaluates customers' experience with product
Operations	Builds up plant to volume targets, refines quality, yield, and cost performance
Post-scale service	
Marketing	Gathers customer feedback
Engineering and Operations	Study warranty data

requirements, such as cooled and/or modified atmosphere storage, require additional facilities, thus additional costs.

- *Engineering personnel* must evaluate the processes needed for manufacturing of the new product(s). If new or innovative processes are necessary, then additional costs and time must be established.
- *Production or manufacturing* screens the impact of new product(s) on current plant production systems. They indicate if required skills, personnel and physical plant are available or must be attracted.
- *Research and development* department plays a vital role in the screening procedure. They control the technology of the development process and indicate technological possibilities and limitations concerning ideas and prototypes.
- *Purchasing department* must draw the balance between availability of proper materials (e.g. ingredients, packaging), costs (e.g. of materials, of transportation to get the materials), reliability of supply, and quality (i.e. complying with specifications of design). Product costs are a major screening tool.
- *Quality department* should consider potential hazards and indicate relevant control points.
- In addition, *packaging expertise* is also required to advise on proper packaging materials with suitable properties for the new product, at an early stage of product development. In practice, packaging considerations commonly enter the process at the end, often resulting in (unnecessary) shelf life problems.

It must be remarked that in smaller companies one person often fulfils several roles, whereas large companies have distinct departments.

According to Forza and Salvador (2001) it is essential to *maintain communication* during the whole design process. Horizontal communication among departments is essential in the design process in order to make trade-offs in translating customer needs into product specifications. Overcoming inter-functional communication barriers, due to different worlds of thoughts, languages, task priorities, responsibilities and physical barriers, is possible in off-site workshops. Here specifications are written jointly, which will allocate communication to take place simultaneously rather than through sequential information exchange on specifications between departments. Forza and Salvador also stated that Quality Function Deployment (QFD) can support communication, because planning and communication routines will help in translating customer requirements into production and manufacturing specifications. The main purpose of information exchange in this context is to prevent one product feature being defined without sufficient information on how this choice impacts on product performance as a whole.

4.6.3 Project management

A project is an interrelated set of activities that has a defined starting and ending point. It should result in a unique product or service. Furthermore, each project is unique even if it is routine. Uncertainties, such as development of new technologies or timing of certain events, can change the character of the project. Finally, projects are temporary activities because personnel, materials, and facilities are assembled to accomplish a goal within a specified time frame. After that time the project team is disbanded (Krajewski and Ritzmann, 2007).

Successful project management involves co-ordination of tasks, people, organisations, and other resources. Three major elements are the project manager, the project team, and the project management system.

The *project manager* has the responsibility to integrate the efforts of people from various functional areas (departments) to achieve specified project goals. Traditional organisational hierarchies tend to slow the progress of projects due to lack of communication, co-ordination, and sometimes motivation. A project manager can overcome these obstacles to project completion. The manager is responsible for establishing project goals and providing the means to achieve them. The project manager must also specify how the work will be done and ensure that appropriate hiring of employees is done and any necessary training is conducted. The project manager must demonstrate leadership and provide the motivation necessary to accomplish the tasks required. Finally, the project manager evaluates progress and takes appropriate action when necessary. In product development projects, the project manager will usually be recruited from the marketing department or the R&D department.

The *project team* in product development will consist of members as described in section 4.6.2, i.e. cross-functional teams. Sometimes there are representatives from outside firms involved in the project team. Size and composition of the team may fluctuate during the life of the project. The team will be disbanded when the project has been completed.

The *project management system* consists of an organisational structure and an information system. The organisational structure defines the relationship of the project team members to the project manager.

- The *traditional structure* is the functional organisation, whereby the project is housed in a specific functional area, presumably the one with the most interest in the project. The project manager must negotiate assistance from personnel in other functional areas. Under this structure, the project manager has minimal control over the timing of the project, but resource duplication across functional areas is minimised.
- At the other extreme is the *pure project structure*, whereby team members work exclusively for the project manager on a particular project. Although this structure simplifies the lines of authority for the project manager, it could result in significant duplication of resources.
- A compromise is the *matrix structure* depicted in Figure 4.13. The activities in a design project are depicted in the upper part of the figure. Figure 4.13 shows how different departments in the organisation may contribute to a variety of project activities, which are logically organised for project purposes. The matrix structure is a combination of the traditional and the project structure. In fact, it is an attempt to pool the advantages of the two structures by using permanent cross-functional teams to integrate functional expertise with a divisional focus. Workers in a matrix structure belong to at least two formal groups at the same time - a functional group and a product or project team. They also report to two bosses - one within the functional department and the other within the project team (Schermerhorn, 2007). The two-boss system is the major disadvantage, because it is susceptible to power struggles between functional supervisors and team leaders, both trying to gain authority.

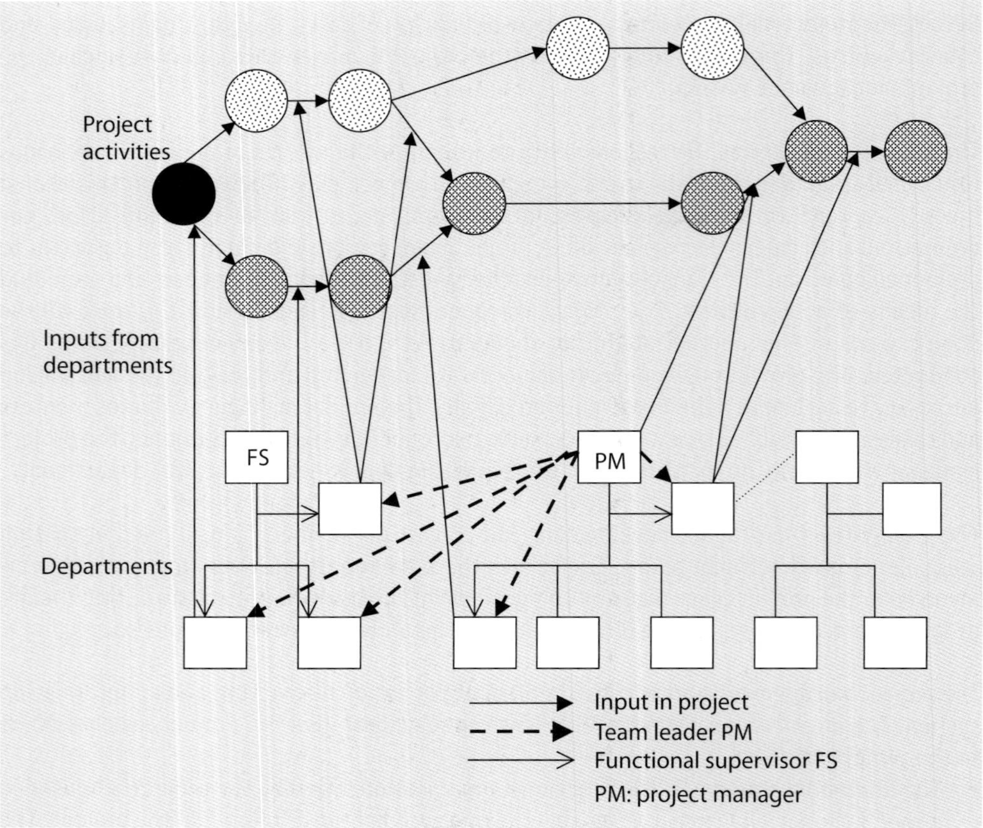

Figure 4.13. Different departments contributing to project activities in a matrix structure.

The project management system should also provide integrative planning and control of the project, accumulation of information related to performance, costs, schedule changes, and projected time and costs of project completion (Krajewski and Ritzman, 2007). *Project planning* is mostly based on network planning methods. A well-known technique is PERT, programme evaluation and review technique, which identifies and controls key activities to complete a project. It ensures that all activities on all paths are executed in proper sequence and in time. The path requiring the longest time is designated as the critical path and it attains special attention in the control process (Schermerhorn, 2007).

Besides project planning, *project control* is very important because of the high probability of changes in both market conditions and engineering. Changes increase product design costs, especially when they are made late in the design process, which results in the *escalator effect*. The escalator effect refers to the process in which the design becomes less flexible and the cost of incorporating engineering changes tends to increase by a factor of 10 as the product moves from concept stage to market launch (Noori and Radford, 1995). Therefore, in design projects, project control based on go-no-go decisions accompanying each phase is very important (Figure 4.14).

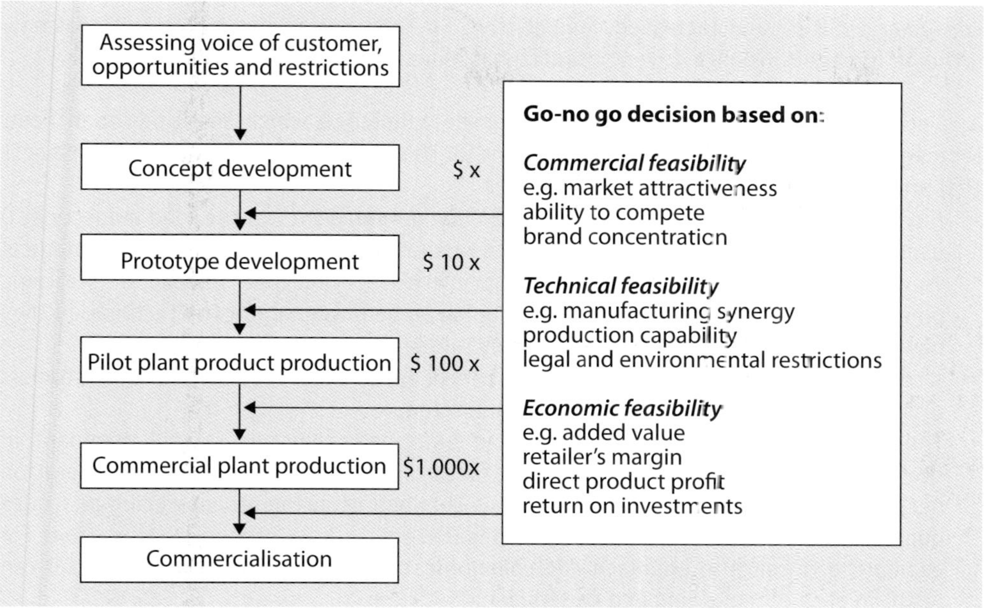

Figure 4.14. Control in the design process (x = costs of modifying the design).

At the end of each design phase an analysis should be made of commercial, technical and economic feasibility. However, one should recognise that information about project results in the beginning of the process is usually rather unclear. Therefore, proper estimates should be made in order to judge expected project outcome in relation to strategic goals and objectives. Less successful projects can be ended then, before too much money and time has been invested.

4.7 Quality design in practice

Quality design processes have become a core activity due to the dynamic and competitive environment in which agribusiness and food companies have to operate. In general, product development can be characterised in two dimensions, i.e. product origin and difficulty. Product origin refers to where the product idea was created, which can be divided into originated from Research and Development (R&D) or from marketing. Difficulty refers to complexity of the new product and technology used. Based on these two dimensions pharmaceuticals are commonly characterised as originated from R&D and highly difficult, whereas foods are typified as originated from marketing and not difficult. Although these characterisations tend to overemphasise extremes, it clearly shows how food product design is generally considered. As a matter of fact, a part of the (new) food products can be classified as originated from marketing and not difficult. However, there is a trend towards more complex food products like ready-to-eat meals and nutraceuticals. In these food concepts the technological degree is high and therefore must be considered as difficult. Moreover, the drive from the market has been intensified in the last years by a more critical consumer attitude and increased competition. So food design has become increasingly difficult, but the consumer driven potential has increased as well.

As a consequence, the food industry faces a major challenge, which might be too difficult for many companies in this sector. The reasons for this may be due to the following aspects that are typical for the food industry:

- *Product design as major activity of 'Research and Development'*. In the food industry R&D usually proceeds with the full design process and considers it mainly from a technological viewpoint. Sometimes even individual priorities are decisive instead of strategic objectives as set by top management. People can be biased about the technology they have developed themselves and which they refuse to abandon.
- A *restricted market orientation*. Many companies have less contact with their customers or consumers and market research is often limited.
- The *functional organisation structure*. The functional structure is still dominant in the food industry and often results in a focus on discipline and lack of co-operation between departments. Cultural differences between departments marketing and sales departments versus technical research departments hinder proper collaboration. For example, difference in language and communication between marketing and R&D can result in misunderstanding and frustration.
- *Process design activities are initiated too late*. Modifications in process design are often carried out just after problems occur during pilot plant production.
- *Lack of proper project organisation*. Product design is commonly considered as a task of one department. Instead it should be considered as a joint project that is properly managed by using, amongst other things, appropriate project planning methods.
- *Restricted collaboration* with other members from the production chain. Although intensive mutual dependencies exist between members of food productions chains, there is commonly low collaboration between suppliers and customers in the area of design. One of the factors is knowledge protection, but a lack of proper chain management also plays a major role.
- *Feedback of design process to strategy of organisation*. Designs regularly have to be modified or design projects cancelled because strategic goals have changed.

This list of potential reasons for bottlenecks in food design processes is not complete but shows that there are many improvement opportunities.

From the techno-managerial approach, solutions for this type of problems should combine the unique technological aspects with appropriate management attention. This should occur in supply, production and distribution. Three points in the design process require in particular special attention, i.e.:

1. During the initiation phase, when consumer demands and needs have to be translated into technical and technological specifications
2. During product and process design (concept and prototype phases), when technological and technical knowledge must be adequately transferred to prototypes.
3. In the process where technical and technological knowledge and experience of production and distribution have to be perfectly translated into production and distribution concepts.

Within the techno-managerial approach, Quality Function Deployment (QFD) can be a strong tool during the initiation phase. QFD is a tool that enables a structured collaboration between people from different disciplines (or departments) in order to design a new product based upon consumer demands. The house of quality and other planning matrices of QFD provide a systematic approach for translating consumer (marketing) language into technological language, so supporting communication between different departments like marketing, sales, engineering and production. Moreover, QFD forces discussions about prioritising quality attributes in terms of consumer demands, future production and costs. In this way it enables pro-active planning and control of the design process so preventing the escalator effect.

The concept, prototype and pilot phase (Figure 4.2) can be strongly supported by modelling techniques. These techniques can predict behaviour of e.g. microbial, enzymatic, chemical or physical processes in fresh and processed food as a function of the applied technological processing and storage conditions. In addition, more complex food-packaging interactions can be modelled. These technological models can provide knowledge and information at an early stage (even before prototype production), which enable managers to consider the consequences of different choices. Based upon these considerations, for example, better choices can be made with respect to packaging concepts where a proper fit between barrier properties and product requirements is necessary. This type of information can also enable a better tuning of time-temperature requirements of sensitive products on the one hand, and logistic planning on the other. So, although the existing models are often simplified reflections of much more complex situations, they can support decision-making at an early stage of product development.

Above mentioned methods and techniques are only valuable and effective when they are properly embedded in the organisation. Consequently, attention should be paid to principles of concurrent engineering and cross-functional collaboration, not only with other departments within one organisation but also between partners in the production chain (like suppliers and customers).

Chapter 5
Quality control

Controlling means not just inspection but also taking corrective actions when the performance is not meeting specifications. The major objective of quality control is to produce a product that complies with targets within set tolerances. Therefore, a profound understanding of the sources of variation is required.

In this chapter relevant technological and managerial factors, which are important for understanding food quality control, will be discussed. After a general introduction (section 5.1) the control process will be detailed in its basic steps (section 5.2), followed by an analysis of the most important control activities linked to the basic technological functions (section 5.3). Then, statistical tools will be explained together with some typical measurement and analysis methods for food control (section 5.4). Finally, food quality control will be discussed in the context of management of the company (sections 5.5 and 5.6).

5.1 Introduction

Variation is a typical characteristic of food products because they consist of biological materials. In particular, raw materials directly delivered from primary production (fruits, vegetables, milk, meat, etc.) to processing companies show considerable variation. The large variation in incoming materials is often just a fact and can be scarcely influenced by primary production. Standardisation is partly realised by modification of recipe and/or processing conditions. But, for instance, for fresh fruits and vegetables, variation remains during the whole production chain including the moment of consumption. As a consequence of the large variation in food production chains, there is considerable uncertainty about the actual properties of raw materials and products. This uncertainty can be partly overcome by control during manufacturing of food products. Quality control can be considered as an important process in the realisation of the quality of food products.

In Figure 5.1 the quality control function is described in its main elements. It includes a decision-making process wherein quality control objectives are translated into product quality within acceptable tolerances by measuring product and/or process properties, by comparing the measurement results with target values and tolerances, and by taking corrective actions, when necessary.

This process is dependent on the ability to measure product properties and processes (both technological processes and food handling processes). For instance, measuring product temperature is simple, but analysing microbial contamination is not so easy. Measuring fat content is rather straight forward, but measuring homogeneity after mixing is more difficult. Measuring food handling processes is always difficult due to lack of objective measurement tools.

But there are typical tools and methods for measuring food product properties and for analysing and controlling these in a statistically reliable way. Besides that, typical organisational measures can support the effectiveness of the control process, by for

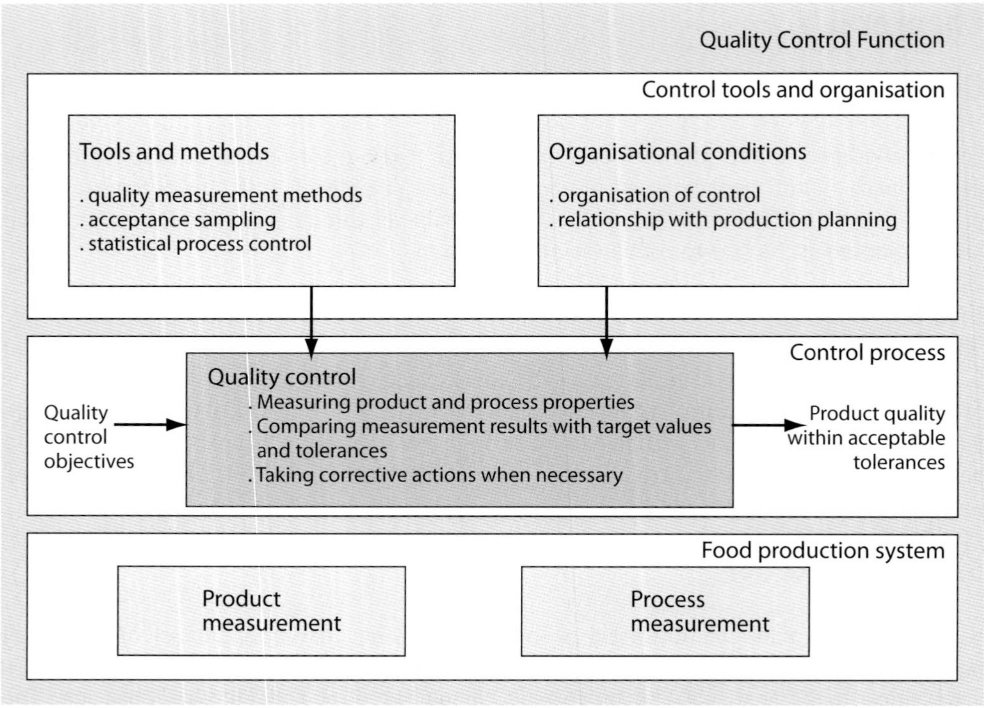

Figure 5.1. Quality control function.

instance bringing responsibilities for control to the work floor. These factors are described in the next sections.

5.2 The quality control process

Quality control is a basic activity of food quality management and the objective is to keep product properties, production processes, and human processes between certain acceptable tolerances. Quality control is the ongoing process of evaluating performance of both technological and human processes and taking corrective actions when necessary (Evans and Lindsay, 2005).

> *The core mechanism of quality control is to maintain desired quality in product and production processes by measuring product properties and processes, comparing actual measurement outcomes with the standard, and taking necessary corrective action to ensure that final quality will meet or exceed customer/consumer needs and legal requirements.*

In this section the activities in the quality control process, as well as the decisions involved are described.

5.2.1 Quality control activities

The above mentioned control activities together form the control circle. Control circles are not only applied at the level of production and processing but also at management level. However, this section is focused on aspects of control circles used in the operational manufacturing process.

Control circles in food production and processing can be described in four basic activities (Figure 5.2), i.e.:

1. *Measuring* a product or process parameter, e.g. sugar content, process temperature.
2. *Testing* to compare measured value/data against the target/norm/standard and specified tolerances/limits.
3. *Regulating* to determine in which direction the product/process must be corrected and by how much.
4. *Taking corrective actions* to ensure that the actual action (e.g. reworking product batch, correcting process temperature) is carried out.

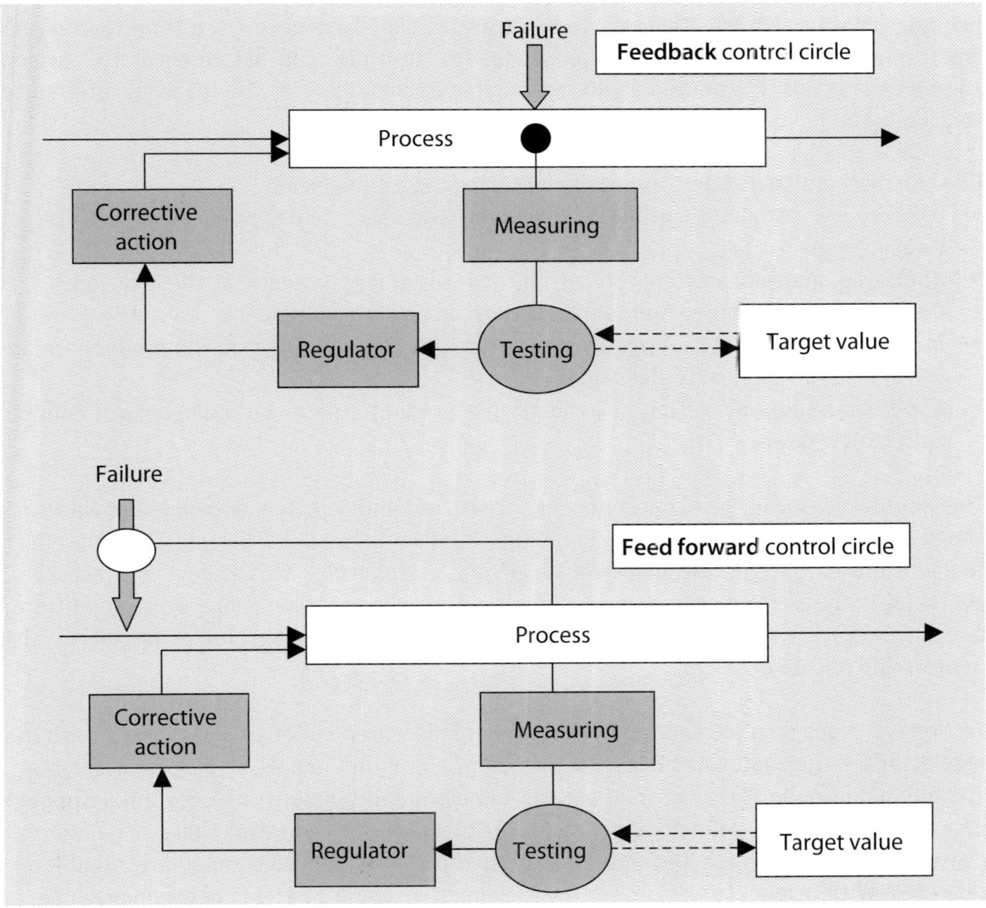

Figure 5.2. Quality control process and activities.

It should be noted that regulating and taking corrective action is often embodied in one handling, person or equipment.

Measuring the product status includes taking a representative sample from a batch (e.g. of raw materials, or process batch) and directly measuring (e.g. by visual inspection, or pH measuring) or analyse (by analytical equipment) the product sample. Measuring the process status refers to checking the actual process parameters (e.g. what is the actual temperature in the cold room, what is the actual pressure during processing). Important concerns in measuring are the appropriateness of the sampling design (e.g. is it statistically underpinned or just ad hoc), and accuracy (i.e. is the right value measured), preciseness (check), sensitivity (signal/noise ratio) and specificity (no interference with other substances) of the measuring or analytical equipment. When measuring is done by people then aspects like accuracy and competences of the food handlers and conditions under which they make visual inspections (enough light, time, support tools) are important concerns.

Measuring can be done automatically (or continuously) or manually. Measuring can be visual or instrumental like pH, time, temperature or flow rate. It can also involve analyses like microbial or sensory analyses. These analyses usually require more time than direct measurements, which can have consequences for the time span before corrective action can be carried out. For details about measurements and analyses the reader is referred to section 5.4.3.

One can distinguish different types of measuring devices, such as:
- *Off-line,* i.e. manual sampling followed by transport to laboratory for analysis or measurement.
- *At-line,* i.e. manual sampling followed by analysis or measurement at the location.
- *On-line,* i.e. automatic sampling followed by automatic analysis.
- *In-line,* i.e. the signal is observed by a sensor (thus no sampling) in the product stream and translated to an external signal.
- *Non-invasive,* i.e. measuring of a signal in a product stream without physical contact with the product.

The way of measuring has consequences for the reliability of the outcome and how fast the process can be corrected. Usually fast measurements are used, like measuring product temperature to control the inactivation of micro-organisms; this is fast and accurate. Microbial analyses are then only executed to verify if the set time-temperature conditions, and the actual time-temperature in the product are sufficient for inactivating the undesirable micro-organism.

Testing is the action of comparing the outcome of measurement or analysis results with the established target and tolerances. Testing can be done by people or may be automated. The outcome can be a number (e.g. amount of colony forming units of a certain pathogen) or a visual result (e.g. colour, appearance). For this part of the control circle, often control charts are used in which the actual results, the target and tolerances are graphically reflected. With respect to the causes of variation, the combined effect of common causes is reflected in the tolerance levels, which can be statistically established. Specific causes can be noticed in the control charts as the out-of-control situations. There are several types of control charts. The choice of a certain control chart depends on *what* must be monitored and the *type* of data. For example, different charts are used for variable (i.e. measured along a continuous scale) and attribute data (e.g. pass or fail). More details about control charts are described in section 5.4.2 on statistical process control.

Regulating is the process of determining which corrective action is required based upon the result of the comparison with the target value and tolerances. So, the extent (great or small) and direction (positive or negative) of the corrective action are set. Commonly, regulators are employees, like operators, supervisors or quality inspectors, but in automated production processes regulators are automated equipment, like PLC's (process line computers). These regulators must be selected based upon the character of the process and the required accuracy. A wide variety of regulators are used in practice. They vary from very simple regulators to complex systems. Some common regulators include:

- the *open-close regulator*, which is the simplest regulator, and has only two fixed positions;
- the *proportional regulator*, in which the magnitude of the action is proportional to the deviation of the process parameter;
- an *optimising regulating system*, in which several process parameters are compared to various targets in order to obtain an optimal regulation;
- an *expert system*, which is a more sophisticated regulator, incorporating expertise and knowledge of the system. For example, in the washing process of re-usable PET bottles, a pattern recognition system is used. Headspaces of bottles are quickly analysed (within seconds) and the patterns of volatile compounds are identified and compared with the information in the expert system. The bottles are removed if the patterns do not fit, e.g. because there are still contaminants in the bottles after washing.

Taking corrective action is the actual handling that is carried out as a consequence of exceeding the target tolerances, i.e. out-of-control situation. The corrective action can be an instrumental (e.g. temperature increase) or a human handling (e.g. removal of the non-conformance products). The accuracy of the corrective action is essential for a successful execution of the control circle. Moreover, it should be clear who is responsible for taking the corrective actions, and how the findings should be documented. Documentation is important as basis for improvement processes and as part of assurance.

There are various forms of control circles, but the basic principle is the same in each. Two common control circles are the feedback circle and the feed-forward control circle, as illustrated in Figure 5.2. In the *feedback* control circle, the corrective action is performed *after* the trouble has occurred in the process. In the *feed-forward* control circle, the process trouble is recognised *before* it occurs. The feed-forward circle anticipates the expected situation based on measurements earlier in the process. For example, the composition (e.g. sugar content) of incoming raw tomatoes can be analysed in order to modify the recipe prior to production to obtain ketchup that complies with the specifications.

For the correct functioning of the control circle it is important that the basic elements are properly tuned. A profound knowledge of the process is required to assess the appropriate measuring parameters that correspond with the quality attributes that must be controlled. Moreover, the control circle must consist of all basic elements/activities and it must be tested *in situ*, prior to use. Besides tuning, another important aspect is the run-time of the control circle. The process stability must be such that during the run-time (i.e. the time between measuring the deviation and performing the corrective action) no unacceptable troubles occur. In other words, the run-time must be short enough to get no troubles in the intervening time. As a matter of fact, the corrective action cannot be always carried out in the same process. Where the action is performed depends, among other things, on the time between measuring or testing and the corrective action, and on the type of process (batch or continuous).

5.2.2 Quality control decisions

The food quality decisions grid (Figure 3.17) shows which decisions can be seen as a result of the quality control activities, i.e. decisions on:

- Out-of-tolerance of product properties and corrective action (influencing food dynamics), e.g. measuring incoming material properties and rejecting the supplied batch when it doesn't meet specifications, or final product inspection and accepting the batch when it meets customer specifications.
- Deviation from requirements on process parameters, equipment and buildings and corrective action (influencing technological conditions), e.g. stopping production when pasteurisation temperature is below standard, or taking samples on surfaces of equipment, testing them on microbes and when necessary deciding on extra cleaning.
- Out-of-tolerance of people actions and corrective action (influencing human dynamics), e.g. visual inspection of hygienic behaviour and, in case of insufficiency, showing them how to do it correctly.
- Deviation from requirements on personnel, procedures, and information systems and corrective action (influencing administrative conditions), e.g. checking if necessary procedures and instructions are there, and when not, giving orders to provide them.

These different types of decisions show the complexity of control processes, dealing with all kinds of variation and with different influences on quality at the same time.

Quality control is not exclusively a task of the quality department; it is the job of everyone in the company. The quality department will however do specialised inspections and tests, and will support departments in building up and maintaining their control systems.

5.2.3 Variation

In Chapter 1 we defined quality as 'Quality is meeting or exceeding customer expectations'. Another definition of quality is as follows: 'Quality is inversely proportional to variability' (Montgomery, 2008). This definition implies that if variability in the important characteristics of a product decreases, the quality of the product increases. Reduction of variation leads to reduced scrap and rework, fewer line stoppages, greater planning efficiencies and what might be more important, happier customers and repeated sales.

Every process is subject to variation. Variation is a natural phenomenon, it is everywhere. We live in a world of variability. No two units of product produced by a manufacturing process are identical. Some variation is inevitable. Even products in a properly maintained production line - with adequately trained and motivated workers, the right raw materials, expert supervisors, and quality control employees who know what they are doing - cannot be perfectly uniform, with no defects and no variation (Hubbard, 1996). Therefore, no process is absolutely predictable, but the performance of any reasonably reliable process can be predicted within limits.

Food production is especially challenged to deal with variation. According to Gould and Gould (2001), one of the greatest variables the food processor works with is the materials used for the manufacture of a given product, including all the raw materials. As an example, fruits and vegetables experience changes in product properties from all kinds of causes like growing area, seasonal effects, harvest maturity and storage temperature (Tijskens *et al.*, 2000). Raw material variability is a challenge in every segment of food processing.

If a product is to meet customer requirements, generally it should be produced by a process that is stable and repeatable. More precisely, the process must be capable of operating with little variability around the targets of the product's quality characteristics (Montgomery, 2008). To properly establish quality control, an understanding of product and process variation and the causal mechanisms that create that variation is needed. To regulate or improve a process we must understand the causes of variation.

As stated before, in any production process, regardless of how well designed or carefully maintained it is, a certain amount of inherent or natural variability will always exist. This natural variability is the cumulative effect of many small, essentially unavoidable causes (Montgomery, 2008). These sources of variation include people, materials, machines and tools, methods and measurements, and the environment. In the framework of statistical quality control, this natural variability is referred to as *'common causes'* of variation or also *'structural variation'*. A process that is operating with only common causes of variation present is said to be *in statistical control*. Shewart (1980) once stated 'A phenomenon will be said to be controlled when, through the use of past experience, we can predict, at least within limits, how the phenomenon may be expected to vary in the future'. Here it is understood that prediction within limits means that we can state, at least approximately, the probability that the observed phenomenon will fall within the given limits. The critical point in this definition is that control is not defined as the complete absence of variation. Control is simply a state where all variation is *predictable variation*. Common causes are an inherent part of the process. Common causes generally account for 80-90% of the observed variation. The extension of common causes can only be reduced by structural improvements in personnel (e.g. by education), materials, machines, tools, methods, measurement and/or environment.

Other kinds of variability are referred to as *'special causes'* or *'incidental variation'*. This variability in key quality characteristics usually arises from three sources (Montgomery, 2008): improperly adjusted machines, operator errors, or defective raw material. Such variability is generally large and usually represents an unacceptable level of process performance. A process that is operating in the presence of special causes is said to be out of control. Special causes of variation account for 10-20% of the variation.

Summarising it can be said that a process can be perceived as some combination of machines, materials, methods and people that transforms an input material into a product (Figure 5.3). This product output has one or more observable quality characteristics. Some of the process variables x_1, x_2,..., x_p refer to common causes and are controllable. The other factors, z_1, z_2, ..., z_q refer to special causes, they are uncontrollable. Special causes can be detected by the use of (graphical) control charts (see Figure 5.15 for an example).

Better understanding of the sources of variation which could lead to deviations from pre-established standards might enable food companies to better control product outcomes. In an industry where quality is synonymous with consistency and conformance, managing variability is a critical first step towards higher-quality products.

With respect to quality control in food production and processing, a number of typical variables are important to elaborate in more detail, i.e.:

- *(Raw) materials* in food processing often form a large source of variation. This can be assigned to the following aspects. Raw materials from plant or animal origin can have a natural variation of 10%. Causes of variation relate to inherent variation of materials (cultivars/breed), seasonal effects, harvesting and breeding practices, storage and transport conditions. Moreover, raw materials are often unstable due to a variety of reactions that occur immediately after e.g. harvesting, disruption or processing (Chapter 2). Furthermore, batches of materials or raw products are often inhomogeneous which can complicate quality control. Also the low concentrations of agents with a high impact (e.g. off-flavours or pathogenic bacteria) put high demands on quality control. Finally, the origin and or history of raw materials is often unknown, which renders the quality control more difficult. As a consequence of these sources of variation, the tolerances should not be specified too narrowly unless it will result in unsafe products.
- Moreover, the large variation in raw materials puts high demands on the *methods and measures* used. Especially, sampling (taking the right number of samples at an appropriate location) and sample preparation are crucial for the control process. Also, the time span

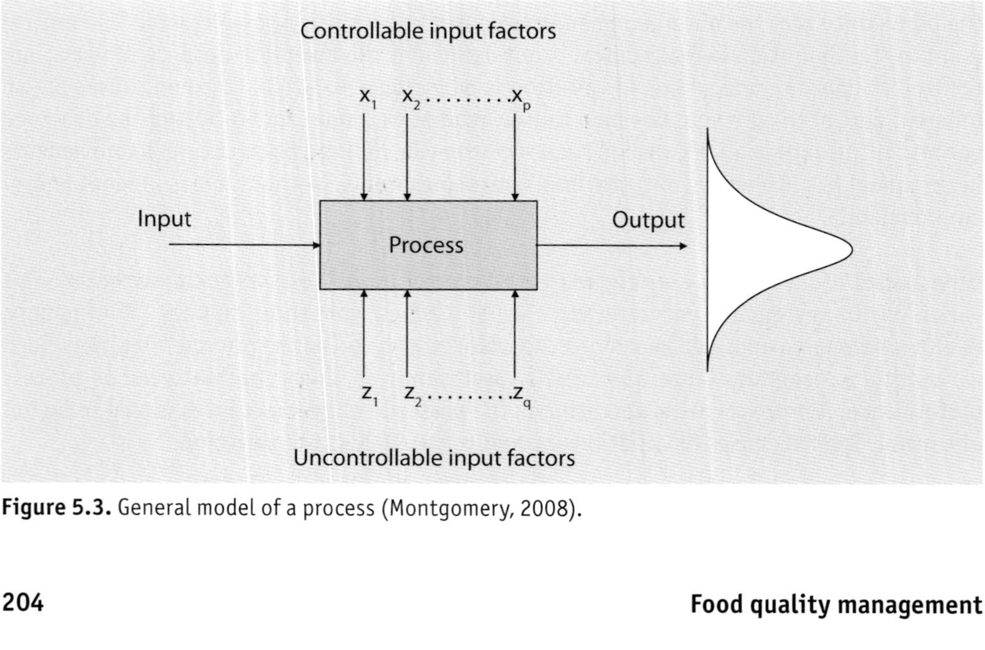

Figure 5.3. General model of a process (Montgomery, 2008).

of some sampling methods (e.g. microbial analyses can take more than three days) makes it difficult to react quickly to disturbances in the process. Furthermore, the methods are often destructive, which means that the products that are actually consumed have not been controlled. Thus the methods and measures, especially the way of sampling, can result in a large source of variation.

- Another important source of variation is the *people*. On the one hand in food production processes personnel with rather low education levels are often employed. On the other hand, hidden safety risks (like contamination from the growth of pathogens due to unhygienic conditions or improper processing) require a rather profound knowledge of microbial processes in agri-food. Also shift-work can be a great source of variation. Mistakes due to misunderstanding or too low levels of knowledge can result in quality problems. Another aspect related to personnel is the fact that in food processing, many control processes include visual inspections. This type of measurement is very sensitive to variation due to the objectivity of the measure and low accuracy.
- Finally, the *machines and equipment* used in food production are often very specific for a certain application. As a consequence equipment is not easily exchangeable. For each type of equipment specific types of controls are required, which can form a potential source of variation. In addition, the *design* of equipment and food processing machines is essential for food safety. Microbial contamination due to poor design and/or improper cleaning puts high demands on the control process and is a major source of variation.

5.3 Quality control in food production

In Chapter 3 about quality management, a distinction is made between product decisions and resource decisions (section 3.2.1). These types of decisions are relevant in quality control. Quality can be controlled via product and resources decisions. It has been hypothesised that both product and resource decisions influence business performance. However, a strong relationship exists between them. Quality control in food production will be analysed on these two types of decisions with respect to the three basic technological functions, i.e.:

- input, the supply control;
- transformation, the production control;
- output, the distribution control.

All these processes will be described from both the product and resource point of view, as shown in Figure 5.4. The essential difference between product and resource control is what is measured and judged. In product control, product properties are measured and tested against standards for these properties; while in resource control, process parameters and/or resource characteristics are measured and compared against standards. Examples of control activities are mentioned in Figure 5.4 and are further explained in the next sections.

5.3.1 Supply control

In a company, the purchasing department is responsible for obtaining the supplies (i.e. materials, parts and services) needed to produce a product or to provide a service. The importance of purchasing is clear considering that in manufacturing up to 60 percent of the costs of finished goods comes from purchased parts and materials. In retail and wholesale

	Supply	Production	Distribution
Product control	• Incoming material control • In storage control	• Product control • In storage control	• Distribution control • Service control • Complaint service
Resource control	• Supplier evaluation • Supplier relationships	• Process control • Process capability analysis • Equipment control	• Distribution channel control • Customer relationships • Market analysis

Figure 5.4. Quality control activities from a product and resource perspective.

companies it sometimes can exceed 90 percent. However, not only are costs important; in food chains in particular, the quality and timing of deliveries of goods have a significant impact on operations. In fresh-product business the quality of the end products are almost completely dependent on supply-quality. Control activities in supply management can be summarised in the purchasing cycle as sown in Figure 5.5.

The purchasing cycle begins with a request from within the organisation to purchase material, equipment, services or other items from outside the organisation. The cycle ends when the purchasing department is informed that a shipment is received in satisfactory

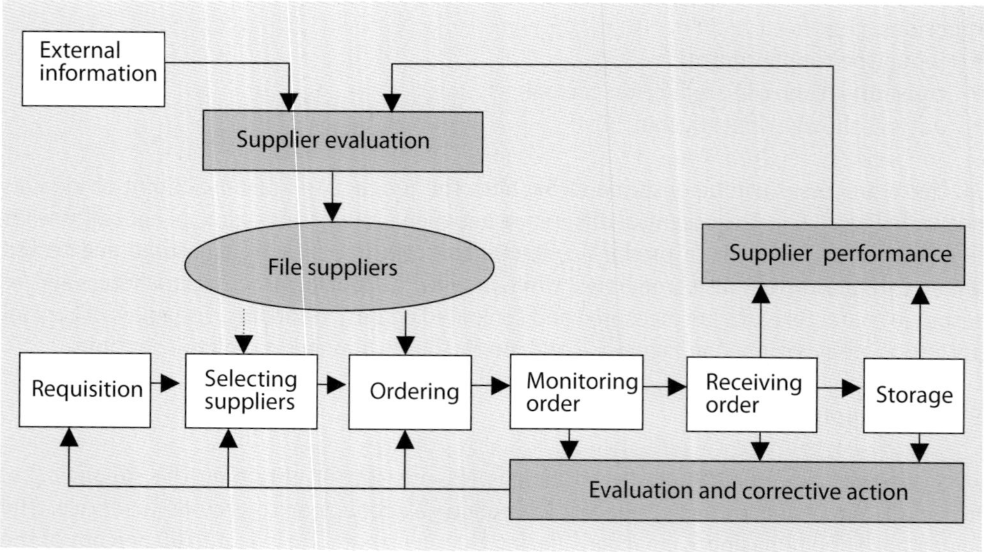

Figure 5.5. Supply control.

condition and that it is properly stored so that quality will continuously meet production requirements. The main steps in this cycle are (Stevenson, 2006):

- *Receiving the request* (requisition). The request includes a description of the item or material desired, the quality and quantity necessary, and desired delivery dates.
- *Selecting suppliers*. The purchasing department must identify suppliers who have the capability to supply the desired goods. If no suppliers are currently listed in the files, new ones must be identified. Vendor ratings may be used to choose appropriate vendors. Rating information can be communicated to the vendor with the idea of upgrading future performance.
- *Placing orders*. If the order involves a large expenditure, particularly for a one-time purchase of, for example, equipment vendors will usually be asked to bid for the order. Operating and design personnel may be asked to assist in negotiations with a vendor. Large-volume, continuous-usage items may be covered by blanket purchase orders, which often involve annual negotiation of prices with deliveries subject to request throughout the year. Moderate-volume items may also have blanket purchase orders, or they may be handled on an individual basis. Small purchases may be handled directly between the operating unit requesting the items and the supplier, although some control must be carried out to keep it in hand.
- *Monitoring orders*. Routine follow-up on orders, particularly large orders or those with lengthy delivery schedules, allows the purchasing department to predict delays and communicate this information to the appropriate operating units. Likewise, the department must inform suppliers about changes in the quantity and delivery needs of the operating units, so they have time to adjust their plan.
- *Receiving orders*. Receiving must check incoming shipments from vendors for quality and quantity. It must report to purchasing, accounting and the operating unit that requested the goods. If the goods are not received in satisfactory condition, they may have to be returned to the supplier for credit or replacement; or they must be subjected to detailed inspection. Again, purchasing, accounting and the operating unit must be notified of this information. In either case, vendor evaluation records must be updated.
- *Storage*. Received goods must be kept in storage before they are used in production. Quality control is needed on stored goods and/or storage conditions due to the specific requirements of the product (for instance temperature control in food storage).

In agribusiness and food industry, quality control in supply management is often based on acceptance sampling, due to cost consideration. However, in some cases full (100%) inspection is required due to specific customer requirements or safety issues. Figure 5.5 shows the feedback loop from quality inspection points to phases where corrective action can be taken. This feedback loop is called the *product-control* activity.

The *resource-control* activity can be summarised as measuring supplier performance, evaluating supply sources and developing supplier relationships. Good suppliers are a vital link in the supply chain (Stevenson, 2006). Late deliveries of materials or parts, or missing or defective items, can cause large problems for manufacturers by disrupting production schedules, increasing inventory costs and causing late deliveries of end products. Substandard supplier services can have similar results.

Choosing a supplier involves taking into account many factors. A company should consider price, quality, the supplier's reputation, past experience with the supplier and service after sale; this process is called vendor analysis. It is very important that a company provides suppliers with detailed specifications of the materials or parts that are needed, instead of buying items off the shelf. In this way the suppliers can comply with the ordered quantities and specific production requirements. The relationship with the supplier depends on the situation. For instance, in the food industry, intensive communication is often required to design packaging materials that comply with the specific requirements of the manufacturer regarding product demands, packaging and distribution process characteristics.

Typical factors a company should take into account when *evaluating a supplier* are:

- *Price*. This is the most obvious factor, along with any discounts offered, although often it is not the most important one.
- *Quality*. What is the quality reputation of the supplier; what quality assurance system does the supplier have; is the supplier able to collaborate in future design processes?
- *Services*. Special services can sometimes be very important in selecting a supplier, like replacement of defective items, sharing purchase-market information, fast response on complaints can be key factor in selecting one supplier over another.
- *Location*. Location has an impact on shipping time, transportation cost and response time.
- *Emergent cases*. Sometimes quality claims due to origin can play a role (food and drinks).
- *Inventory policy of supplier*. When customer demands are fluctuating (for instance seasonally) guarantees are needed about in-time deliveries.
- *Flexibility*. The willingness and ability of a supplier to respond to changes in demand and to accept design changes can be important considerations.

Periodic *supplier audits* are a means of controlling supplier's production capabilities, quality and delivery problems and resolutions, as well as suppliers' performance on other buyer criteria. If an audit reveals problem areas, a buyer can address these before they will result in serious problems. Among the factors typically covered by a supplier audit are management style, quality assurance, materials management, the design process used, process improvement policies, and procedures for corrective actions and follow-up. Supplier audits are also an important first step in supplier certification programmes.

Supplier certification is a detailed examination of the policies and capabilities of a supplier. The certification process verifies that a supplier meets or exceeds the requirements of a buyer. This is generally important in supplier relationships, but it is particularly important when buyers are attempting to establish long-term relationships with suppliers. One advantage of using certified suppliers is that the buyer can eliminate much or all of the inspection and testing of delivered goods. Although problems with supplied goods or services might not be totally eliminated, there is much less risk than with non-certified suppliers. Many companies rely on standard industry certifications such as ISO 9000 - perhaps the most widely used international certification (see also Chapter 7).

Nowadays, *supplier partnership* has become increasingly important. In the past, too many firms regarded their suppliers as rivals and dealt with them on that basis. The relationship between customer and supplier has been seen as an arms' length relationship. It was seen to be necessary to keep a distance, neither becoming too dependent upon a supplier nor allowing a supplier to become too dependent upon any customer. This view led naturally to multiple sources as the norm because a single source, i.e. having one supplier for each good, placed too much power in the hands of the supplier.

The concept of managing the total supply chain, however, stressed the benefits of a mutual dependence between supplier and customer. This is totally different from the traditional approach. In the mainstream activity, the survival of each partner is correctly perceived as being dependent on the other. This implies a long-term commitment, not simply to do business with each other but also to find new business for each other. Suppliers support their customers even at the cost of short-term difficulties for themselves. Benefits in terms of profitability or of quality improvement are there to be shared, based upon an open relationship. This is therefore the notion of partnership based upon a perception of mutual advantage, or the so-called win-win synergy.

Juran (1992) has developed a framework to distinguish adversarial (rivals) and teamwork relationships with suppliers (Figure 5.6). Nowadays much effort is put in building sustainable long-term supplier relationships to realise food quality (e.g. White, 2000; Rábade and Alfaro, 2006). Also, global sourcing (i.e. retrieving raw materials from all over the world) puts increasing demands on supplier relationships when controlling resources.

For more information about supplier evaluation, performance and relationships the reader is referred to Bossert (2004), Theuvsen and co-editors (2007), and Gordon (2008).

Dean and Evans (1994) observed that in the teamwork concept suppliers must be recognised as crucial to organisational success, because they make it possible to create customer satisfaction. Neither the quality nor the cost of the organisation's product can reach competitive levels and continuous improvement without the contributions of suppliers. Customer-supplier relations, therefore, must be built on trust rather than suspicion. Besides, from the obvious teamwork implications, monitoring supplier or customer behaviour does not add any value to the product. Trust is developed over time through a pattern of success by all parties to fully and faithfully deliver what was promised.

Element	Adversary concept	Teamwork concept
• Number of suppliers	• Multiple, often	• Few, often single source
• Duration of supply contracts	• Annual	• Three years or more
• Criteria for contracts	• Conformance to specifications	• Fitness for use
• Emphasis of surveys	• Procedure, data systems	• Process capability; quality improvement
• Quality planning	• Separate	• Joint
• Pattern of collaboration	• Arms length, secrecy, mutual supervision	• Mutual visits, disclosures, assistance

Figure 5.6. Adversary versus teamwork concept of Juran (1992).

5.3.2 Production control

The purpose of quality control of production activities is processes to run in an desired manner. Companies accomplish this by measuring process outputs using statistical techniques (see also 5.4.1 and 5.4.2). If the results are acceptable no further action is required; unacceptable results call for corrective action. Product control involves *acceptance sampling* procedures. In addition, more and more companies give emphasis to designing quality into the process, thereby greatly reducing the need for inspection of final products (Stevenson, 2006). In fact, attention is nowadays shifted towards controlling the process, which is referred to as (statistical) *process control*. In process control, feed-forward mechanisms are preferred rather than feedback control. A feed forward system will prevent defects and variation. As a matter of fact, product inspections will be still necessary when applying process control, but inspection results then are intended to verify that the process is running in conformity to the standard.

Production control activities are summarised in Figure 5.7, wherein both product and resource control are described. The main steps are (steps 1 to 4 are product control and steps 5 and 6 are resource control):

Production planning is usually not recognised as a quality control activity. However, it should be noted that it has a great influence on quality performance. Production planning combines information about equipment (availability, quality performance, and cost), and customer orders, including product specifications, into a production schedule. Many problems with product quality can be introduced here, when costs and delivery time result in production schemes in which time pressure makes it impossible to meet quality standards.

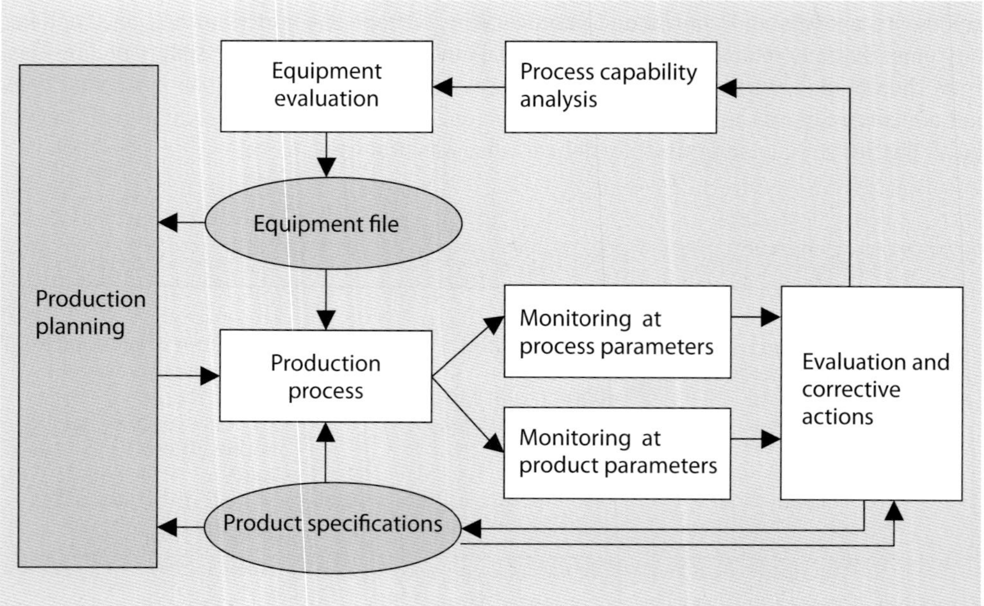

Figure 5.7. Production control.

In the *production process* quality activities are related to the process-organisation. Different types of process-organisation can be classified, such as:

- *Job-shop production*: a highly skilled workforce using general-purpose equipment makes a wide variety of customised products. These processes are referred to as a jumbled-flow process because there are many possible routings through the process.
- *Intermittent flow*: a mixture of general-purpose and special-purpose equipment is used to produce small to large batches of products. Often specifications of the product are new or specifically for one batch (example: a catering firm)
- *Repetitive flow*: several standardised products follow a predetermined flow through sequentially-dependant work centres. Personnel are typically assigned to a narrow range of tasks and work with highly specialised equipment. Food manufacturers often can be typified as repetitive flow producers.
- *Continuous flow*: commodity like products flow continuously through a linear process (example: a sugar refinery) (Noori and Radford, 1995).

Quality activities are clearly different in each process-organisation. In the job-shop production professional skills are very important. During production the personnel carry out quality control, by checking their own work. Occasionally, quality is not clearly defined because unique products are made in an experimental stage. On the other hand, in continuous flow production products are specified in detail and there is much knowledge about the product and the production process. Quality control is for a great part incorporated in the process and automatically worked out. Nevertheless, operators control the process and bring with their experience and insights the finishing touch to the process, being responsible for the important last percentages of quality.

Measuring in the production process can occur at three points: before, during and after production. The logic behind checking conformance before production is to assure that inputs are acceptable. The logic of checking conformance during production is to make sure that the conversion of inputs to outputs is proceeding properly. The logic of checking conformance of output is to make a final verification of conformance before passing goods to customers (Stevenson, 2006).

Questions that arise when developing measuring procedures are: how much to inspect, how often and at what points in the process inspection should occur. *Low-cost, high volume* items (continuous flow production) often require little inspection because cost associated with defectives are quite low and the processes that produce these items are usually highly reliable, so that defectives are rare. Of course when there is a safety risk, as in food production, process conditions are controlled very intensively, e.g. the intensive cleaning and disinfection to prevent microbiological problems. When a clear relationship between product requirements and process parameters exists, then the production processes can be controlled by measuring these process parameters. For instance control of temperature and pressure in food processes are often closely related to product requirements.

In *high-cost, low-volume* production more intensive inspection is necessary due to the high costs associated with passing defectives. In general, inspection costs must not exceed the decrease in failure costs. Concerning the question where to inspect, some typical inspection points are:

- raw materials and purchased parts;
- finished products;
- before a costly operation (for instance before adding expensive ingredients);
- before an irreversible process (for instance before filling and labelling food products in cans).

Evaluation and corrective action means that product and process parameters are checked for consistence with standards and tolerances. Management must establish a definition of out-of-control situation. Even a process that is functioning as it should, will not yield output that conforms exactly to a standard simply because of natural variations inherent in all processes (especially for food processes with biological materials). The main task of quality control is to distinguish random from non-random variability, because non-random variability means that the process is out of control. When a process is judged out of control, corrective action must be taken. This involves identifying the cause of non-random variability (for instance work equipment, incorrect methods or materials, failure to follow specified procedures) and correcting it (Stevenson, 2006).

Process capability analysis is important when evaluation indicates an out-of-control situation in the process. Capability analysis involves determining whether the inherent variability of the process output falls within the acceptable range of variability allowed by the design specifications for the process output. The principle of process capability study is already described in section 4.5.5, so the focus here is on possible solutions for out-of-control processes (situations). Stevenson (2006) mentioned a range of solutions:

- redesign the process so that it can achieve the desired output;
- use an alternative process that can achieve the desired output;
- retain the current process but attempt to eliminate unacceptable output using 100% inspection; and
- examine the specifications to determine if they are necessary or could be relaxed without negatively affecting customer satisfaction.

Equipment evaluation includes assessment of possibilities of improvement by redesign or modification of the equipment and on changing maintenance (and clearing) programmes. Redesigns and modifications are mainly intended to improve reliability and, in the food industry, hygiene. Hence the connection between quality control and quality design. In Chapter 4 quality design techniques for reliability analysis (FMEA) and hygienic design have been described in more detail.

Maintenance includes all activities involved in keeping a system's equipment in working order. Maintenance is concerned with avoiding undesirable results due to system failure. Even when the results are not catastrophic, they can be disruptive, inconvenient, wasteful and expensive.

Maintenance falls into two categories: preventive maintenance and breakdown maintenance (Heizer and Render, 2006). *Preventive maintenance* involves performing routine inspection and servicing, and keeping facilities in good repair. Preventive maintenance activities are intended to build a system that will find potential failures. It makes changes or repairs that will prevent failure. Preventive maintenance is much more than just keeping machinery and equipment running. It also involves designing technical and human systems that will keep the productive process working within tolerance: it allows the system to perform. The emphasis is on understanding the process and allowing it to work without interruption. *Breakdown or corrective maintenance* is remedial; it occurs when equipment fails and then must be repaired on an emergency or priority basis. Figure 5.8 indicates that failure occurs at different rates during the lifetime of equipment; it may follow different statistical distributions.

A high failure rate, known as infant mortality, exists for most equipment in the initial (starting) phase. Once the machine or process is settled, a study can be made of the MTBF (mean time between failures) distribution. When the distributions have a small standard deviation, then it is a candidate for preventive maintenance. Whether preventive maintenance is applicable depends on the relationship between preventive maintenance costs and the costs of breakdowns. With modern technical instruments for diagnosis of wear-out (measuring metal parts of oil, ultrasonic measuring of vibrations, infrared thermograph) prevention costs can be kept low, whereas breakdown costs tend to become higher due to safety risk and losses in goodwill and image. When reliability cannot be achieved and prevention maintenance is not appropriate or does not work, management can enlarge or improve the repair facilities. Remedial maintenance can then be performed in order to put the system back into operation.

5.3.3 Distribution control

Whereas purchasing deals with the incoming flow of materials, distribution deals with the outgoing flow of materials (Krajewski and Ritzman, 2007). Distribution control involves management of the flow of materials from manufacturers to customers and from warehouses to retailers. It also includes the storage and transportation of products. Distribution broadens the marketplace for a firm, adding time and place value to its products. Quality

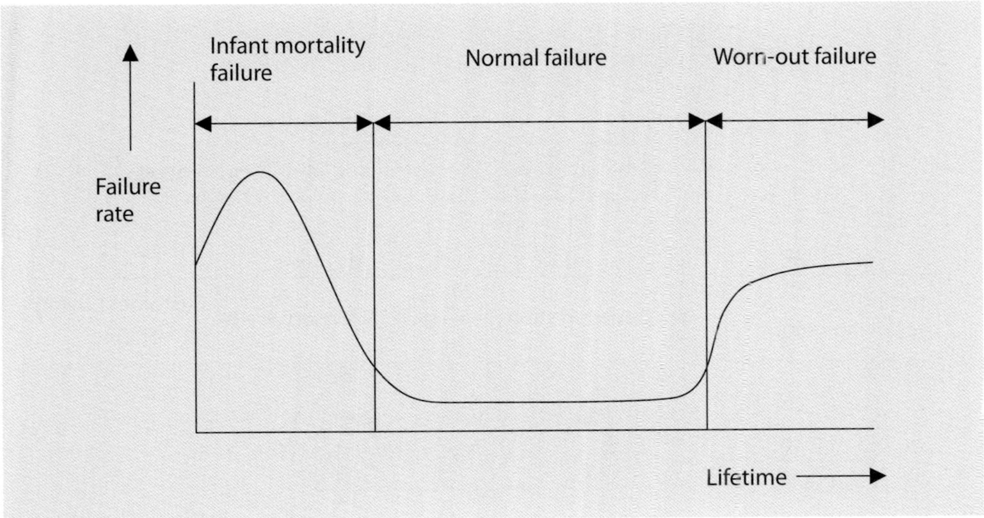

Figure 5.8. Lifetime failure rates.

control in distribution management concerns decisions around transportation and storage of products and measuring the whole process including consumer usage of the products (Figure 5.9).

The main steps in distribution control include product and resource decisions. They are:

- *Selling, receiving orders*. In the food industry, selling is commonly established in annual contracts and within one contract many delivery-orders are agreed. Quality in selling means that no deals are made that cannot be sustained in the course of the year. Detailed agreements about delivery time, price and quality specifications are a part of this process.
- *Transportation and storage*. In food distribution, transportation and storage can be complex processes within a long supply chain, often requiring international transport movements. The major quantity of food products follows a 'common' routing from manufacturers to retailers' distribution centres and from there in combined freights to retail stores. *Product control* during transportation and storage concerns maintaining quality of the products by monitoring the product and taking corrective action when necessary. *Resource control* involves transportation equipment, distributor organisation and storage conditions, both in distribution centres and in retail stores. In the food industry temperature and humidity conditions are very important control parameters, as well as hygienic conditions. Distributors are often obliged to introduce quality assurance systems in their organisation to ensure acceptable quality levels.
- *Buying and consumer usage*. In Figure 5.9, the common situation in the food business is described where retailers (storage) are considered as customers and the individual person buying from the retail store is regarded as consumer of food products. Quality

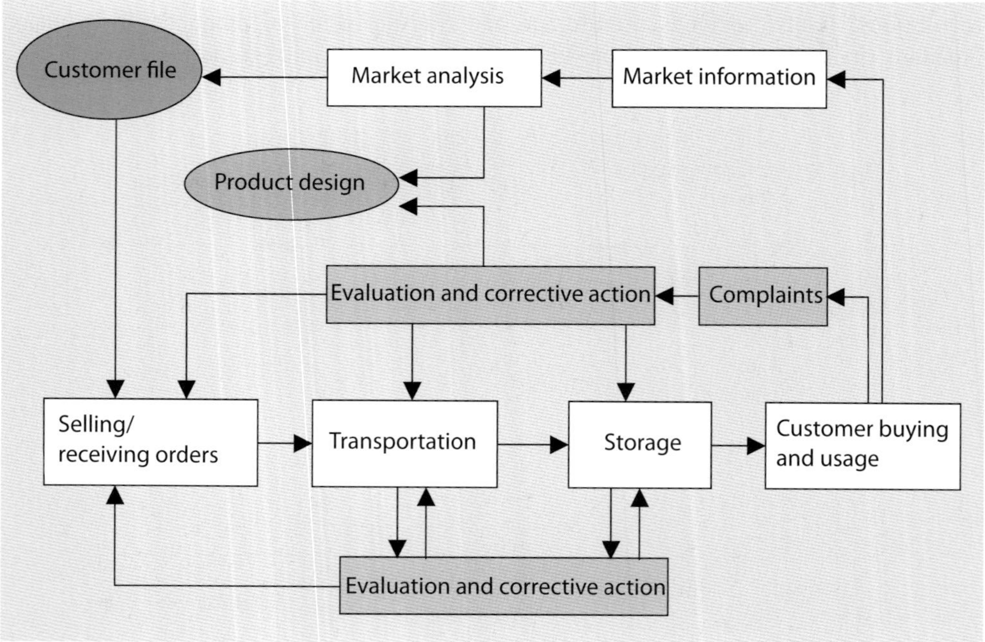

Figure 5.9. Distribution control.

control in the stage of buying and consumer usage is often restricted to, for instance, information on packages, in leaflets and magazines.

- *Complaints, market information and analysis*. Complaints and market research provide lots of information about quality and quality performance. This information can be used for quality improvement. Firstly on a short-time basis, when consumer problems have to be solved immediately and secondly in the long term, where this information can be a starting-point for product redesign. In Chapter 4 (customer-oriented product design) the process of turning market information into product design has already been described. An example of a complaint procedure is described in Figure 5.10 (Korteweg, 1991). When designing a complaint procedure, use can be made of the ISO standard 10002:2004, which provides requirements on developing a complaint procedure (ISO, 2004).

The advantages of a controlled complaint procedure are insight into complaints per product and into cause and effect relationships, and faster solutions of complaints. Essential elements in a complaint procedure are:

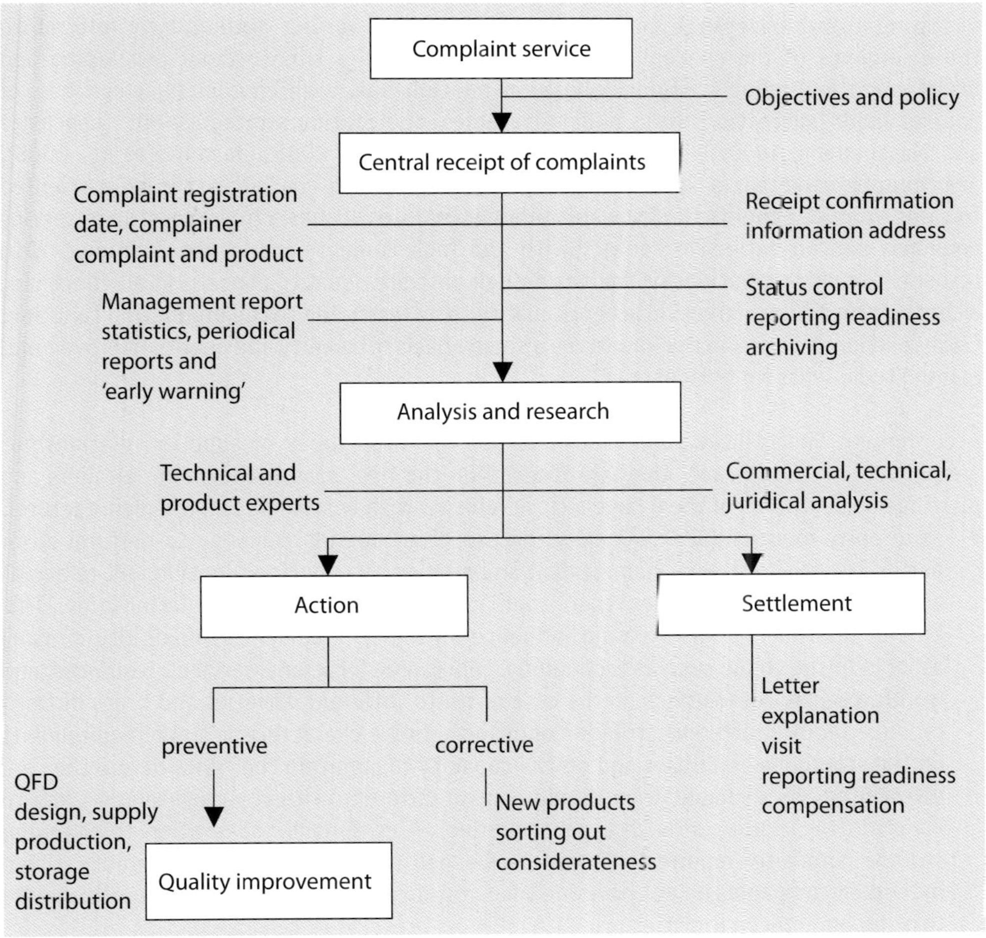

Figure 5.10. Complaint procedure.

- *Central receipt of complaints*, where complaints are registered, controlled and reported. This is also the central point for external information.
- A *procedure for analysis and research*, ensuring that complaints are analysed from a technical, commercial and juridical viewpoint.
- A *procedure for adequate action taking*. Firstly, complaints should be settled in a customer/consumer friendly way. Secondly, corrective action could be necessary, for instance delivering new products, or return of products and sorting out bad materials, making a deal that is satisfying for the customer. Thirdly, preventive action should be taken to improve quality, so making quality complaints a basis for organisational learning. Of course a complaint procedure must have the linkage to recall procedures, which are an element of various quality assurance standards and guidelines that are implemented in a company's quality management system (see also Chapter 7).

Considering distribution, relationships are very important. It automatically addresses a discussion about *customer and supplier relationships*.

In the food business customer relationships are becoming more important, particularly in the relationship between manufacturer and retailer. Various authors have referred to the emergence of the concept of relationship marketing. This is about developing and maintaining mutually satisfying long-term relationships with customers, which has, of course, major implications for product strategies, distribution strategies, etc. (Lancaster and Massingham, 1999; White, 2000; Rábade, and Alfaro, 2006; Clements *et al.*, 2008). *Relationship marketing* is the opposite of *transaction marketing*. The first is more oriented towards product benefits than towards product features. It has a higher customer service emphasis and an extensive contact with and high commitment to the customer. With respect to quality, relationship marketing demonstrates quality concern in all functions, whereas transaction marketing accepts quality to be restricted to the production function. Close relationships are, in the long term, a steady basis for quality and quality improvement, leading to benefits for both parties.

According to Simchi-Levi and co-authors (2000), three types of supplier relationships are particularly significant. They are focused on the final part of the chain, retailers and distributors, and some of them have a close relation with customers and the manufacturer.

- *Third party logistics* (3PL) is simply the use of an outside company to perform all or part of the firm's material management and product distribution function. 3PL relations are typically more complex than traditional logistics supplier relationships. This kind of relationship has a strong influence on the development of flexibility in many aspects of the chain, such as technology, geographic locations, services, resources, and workforce size. For instance, in the case of fruits, different varieties and types of fresh produce require different storage conditions. These characteristics put requirements on the storehouse facilities and on the capacity to maintain the fruits before they are distributed. For example, they should control different critical parameters in order to maintain the quality of each type of product. This dynamic adaptation to changing storage parameter requirements can be reached with technological flexibility. Often individual companies do not have the time, resources, or expertise to constantly update their technology on logistics.

- *Retailer supplier partnerships* are built as margins become tighter and when customer satisfaction becomes more important. Then it makes sense to create cooperative efforts between suppliers and retailers in order to leverage the knowledge of both parties. The types of retailer-supplier partnership can be viewed as a succession. At one end there is information sharing, which helps the supplier plan more efficiently, and at the other end there is a consignment scheme, where the supplier completely manages and owns the inventory until the retailer sells it.
- *Distributor integration* (DI) is interesting because distributors have a wealth of information about customer needs and wants, and manufacturers could use this information when developing new products and product lines. On the other hand, distributors typically rely on manufacturers to supply the necessary parts and expertise. Modern information system technology provides solutions in which distributors are integrated so that the expertise and inventory located at one distributor is available to the others. Distributor integration (DI) can be used to address both inventory and service related issues. In terms of inventory, DI can be used to create a large pool of inventory across the entire distributor network, lowering total inventory costs, while raising service levels. Similarly, DI can be used to meet customers' specialised technical service requests by steering these requests to the distributors. DI can also be used to improve distributors' technical ability and ability to respond to unusual customer requests (Simchi-Levi *et al.*, 2000).

5.4 Tools and methods for quality control

As shown in the previous section, quality control is usually carried out at three major points in the food production processes, i.e. at receipt of the (raw or semi-processed) materials, during processing, and prior to and during distribution of the (end) product.

On receipt of the incoming materials, quality control is carried out by product inspection. At this stage it is decided whether batches are rejected or accepted for processing. For this purpose different methods are available like spot-check procedure, hundred-percentage inspection or acceptance sampling.

During processing or manufacturing quality control is executed by measuring product or process conditions at critical points in the process. For a proper functioning of the process, the principles of statistical process should be applied, using control charts to monitor the process.

Before distribution and storage, final inspections are carried out to check the actual quality of the manufactured or processed product. For this purpose, similar techniques to those used for incoming materials control can be employed.

In the last few decades, a shift can be observed from quality control by acceptance sampling towards controlling the supplier by auditing or certification. The supplier is now responsible for the quality of raw materials and products as delivered to the customer. Moreover, a significant shift has occurred from final inspection to quality control by (statistical) process control. Statistical methods can help in taking correct decisions with respect to, for example, acceptance or rejection of a batch of raw materials. Some regular statistics used in quality control are acceptance sampling and statistical process control, which are briefly explained. For details about statistics behind acceptance sampling and statistical process control, the reader is referred to specific handbook like Banens and co-authors (1994), Otten and Verdooren, (1996), and Schilling and Neubauer (2009). Furthermore, an overview and considerations are given on major measurements and analyses used in quality control of food production.

5.4.1 Acceptance sampling

Usually a quality control activity occurs on delivery of incoming raw materials, i.e. delivery control. The aim of delivery control is to decide whether a lot or batch will be accepted or rejected based upon its compliance with set specifications. Several inspection methods are commonly used in agribusiness and the food industry, such as spot-check procedure, 100% inspection, and acceptance sampling.

In the *spot-check procedure*, a fixed percentage of the batch is selected for inspection. For example, 10% of each lot or a periodical selection of every 10th product are selected for inspection. This method has no statistical basis. It gives no information about the risk of making an incorrect decision. The spot-check procedure is often used to check quantities (e.g. number of packaging materials delivered) and not as decision tool for quality.

Hundred-percentage inspection is in fact a sorting method and will in theory eliminate all non-conforming products from a batch or lot. This inspection method is often used for grading of size and/or shape of agricultural products. In addition, 100% inspection is also applied for products with very critical safety requirements or with high failure costs. The limitations of 100% inspection are:

- The accuracy is typically 85%, because the monotony and repetition can create boredom and fatigue of inspectors.
- It is only applicable for non-destructive testing.
- It is often costly and impractical.

An alternative to 100% inspection is acceptance sampling. To decide which inspection method should be applied, the cost of inspection and the cost of failure should be considered. The cost of failure depends on the implications of non-conforming products for customers and the likelihood of failure occurring.

Acceptance sampling is commonly used at delivery of incoming materials but can also be applied before materials are submitted to an expensive operation, handling or processing step. The method is based on statistical principles and provides a judgement of the risk of a decision that is taken. The risk refers to on the one hand, *producer's risk* (α), i.e. the probability that a batch is rejected whereas the actual batch quality was acceptable and on the other hand *consumer's risk* (β), i.e. the probability that the batch is accepted whereas the batch quality level is not sufficient.

Acceptance sampling includes the following steps:

- Inspectors take a statistically determined random sample (n) from a batch or lot (N), i.e. *sampling*.

- Then, the number (or amount, or level of, etc.) of non-conforming items is *determined* by visual inspection, instrumental measurements or analyses.
- Subsequently, the results will be *compared* with acceptance criteria (c).
- Finally, it will be determined if the batch or lot is accepted or rejected, i.e. *batch or lot sentencing*.

Application of acceptance sampling involves several aspects that must be considered, as described below.

Location and type of data

First, the location of acceptance sampling should be determined. Besides at delivery control, it can also be used e.g. after changes in responsibility or before high added-value processes. Furthermore, the type of data that will be used as acceptance criteria should be assessed. Criteria can be either attribute data, whereby the product is inspected on a pass (i.e. conforming) or fail (i.e. non- conforming) basis, or variable data, whereby the measured value of the inspected parameter is recorded and used as the basis of evaluation.

Sampling design and operating characteristic (OC) curve

Design of a sampling plan is another important aspect of acceptance sampling. It includes two key questions i.e. how many should be sampled (n) and how many defectives are allowed in the sample (c). Sampling plans can be designed by using distribution tables (e.g. modified Thorndike chart) or by national or international standards tables. ISO 2859 provides sampling procedures for inspection by attributes to support companies in designing their sampling plans (ISO 2859, 2006) The modified Thorndike chart reflects the relationship between the average number of defectives (x-axis) and the probability of occurrence of 'c' or fewer defectives (y-axis) for different acceptance numbers (c). The design of a sampling plan can influence the probability of acceptance. For a specific sample size (n) and acceptance number (c), the modified Thorndike chart can be used to determine the probability of accepting the batch for a range of values of percentage defectives. Plotting this relationship generates the typical operating characteristic (OC) curve as illustrated in Figure 5.11. The OC-curves show that the probability of accepting a batch (e.g. with 30% defectives) is much higher in a sampling plan with a small sample size (n=5) and high acceptance criteria (c=2), than in a plan with a larger sample size (n=20) and more severe acceptance criteria (c=0). In other words, the chance that the batch is unjustly accepted is higher in the first than in the second sampling plan.

Quality levels

Another aspect is the quality level against which conclusions are drawn with respect to acceptance or rejection of the complete batch. Two common types of quality levels are:

- The *acceptable quality level* (AQL) is the maximum defect level in the complete batch that is still acceptable to the customer. The government can assess the AQL (i.e. legal levels), but usually in practice, both the supplier of raw materials and the purchaser reach agreement on the AQL level.

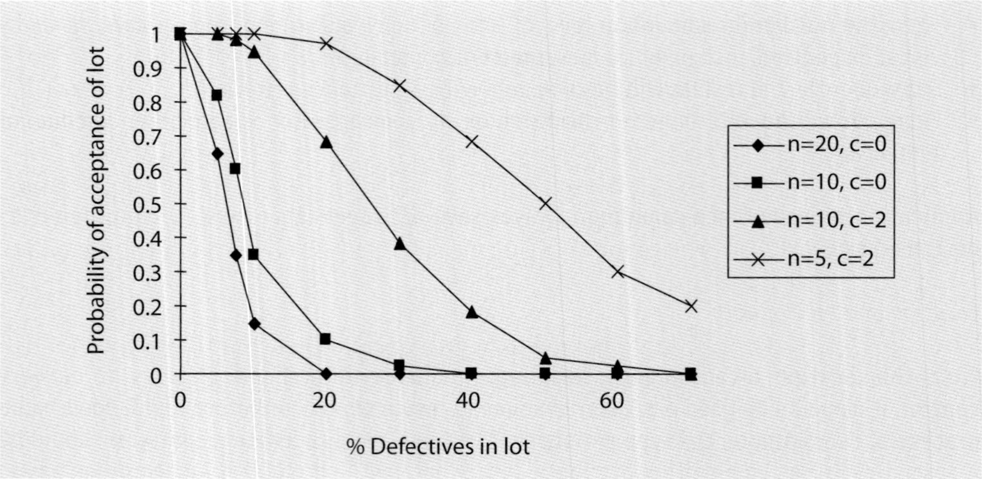

Figure 5.11. Operating Characteristic (OC) curves of different sampling schemes.

- The *limiting quality level* (LQ) or *lot tolerance percent defective* (LTPD) is the defect level in the complete batch that is considered to be unacceptable to the customer. This is the level, assessed by the government or purchaser, at which the batch should always be rejected. The chance that a batch with defects at the LQ level is unjustly accepted must be below 0.1.

If the accepted quality level (AQL), the limiting quality level (LQ), the producer's risk (α) and consumer's risk (β) are agreed, then a final OC-curve can be developed using the Thorndike chart. Figure 5.12 gives an illustration of the AQL and LQ level in an OC-curve.

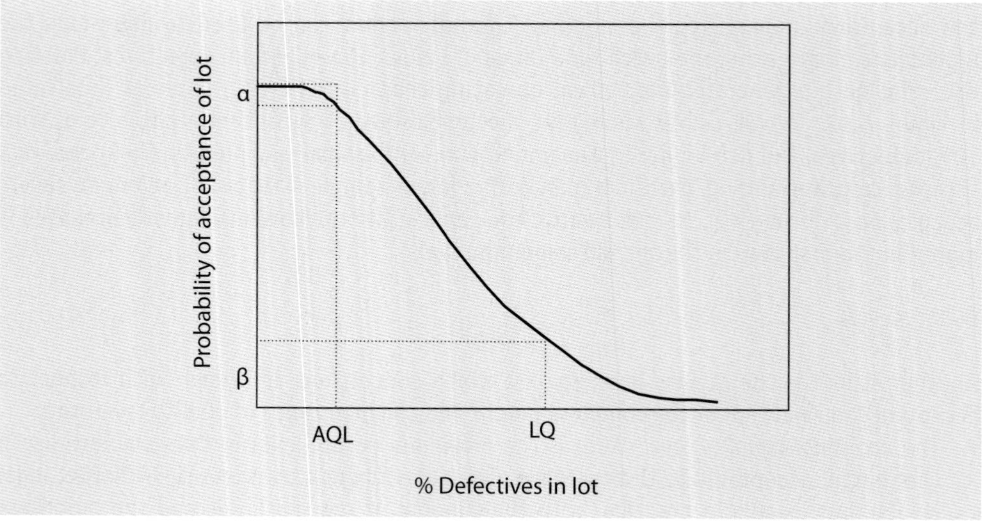

Figure 5.12. Operating Characteristic (OC) curve with acceptable quality level (AQL) and limiting quality (LQ) level.

The OC-curves described above are only suitable for single sampling using attribute data (i.e. inspection on pass or fail basis) and require rather a lot of samples. However, in many practical situations it is desirable to reduce the costs of acceptance sampling. For example, when the process under inspection had previously shown that it is capable of producing materials well within the acceptable quality level and outside the limiting quality level. Moreover, the OC curve is only suitable for attribute and not for variable data. For such situations other types of sampling plans are available, which require smaller sample sizes or sampling plans that can be used for variable data. Some commonly used sampling plans are briefly described below.

A common type of sampling which requires fewer samples is called multiple sampling where up to a maximum of seven samples can be taken. In this sampling plan:

- More but smaller sample sizes (n) are selected and analysed.
- The number of defectives in the initial sample is determined and compared with the acceptance number (Ac).
- If it is less, then the whole batch is directly accepted (using fewer samples than at simple sampling thus less product loss).
- If the number of defectives exceeds the rejection number (Re), then the batch is immediately rejected.
- In the intermediate range (Ac < number of defects < Re), a new sample is selected and analysed.
- The numbers of defects in the first and second sample are summarised and this total number must be compared with the Ac and Re of the second sample.
- Based on this comparison the batch is directly accepted or rejected, or a further sample is taken.
- At the final (maximal 7th) sample, the decision to accept or reject the sample is based on the cumulative number of defects detected in all samples.

Another sampling plan that can be applied to reduce sampling costs is *sequential sampling*. This is an important technique commonly used to monitor ongoing quality of production. The principle of this method is as follows:

- A sample is taken (n=1) and immediately analysed.
- Subsequently a new sample is taken and analysed.
- The cumulative results of both samplings are judged against the acceptance and rejection lines as illustrated in Figure 5.13.
- A batch is *accepted* if the cumulative number of defects passes the line of the accept region. A minimum number of samples are required to pass this line, i.e. minimum sample size for acceptance.
- A batch is *rejected* if the cumulative number of defects passes the line of the rejection region.
- If the value of the cumulative number of defects remained between both lines of acceptance and rejection then no decision can be made, and a multiple sampling plan must be selected.
- The graphical acceptance and rejection lines (sampling design) can be constructed by using the values of ($1-\alpha$, AQL) and (β, LQ) or by using national or international standards such as BS 6001.

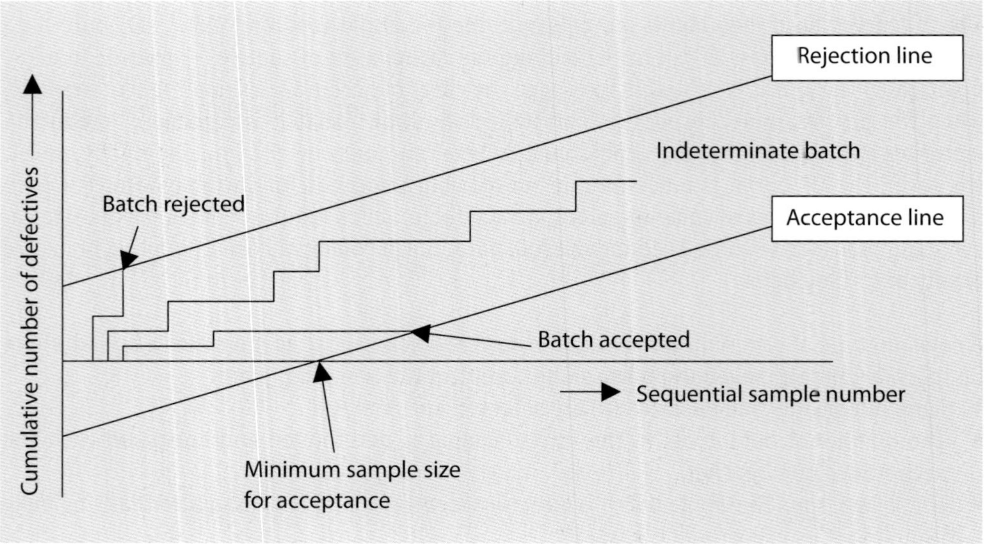

Figure 5.13. Sequential sampling.

The sampling schemes discussed above are suitable for *attribute data* only, which means for pass-failure decisions. However, in case of *variable data* (assuming normal distribution of data), other statistical techniques should be applied with:

- x as the normally distributed variable;
- μ as the mean of the random variable x;
- σ^2 as the (known) variance of x;
- $x_1, \ldots, x_n$ as the a-select sampling with ξ as sample mean.

Single, multiple and sequential sampling plans can be constructed for both attribute data and variable data. For example, in the case of a single sampling plan it is characterised by sample size (n), acceptance criteria (c), and the sample mean x is compared to acceptance criteria. If a lower μ is associated with a better quality (e.g. the lower the bacterial count the better) then the batch will be accepted if sample mean $\xi < c$ (acceptance criteria), it will be rejected if $\xi \geq c$.

If a higher μ is associated with a better quality (e.g. protein content in milk) then the batch will be accepted if sample mean $\xi > c$ (acceptance criteria). It will be rejected if $\xi \leq c$. The values for (n) and (c) of a simple sampling plan can be statistically calculated, using 1-α (α= producer's risk) and β (=consumer's risk) as boundary conditions.

An overview of different sampling methods used for acceptance sampling is given in Table 5.1. For more detailed information about statistical details and construction of multiple and sequential sampling plans the reader is referred to Banens and co-authors (1994), Otten and Verdooren (1996), and Schilling and Neubauer (2009).

Table 5.1. Overview of sampling plans used for acceptance sampling.

Sampling	Characteristics
Single sampling	One attribute data Relatively large sample size (n) Acceptance criteria (c) Critical level (m) Use of OC-curve to assess probability of accepting batch at certain (n, c)
3-Class single sampling	One attribute data Relatively large sample size (n) Critical level divided in 3 classes: $x<m$, $m<x>M$, and $x>M$ If the analysed data $x<m$ then the batch is accepted A restricted number c (i.e. acceptance criteria) is allowed in the middle class $m<x>M$, whereas at $x>M$ the batch is rejected A 3-dimensional OC curve can be used to assess probability of accepting batch at certain (n, c, $P_{acceptance}$, $P_{rejection}$)
Multiple sampling	One attribute data More but smaller sample sizes, n1, n2, n3 ...n7 More acceptance criteria $0 \leq c1 < c2 < c3 \ldots < c7$ Ac: acceptance number Re: rejection number
Sequential sampling	Attribute data Samples taken one by one (n=1) Number of defects are accumulated and compared with rejection and acceptance line If accumulated values remain in intermediate area then other sampling plan
Simple sampling with variable data	Normally distributed variable data Known variance σ^2, expectation μ and sample mean x For e.g. $\mu_0>\mu_1$ (i.e. higher μ if better quality) if $\xi > c$ then batch accepted if $\xi \leq c$ then batch rejected Calculation (n) and (c) for $\mu_0>\mu_1$ with boundary restrictions if $\mu \geq \mu_0$ then chance of acceptance $P_{acc} \geq 1-\alpha$ if $\mu \leq \mu_1$ then $P_{acc} \leq \beta$ with $P_{acc} = P[\xi > c]$

5.4.2 Statistical process control

Statistics are also applied to control manufacturing processes. Statistical process control (SPC) is a specific approach to monitor the operational process in order to identify special variations. Specific variations are out-layers, which exceed the common variation. Common variation is inherent to the process methods, materials, environment and personnel used.

In this section, the procedure for SPC is described and the different types of process control charts that can be used are explained in more detail.

Procedure of statistical process control

The aim of *statistical process control* (SPC) is to monitor the process and to distinguish normal variation from special variation. Normal variation is due to natural variation, which is inherent in the process. It includes all those factors that are not controlled, such as the relative humidity or the temperature of the environment. Special variations represent unusual variability in the process e.g. due to occasional extreme large seasonal differences in raw materials. Prior to the use of statistical process control, a number of aspects must be considered as illustrated in Figure 5.14. The procedure for SPC application includes the following steps:

- *Understanding and defining* the *process*. Firstly, the process must be described (e.g. by developing flowcharts); subsequently the relevant (or critical) process parameters and the available control measures must be identified. Several techniques can be used for this purpose:
  - *Failure mode effects and analysis* (FMEA) for processes, is a systematic procedure to identify failures in terms of cause, effect and controls. Moreover, an indication of the risk is obtained by calculation of the risk priority number (RPN) of the particular failure mode (or critical point).
  - *Hazard analysis critical control points* is a similar approach to FMEA but it focuses on safety failures (i.e. hazards) with respect to the production of agri-food products. HACCP analysis reveals critical control points and identifies the required control measures (see Chapter 7).
  - *Pareto analysis* is a method for identifying the most important processes with respect to quality failures (see Chapter 6).

  An interdisciplinary team should be used to carry out this part of SPC to ensure incorporation of all relevant knowledge. The team should include personnel from various areas, like production, purchase, quality inspection, and maintenance.
- *Analysis* of the *process* is required to assess the normal variation of the process; what is the inherent variability? Firstly, the type of data attribute or variable, should be assessed. Subsequently, the data is collected and the level of the process parameter (i.e. mean or centre) and its spread (i.e. range or standard deviation) must be determined. For SPC, the data should have a *normal distribution* and therefore the distribution of variation has to be established. For this purpose several methods are available including *skewness and kurtosis*, *probability plots* and *chi-squared tests*.
- *Assessment* of *process capability* is required to determine the relationship between natural process variation and the specified tolerances. Thus the manufacturing process should be capable of making products within the set limitations (e.g. legislative limits or norms). The relationship can be presented by the following index:
- *Process potential index* C_p (or *process capability*), which is the quotient of tolerated variation (= upper specification limit (USL)-lower specification limit (LSL)) and the actual process variation, which is 6σ assuming a normal distribution.

  $C_p = (USL-LSL)/6\sigma$

  $C_p = 1$ means that the process is capable, because the actual process variation is equal to the specified tolerances;

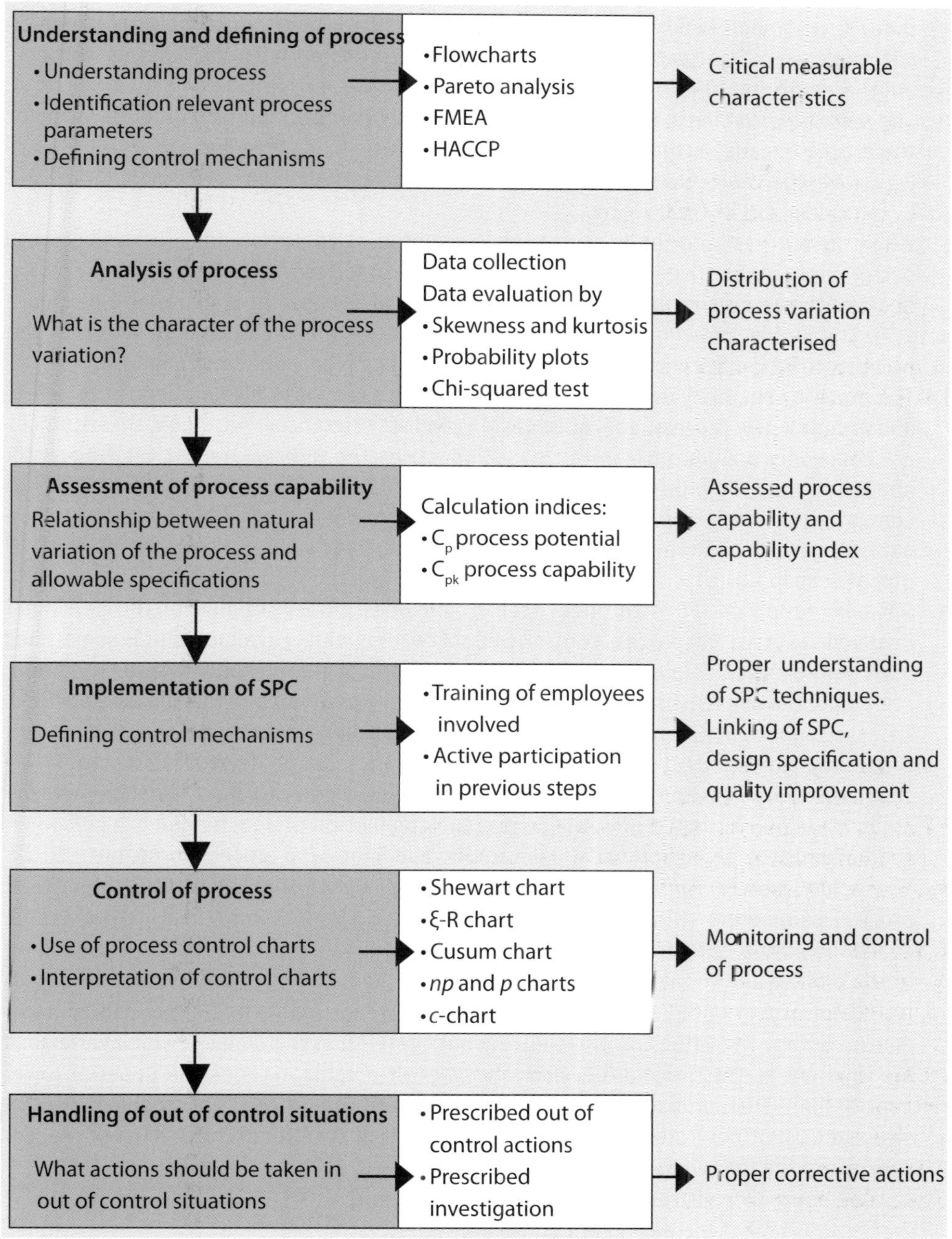

Figure 5.14. Procedure for the application of statistical process control (SPC).

$C_p > 1$ means that the process is very capable;
$C_p < 1$ means that the process is not capable (see Figure 5.15);
The process capability can be improved by either reducing the actual process variability (6σ) or by increasing the specified tolerances (USL-LSL). However, the process potential

index C_p considers only the variation. The extent to which the variability is centred within the required specification range is not considered with C_p. For this purpose, another index has been developed.
The *process performance index* C_{pk} (or *process capability index*) indicates the deviation of the process means in comparison to the target value and is given by:
C_{pk} = (average value-nearest tolerance)/ half the effective range
$C_{pk} = (x - LSL)/3\sigma$ or $(USL-x)/3\sigma$
When the process is correctly centred then the process potential index (C_p) and process performance index are equal ($C_{pk} = C_p$). However, in practice the C_{pk} is smaller that the C_p, because the process is not operating in the centre of the specification range (see Figure 5.7). The indices often used in practice are C_p is 1.67 or even 2.00 to be sure that the process variation fits within the specifications. A common value for C_{pk} is 1.33.
The process capability indices can be used to select raw material suppliers, or to select and accept a new process, and/or to design product tolerances. Although, these indices are commonly applied in (capital) good industries the indices are not yet frequently applied in the food industry. This might be due to the fact that the use of statistics to control processes is not yet very common in the food industry. Moreover, large (often natural) variations in raw materials and ingredients may complicate the assessment of process capability.

- *Implementation of SPC* is the next step. It includes proper training of the personnel involved in statistical process control. SPC software is widely available but is sometimes used before the principles of statistical process control are fully understand. Therefore, preferably the employees should be involved from the beginning of the SPC preparations, i.e. starting with understanding the process, the analysis stage and assessment of process capability indices. By the time the SPC group reaches the implementation stage, they should be familiar with the techniques and aware of the background. Furthermore, the application of SPC should not only be just the application of a technique but personnel should know the philosophy of understanding and improving process variation.
- The *actual process control* includes the ongoing monitoring and control of critical process parameters using process charts (Figure 5.15). The graphical nature of the charts helps the operators, supervisors or managers to obtain a better understanding of the process and to take corrective action if required. The interpretation of the charts is a major aspect of SPC. All process charts have a central line and upper and/or lower control limits. The values for the control limits determine the range in which variations are inherent to the process, i.e. *normal causes*. If a value exceeds the upper or lower control limit, than probably a *special cause* disturbed the process, i.e. an out-of-control situation. The three rules commonly used in practice, for the interpretation of process charts, are the following patterns:
  1. One point exceeds the upper or lower control limit;
  2. Six subsequent increasing or decreasing values;
  3. Ten subsequent values above or below the central line.
- Finally, the approach to *handling out-of-control situations* should be clearly defined. What are the corrective actions in case of an out-of-control situation and/or which investigations must be initiated? This last stage of SPC is sometimes not clearly defined. The data of SPC should also form part of the management review of the overall process effectiveness.

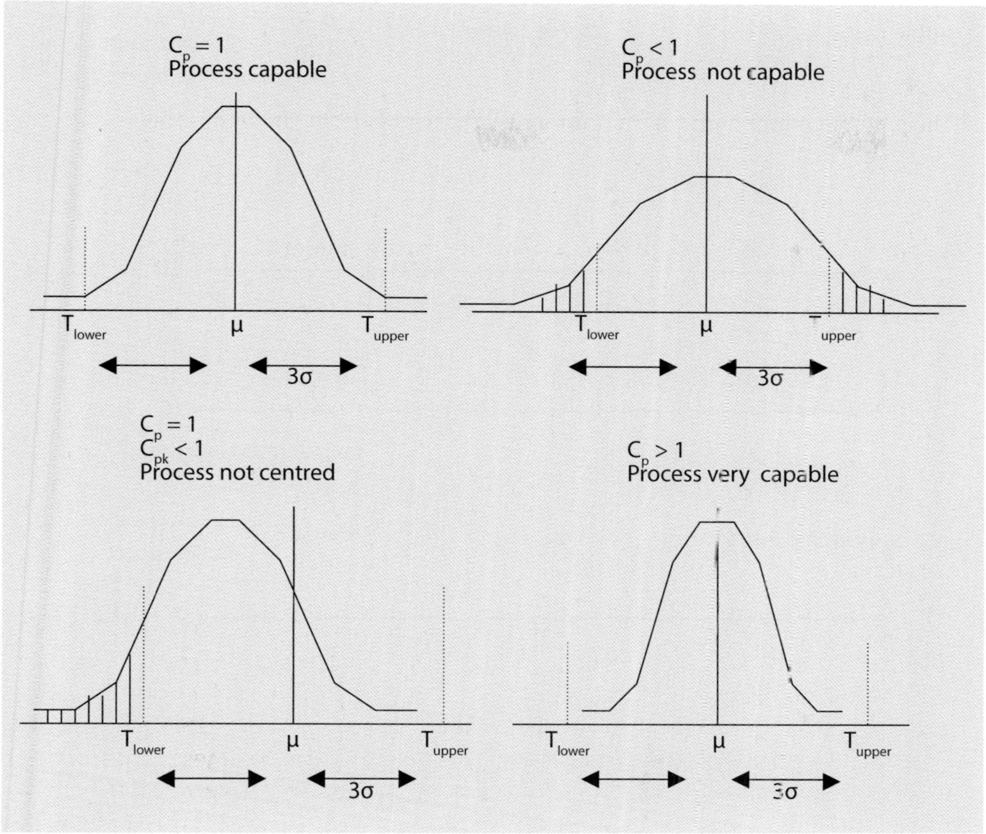

Figure 5.15. Illustration of process potential index (or process capability C_p) and process performance index (or capability index C_{pk}).

Use of control charts

Process control charts can be used to monitor the performance of (production) processes. The choice of a control chart depends on the type of data that can be measured or analysed. This can be attribute (pass or fail) and variable data (e.g. amount of component, temperature). Some charts are only suitable for the detection of out-of-control situations (e.g. Shewart chart), whereas other charts can be used to detect structural but small shifts in data (e.g. Cusum chart). Below a brief overview is given of the control charts used in industry (Figure 5.16).

- *Charts for variable data*. Typical control charts for variable data include the *Shewart chart*, the x-R chart, and the Cusum chart. The Shewart chart plots the control variable against the time (t) or the sequence sampling number (n), as illustrated in Figure 5.16. The control variable can be the sample mean (x), the median (m) or the sample range (R). The Shewart chart consist of the following lines:
  - The *central line* (CL) which is located at the target value. The central line can be the estimated sampling mean (x), or the median of the sampling (M), or the estimated process variation.

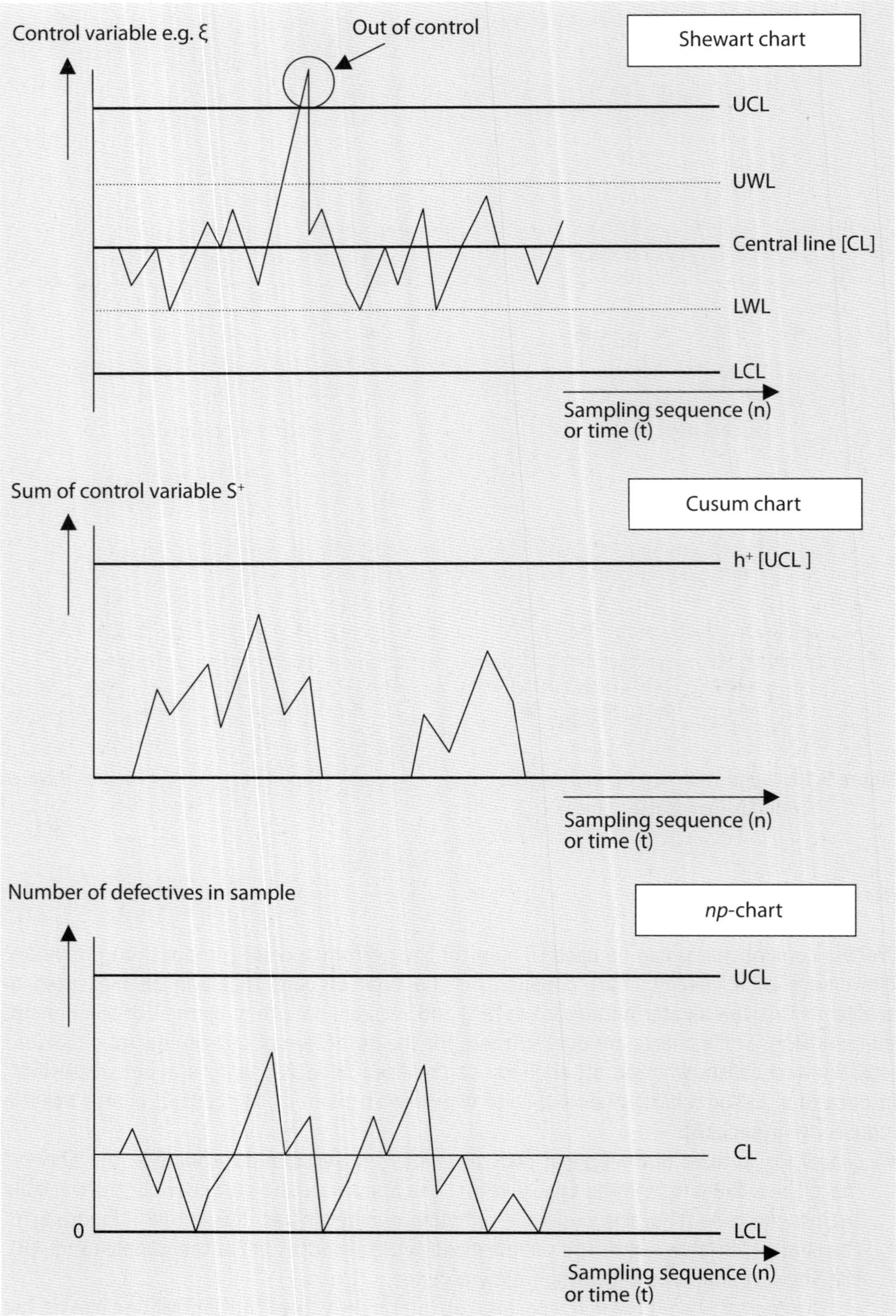

Figure 5.16. Schematic presentation of different process charts: Shewart chart, Cusum chart and *np*-Chart.

  - The *lower warning limit* (LWL) and the upper warning limit (UWL) which are often calculated as follows CL ± 2σ (σ = standard deviation of the control variable).
  - The *lower control limit* (LCL) and the upper control limit (UCL) which are often calculated as follows CL ± 3σ.

  Generally, action is taken if one datum point exceeds the UCL or LCL or when two subsequent datum points exceed the respective upper or lower limit line. The Shewart chart described is a two-sided control chart. However, in some situations there is only an upper or a lower control limit. For example, legislative regulations often put demands on the maximum level of e.g. pesticide residues or residues of veterinary drugs. In such cases the Shewart chart only has an upper control limit, i.e. one-sided. A one-side lower control limit can be applied when a certain composition is claimed, for example the amount of vitamin C; it should not be lower than the LCL.

  The *Cusum chart* (cumulative sum) was designed to identify small but systematic shifts in a process. In contrast to the Shewart chart, the decision to take action is based not only on the last sampling but includes all sampling data. The Cusum chart incorporates all previous data by plotting cumulative sums of the deviations of sample values from a target value.
- *Charts for attribute data.* In the case of attribute data, there are two options, e.g. good or bad, pass or fail. In this perspective, a distinction must be made between defects and defectives. Defect refers to a single non-conforming quality characteristic, whereas defective refers to an item that can have several defects. Attribute data are often visually evaluated or counted and have a broad application in the (agri-food) industry. A common attribute chart is the *p-chart*, also called the fraction defective chart. It monitors the fraction of defectives produced in a lot. Like the Shewart, the *p*-chart consists of a central line, an upper, and a lower control limit (i.e. zero defects). The np-chart is a control for the absolute number of defectives in the sample. The np-chart can only be used if the sample number is constant, and is commonly used as an alternative for the *p*-chart. The absolute number of defectives is more meaningful than the fraction of non-conforming products and therefore easier for production personnel to understand.

For more details about the statistical background and construction of charts the reader is referred to the literature (Banens *et al.*, 1994; Otten and Verdooren, 1996; Montgomery, 2008).

5.4.3 Quality analysis and measurement

As previously mentioned, the typical aspects of food production require knowledge of appropriate sampling techniques. Important aspects that must be considered before sampling or measuring are:

- How and where should samples be taken; thus how to take a representative sample from, for example, an inhomogeneous product, which changes rapidly after harvesting or processing.
- Which analyses or measuring method should be applied; is it destructive or not, what is the reaction time, does it change the content of the quality attribute, etc.
- On what basis are batches or products rejected or accepted; what are the selection criteria.

Quality analyses and measurements are an important facet of quality control processes. They are applied to evaluate or quantify relevant product properties, quality attributes, and/or to control processes. Different sample types can be distinguished (Pomeranz and Meloan, 2000). Incoming raw materials samples are analysed to check whether they are in compliance with the specifications set by the suppliers. New suppliers submit *buying samples* of raw materials to test them for practical use. *Process control samples* must often be rapidly measured or analysed (e.g. temperature, pressure, pH) to adjust the process in order to obtain uniform quality products. Finished products are analysed to check whether the agri-food products meet legal requirements, comply with product specifications, are acceptable to their customers and/or have the desired shelf life. Complaint samples are submitted by customers or consumers and are analysed to detect process mistakes. Finally, a competitor's samples are analysed to obtain information for product development.

Food samples can be controlled by direct measurements, e.g. measuring of pH or visual inspection of colour. However, if samples cannot be directly measured, then an analysis procedure is needed before measuring, including (1) sampling (2) sample preparation and (3) actual measurement or analysis (Figure 5.17). The different aspects of the analysis procedure are described below, keeping in mind that they are often a major source of variation.

Sampling can contribute considerably to the total error as compared to the actual analysis or measurement. Ideally, the sample should be identical to the bulk and must reflect the same intrinsic properties, e.g. same textural properties, or equal concentration of toxins, or same taste. According to Pomeranz and Meloan (2000) typical causes of variation in sampling of agri-food products include:

- Irregular shapes, e.g. round particles flow more easily into the sample compartments than angular particles of similar size.
- During or after sampling, changes in composition may occur, such as water loss, evaporation of volatile compounds or enhanced enzymatic reactions by mechanical injury.
- The non-homogeneity of many agri-foods both between products and within the product. For example, compounds can be located in specific compartments or they can be unevenly distributed.

When taking the samples the following considerations should be taken into account:

- Purpose of inspection, e.g. for acceptance sampling or process control.
- Nature of the material to be tested, e.g. homogeneity, the history, cost of the raw material.
- Character of the test/lab procedure, e.g. (non)destructive, importance of test.
- Nature of lot/batch/population, i.e. sizes and number of units involved, how are sub-lots treated?
- Level of assurance required, cost versus information obtained.
- Characteristic of sampling parameter, i.e. attribute or variable (Gould and Gould, 2001).

Sample preparation is a critical step in the analysis of agri-food products. Major aims of sample preparation are minimising undesirable reactions, e.g. enzymatic and oxidising reactions, preparation of homogeneous samples, prevention of microbial spoilage of samples, and/or extraction of the relevant compounds. Common preparation methods that are used in food control are:

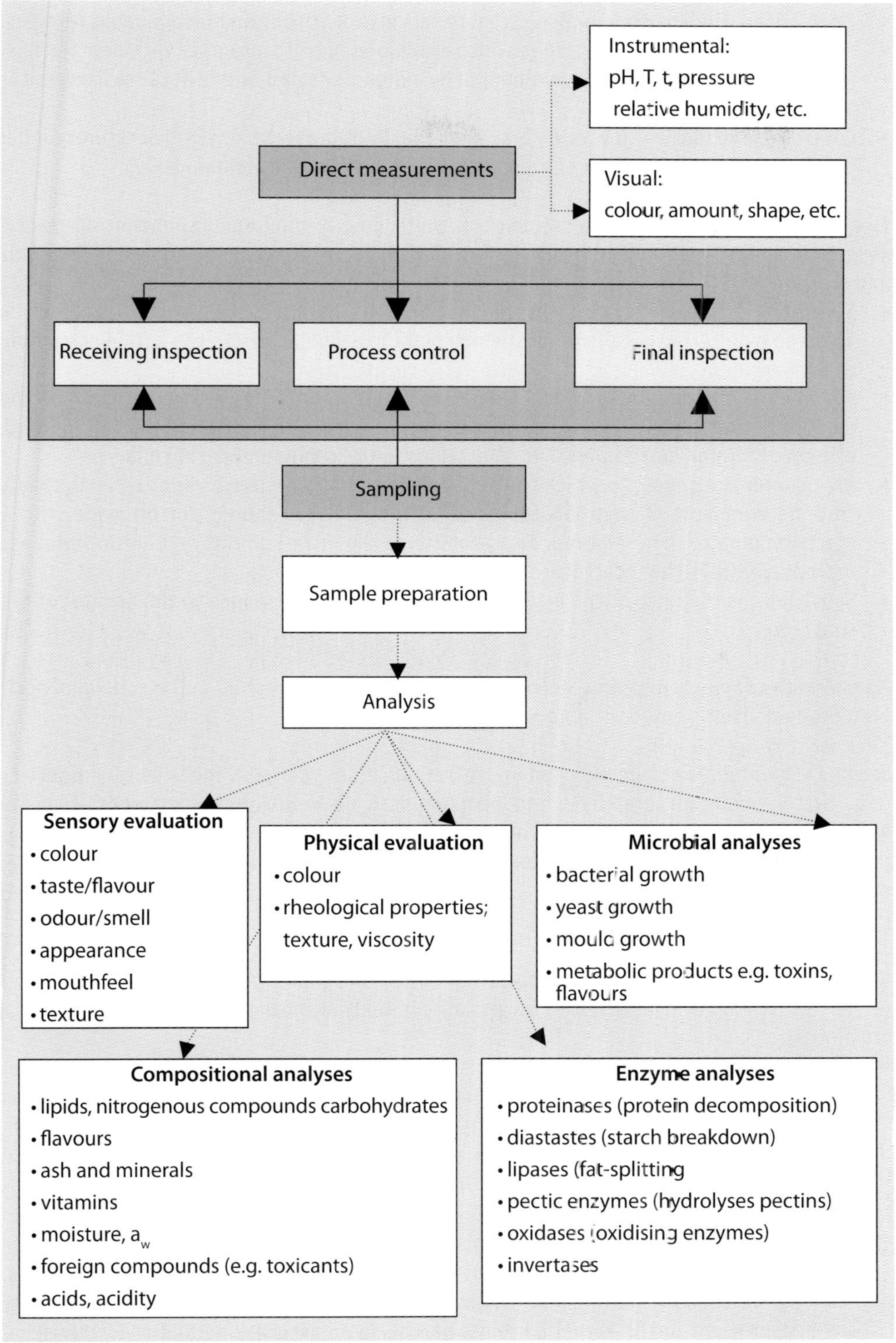

Figure 5.17. Overview of food analyses for quality control.

- Mechanical grinding of dry and moist materials in order to obtain homogeneous samples.
- Enzymatic and chemical treatment, which are also used to disintegrate various materials.
- Enzyme inactivation by heat treatment or by inorganic compounds that cause irreversible enzyme poisoning.
- Controlling oxidative and microbial spoilage by storage at low temperatures under nitrogen and/or by addition of preservatives (Pomeranz and Meloan, 2000).

Finally, the actual *analysis* or *measurement* is performed. The reliability depends on several aspects including specificity, accuracy, precision and sensitivity of the analysis (Pomeranz and Meloan, 2000)

- Specificity is the ability to measure what actually should be measured. Specificity is affected by interfering substances, which react in a similar way to the actual compound to be measured.
- Accuracy is the degree to which a mean estimate approaches a true estimate of the measured substance. Deviation may be due to inaccuracy inherent in the method, effects of foreign compounds, and alterations in the compound during analysis.
- Precision is the degree to which the determination of a substance yields an analytically true measurement of that substance. As a rule analyses should not be made with a precision greater than required. Repeatability is the intra-laboratory precision whereas reproducibility is the total inter-laboratory precision.
- Sensitivity is the ratio of the magnitude of instrumental response to the amount of the substance.

An overview of typical methods, analyses, and measurements applied in the control of agri-food products is presented in Table 5.2.

Sensory evaluation is commonly used in control processes to assess the sensory properties (e.g. taste, smell, texture) by using people as an observation 'instruments'. Sensory analyses are done with intermediate products during processing (as verification of process performance) and final products (as a final production check, and to judge whether products can be released). The principles of various sensory methods have been described in section 4.5.1.

Physical analyses typically include analysis of colour and rheological properties. The latter refers to viscosity and elasticity but also direct texture measurements (e.g. meatiness, crispiness).

- *Rheology* is concerned with the relationship between stress (i.e. compressive, tensile or shear) and strain, which is measured by deformation, as a function of time. The rheological properties of agri-foods can be determined by either an analytical or an integral approach. In the analytical approach the material properties are related to basic rheological parameters (deformation-time curves). In the integral approach, empirical relationships between stress, strain and time are determined. Empirical tests are commonly based on experienced relationships between (poorly) defined properties and textural quality. In imitative tests various properties are measured under conditions similar to practice e.g. during processing, handling or consumption. Examples of rheological measurements are measurement of the kernel hardness of wheats to estimate milling characteristics, viscosity measurement of dough to obtain information on changes in the bread-making

Table 5.2. Overview of typical methods and measurements for analyses of agri-food products.

Quality aspects	Methods and measurements
Sensory evaluation	
Colour, appearance, flavour, odour, taste, mouthfeel, texture, off-flavours	Consumer acceptance (preference) tests Panel difference methods such as: • paired comparison • triangle taste panel, triangular flavour score panel • ranking test • numerical scoring • descriptive terms and quantitative descriptive analysis Visual inspection using e.g. reference colour charts
Physical evaluation	
Colour	CIE tristimulus system (X,Y,Z); simplified tristimulus colour systems: • Munsell system: hue, value, chroma • Hunter colorimeter: L= lightness, a= red, green, b= yellow, blue • spectrophotometric analyses of extracts • chemical analysis of colour pigments
Rheological properties: • Viscosity, elasticity • Texture e.g. mealiness, crispiness	Major rheological devices for analysis of viscosity and elasticity Rotational viscosity meter Capillary viscosity meter Shear press device Wide variety of empirical devices Methods and devices used for textural properties Tenderometer Shear press device Texture metre Penetration measurements
Microbial analyses	
Growth of bacteria, yeast and moulds Metabolic products	Physical methods, e.g. impedance, micro calorimetric, cytometric Chemical methods, e.g. heat stable nuclease, ATP measurement Characterising and finger printing, e.g. serotyping, DNA probes, DNA amplification Immunologic methods, e.g. fluorescent antibody, serology, ELISA
Compositional analyses	
Lipids, nitrogenous compounds, carbohydrates Flavours Ash and minerals Vitamins Moisture, a_w Foreign compounds, etc.	Gas-liquid chromatography Mass spectrometry High-performance liquid chromatography Thin-layer chromatography Column, size exclusion and ion exchange chromatography Infrared spectroscopy Fluorimetrics Visible/ultraviolet spectroscopy Flame photometry, atomic absorption Coulometrics Conductivity Electrophoresis Nuclear magnetic resonance Radio-immunoassay activity
Enzyme analyses	
Enzyme activity Metabolic products	Determination of substrate Assays for determination of (specific) enzyme activity Inhibition studies

process, measuring meat tenderness to evaluate consumer acceptance of meat, and texture evaluation of fruits and vegetables to determine ripeness.
- *Colour* is an appearance property which is attributed to the spectral distribution of light, whereas gloss, transparency, haziness and turbidity are material properties which contribute to the manner in which the light is reflected and transmitted. Colour analyses are applied e.g. to measure the amount of synthetic colorants added, to determine the ripeness of fruits and vegetables, or to determine colour changes during processing and/or storage.

Microbial analyses refer to methods for detecting and characterising micro-organisms. These methods can aim at detecting metabolic activity for certain substrates, measuring growth response, analysis of some parts of cells, and/or detecting produced toxins. Jay and co-authors (2005) distinguished between the following different methods:
- *Physical methods* which are based on measurement, e.g. conductivity, enthalpy changes, or flow sorting which correspond to metabolic activities of the micro-organisms.
- *Chemical methods* to determine metabolic products, or endotoxins or typical bacterial enzymes.
- *Characterisation* and *fingerprinting* methods to identify the micro-organisms.
- *Immunology methods* are used for detection and quantification of micro-organisms and/ or their metabolic products in foods.
- *Assessing* the *genetic profiles*, by using DNA techniques.

Composition analyses are aimed at analysing the composition and concentration of compounds in raw materials/food products. A wide range of analytical techniques is available to determine the composition or a specific compound. Typical techniques are listed in Table 5.2.

Enzyme analyses are aimed at assessing the actual enzyme activity in food matrices. Enzymes are heat-labile proteins, which act as specific catalysts in a wide range of desirable and undesirable, chemical reactions in agri-food products. Enzymes occur naturally in living plants, animal tissues and microbial cells, i.e. endogenous enzymes. Particularly in raw (unheated) materials, these endogenous enzymes can influence textural properties, flavour and off-flavour formation, and decolouration processes. Moreover, enzymes are used as aids in agri-food processing (i.e. exogenous enzymes) for a wide range of applications, such as accelerating fermentation processes, improving cheese flavour, tenderising of meat texture, and bleaching of natural colour pigments. Typical applications of enzyme analyses as part of quality control include:
- Determination of the *quality status and/or the history* of the food product. For example, in milk the presence of bacterial dehydrogenase is an indicator for unhygienic conditions, whereas high catalase levels suggest that the milk might be derived from sick cows (mastitis). Improperly stored grains (high humidity and elevated temperature) show increased fat acidity which is measured by determining the lipase activity.
- *Monitoring or determination* of the *efficiency* of *heat treatment*. For example, peroxidase activity is determined in fruits and vegetables to establish the efficiency of the blanching process. Likewise, phosphatase activity is measured to establish the extent of the pasteurisation process of milk.

Enzyme activity can be determined in several ways including:

- Measurement of rheological changes in substrates. For example, α-amylase (i.e. starch liquefying enzyme) activity can be measured by monitoring changes in viscosity of the starch.
- Analysis of the degradation products of enzyme action. For example, peptidase activity can be monitored by measuring the free amino acids released by peptidase activity.
- Monitoring specific effects of enzyme activity, such as coagulation of milk caused by proteolytic enzymes, which indicates the quality of milk for cheese manufacturing.

Readers interested in analyses and measurements in food are referred to Pomeranz and Meloan, 2000), Gould and Gould (2001); more specifically about sensory analyses (Moskowiz *et al.*, 2006), microbial analysis (in Jay *et al.*, 2005; Maraz *et al.*, 2006), physical (colour, texture) analysis (Bourne, 2002, Soccaciu, 2007).

Direct measurement and the analysis procedure can be applied at acceptance of incoming materials, during processing and at the final inspection of products. The choice of control tests is dependent on the availability of appropriate measurements or analyses, which reflect the actual quality attribute (i.e. accuracy of the test). The time span between sampling and the outcome of the analysis or measurement can also be important. If a quick response is required to adjust a process then fast methods are necessary, which may be not as specific as a more lengthy analysis. Also the costs can play a decisive role in the selection of test methods.

5.4.4 Visual inspection

Visual inspection is commonly used in the control of (agri) food products, because in many cases it is the only available and acceptable inspection method, for technological or financial reasons. However, it is a rather variable and sometimes even unreliable way of measuring product properties, due to the subjective nature of visual measurement and low accuracy.

Gould and Gould (2001) have given a description of *visual measurement*, and used the term subjective evaluation. They described this term using the example of colour. They stated that subjective colour evaluation is any process wherein the human element is involved as an eventual deciding factor in the determination of colour quality for product acceptability.

Subjective methods to evaluate quality are based on the opinions of the investigators. It is usually a psychological reaction, which is a result of past training, experience of the individual, influence of personal preferences, and powers of perception. Advantages of subjective methods are speed and relative ease with which the measurements can be made. Disadvantages are human error, variance between individuals, and non-standardised conditions.

The visual quality inspection task has been described as consisting of the following subtasks: (1) orient the item, (2) search the item, (3) detect any defect, (4) recognise/classify the defect, (5) decide the status of the item, (6) dispatch the item, and (7) record the information about the item. A simpler description of the visual quality inspection task is to search, recognise a defect, and make a decision on the property's acceptability within the quality limits.

From theory it can be derived that visual observation is a rather complex process. The *observation complexity* is dependent on the observer, the product properties or sensory stimuli information, and the characteristics of and conditions set through the organisation.

The starting point for the whole decision-making process is the incoming visual information. How well people measure is dependent on the complexity of the visual information an individual must process, which can be influenced by product complexity and process complexity. *Product complexity* relates to the extent a product has dissimilarities and distinctive parts. Therefore, the level of dissimilarity is of importance for determining the complexity of an observation, but the shape of an object also affects the complexity of an observation. *Process complexity* has to do with the conditions set through the organisation. Because visual measurements are subjective, some objective methods can be used to facilitate these. Typical process factors determining complexity of visual measurements are the availability of facilitative methods (like graphic charts, colour charts), the right illumination, the available viewing time, the presentation and frequency of the stimuli and the background.

Visual measurement results in an assessment of *external product properties*, like colour or appearance. However, people immediately tend to derive from this their judgement of *internal product properties* as well. An example from fruit may illustrate the effect of visual measurement on judging internal product properties. The product appearance of fruit is characterised by the external fruit attributes: size, shape, form, colour, condition and absence of defects. Internal fruit attributes are characteristics that cannot be measured from the outside, e.g. texture and flavour. During visual measurement, employees have to judge the internal fruit properties such as texture and flavour based upon visual characteristics, without tasting the product. *'Visual' observable texture* makes the observer think that the product is, say, crisp (as in biscuits, lettuce or bacon), chunky (chocolate), or tough (meat). These dynamic visual texture properties become apparent when the product is moved or distorted (by feeling/touching the product). Examples of this include strength (of a jelly from its wobbliness), toughness (of meat from its visually perceived behaviour when prodded with a fork), slimy (like yoghurt in the way it pours). *'Visual' observable flavour* is the overall appearance of a product which makes us think that a product has a specific flavour. Observable colour creates expectations about the actual flavour. For example, it has been demonstrated that:

- yellow colour increases sweet, sour and bitter thresholds;
- green colour lowers sweet thresholds, but raises thresholds for sour and bitter;
- red colour raises sour and bitter thresholds (Maga, 1974).

So external product properties can be associated with internal product properties, and will affect the visual measurement performance.

In visual measurements human capabilities are a critical factor. Although the eye is extremely sensitive to small differences in e.g. colour, the memory of colour is not very reliable in the absence of control standards. All human biases, personality differences, and associated frailties subject colour judgement to various interpretations. Equally important in understanding the limitations of visual colour measurement is the near impossibility of describing colour in subjective terms, e.g. how red is tomato red? Although the human eye has some definite weaknesses when used to evaluate colour, subjective determinations can provide meaningful results. The eye is capable of distinguishing small colour differences when it is able to compare the sample to product colour standards. Thus, a standard colour chart, plate, or model for comparison is needed if meaningful subjective colour evaluation is to be made (Gould and Gould, 2001).

5.5 Managing the control process

Clearly merely extending control activities does not automatically imply an increase in quality performance. Effectiveness of control depends heavily on the relationship with planning of the production chain and the embedding of it in the organisation. In this section the relationship with logistics planning will be discussed. Then, embedding control in the organisation will be analysed as well as costs and benefits of control.

5.5.1 Quality controlled logistics

In the food industry production and distribution planning is included in logistics management. *Logistics* refers to the movement of materials within a production facility, and to incoming and outgoing shipments of goods and materials. For the food industry logistics management is rather complex and demands on food logistics are increasing (Van der Vorst and Beulens, 2002). The complexity involves the high number of transport activities for many product types from different supply points. From these supply points, products are transported via storage and production facilities and distribution centres to a large number of selling points. The increasing demands refer to delivery time, which must be reliable and as short as possible. Moreover, in-home stock levels of production companies and retailers must also be preferably as low as possible. Sometimes these demands are parallel to quality requirements, such as for fresh products which require small stock levels because they have a restricted shelf life. Nevertheless, there are sometimes conflicting requirements. For example, availability of raw materials is often linked to optimal harvest time and restricted seasons. However, once the raw material is harvested it is a bulk volume which must be stored and often cannot be processed immediately. Another conflicting requirement is that the production must be fast at low costs. However, sometimes overly fast production, in order to comply with delivery requirements, results in high material loss during operation.

Modern logistic concepts like *JIT* (just in time delivery accompanied by low inventories) and *ECR* (efficient consumer response, a system based on point-of-sale information) enable a company to comply with high logistics demands. However, it requires high and, especially, constant, quality levels of food products. Such high demands on logistics and quality stimulate the food industry and retail sector to develop control systems, which exceed companies' internal systems. This is one of the major reasons for companies to switch to chain co-operation and chain control. The *supply chain concept* is well known, and consists of an interconnected set of linkages between suppliers, warehouses, operations and retail outlets. An essential aspect of this concept is that not only the direct supplier or customer is controlled but also previous or next partners in the production chain, for example when a producer of orange juice is involved in cultivation of the oranges in order to guarantee a high quality standard. Although the supply chain concept seems very attractive on paper, in practice only a few initiatives succeeded in developing chain co-operation structures. This is assumed to be due to lack of *mutual trust* between partners.

In many food supply chains not only good *communication* between chain partners is crucial, but within each of the chain links quality decisions and logistics decisions need to be integrated. For instance, in fruit and vegetable production chains, different quality attributes (appearance, texture, flavour) and different parameters (temperature, relative humidity, air composition) have to be controlled in the chain in order to maintain the quality of the product until it reaches the final consumer. Keeping intact fruits and vegetables within their optimal ranges of temperature and relative humidity is the most important factor in maintaining their quality and minimising postharvest losses. But logistic decisions heavily determine these conditions by defining storage time and time of shipping goods. The distribution chain however, rarely has facilities to store each product under ideal conditions and requires decision makers to compromise on the choice of temperature and relative humidity. These choices can lead to physiological stress and loss of shelf life and quality (Knee, 2002).

For this type of problem a new concept has been introduced, which is strongly dependent on information exchange and collaboration in the supply chain (Van der Vorst *et al.*, 2007). It is called *quality controlled logistics (QCL)* and provides the means to optimise product quality and availability in market outlets and minimises shrinkage. Using time-dependent quality information and quality decay models, it is possible to predict product quality in much more detail, which enables the control of supply chain processes and direct specific products batches – under specific environmental conditions – to specific market segments. Operations management in food supply chains usually takes quality as a given fact; if product quality is approached as a dynamic issue and time-dependent quality information is used, more degrees of freedom emerge to improve supply chain performance significantly.

Quality controlled logistics makes use of the variation in product quality, developments in technology, heterogeneous needs of customers and the possibility to manage product quality development in the distribution chain. Using the definition of logistics management of the Council of Supply Chain Management Professionals (CSCMP), QCL can be defined as follows: *'Quality controlled logistics is that part of supply chain management that plans, implements, and controls the efficient, effective flow and storage of food products, services and related information between the point of origin and the point of consumption in order to meet customers' requirements with respect to the availability of specific product qualities in time by using time-dependent product quality information in the logistics decision process'* (Van der Vorst *et al.*, 2007).

Figure 5.18 shows the essence of the QCL concept. It concerns product differentiation and maximisation of added value created in the food supply chain network (FSCN) by the timely separation in harvesting, storage, processing, and transportation stages, and pro-active control of goods flows. After each stage quality and quantity of the products are measured. The information from these measurements is used for quality control, both for immediate control (e.g. should products be rejected due to bad quality?), but also for prediction

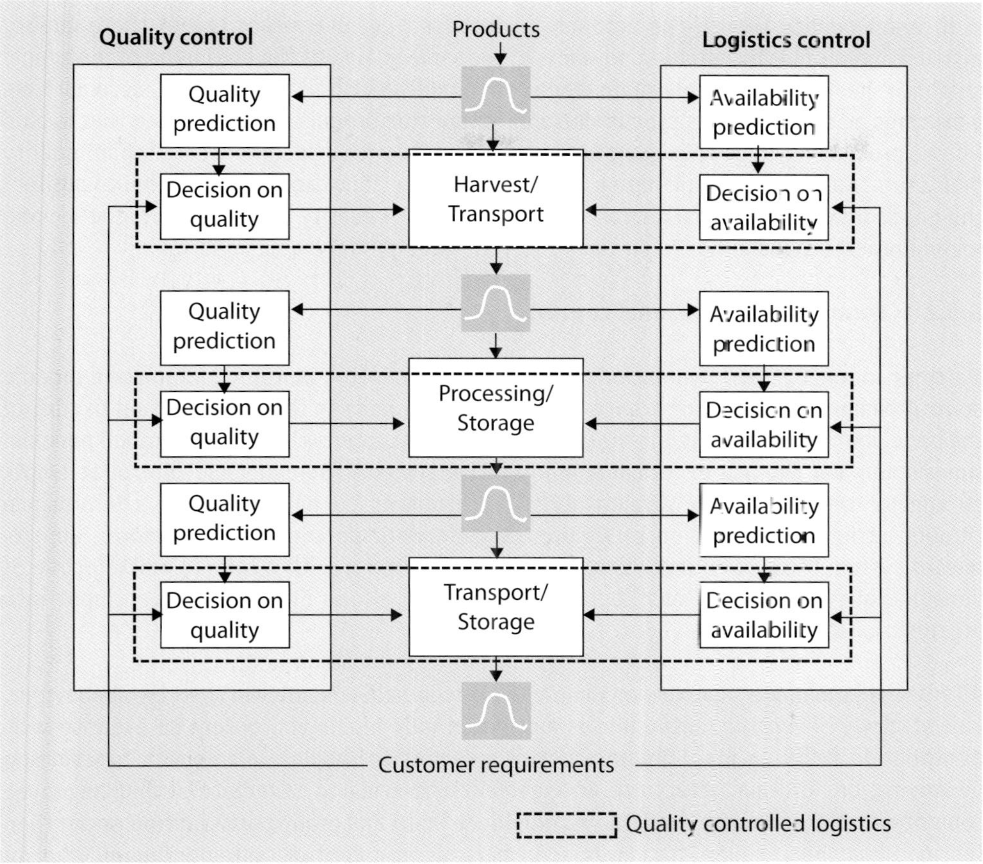

Figure 5.18. Quality controlled logistics (Van der Vorst *et al.*, 2007).

of future quality and required action (e.g. should future process conditions be changed depending on actual quality?). The information is also used for logistic control, again both for immediate control (e.g. how many product batches have the required quality levels to meet customer orders?), and for prediction of the availability over time of different batches with a certain quality level and required action (e.g. should batches be rearranged due to differences in the expected final quality?). When decision-making processes within quality control and logistics control are connected and jointly executed this is called quality controlled logistics.

Appropriate strategies for logistics management can be developed based on scientific insights into the dynamic product quality behaviour profiles throughout the supply chain and an understanding of the impact of technological and managerial conditions. More specifically, QCL starts with obtaining detailed knowledge on customer requirements in the different market segments. At the harvest (or breeding) stage, products are collected and clustered based on variation in the quality parameters. It is well known, for example, that one stable with pigs or one tree with apples deliver products with different quality levels. For example, exposure to the sun means that apples on the outside of the tree have a different quality from those products inside the tree. QCL makes use of these quality distribution profiles by batching products of the same quality at the beginning of the supply chain. In the following supply chain stage comparable decisions have to be made, each time a match is made between customer demand for specific products and the price that is paid for the products with the available supply of products with a specific (variation in) quality prediction. Subsequently one has to determine what actions can be taken either to redirect the goods flows to other markets or try to influence the quality level of the products using technological equipment.

5.5.2 Forms of organisational control

Managers normally have two options with respect to control, i.e. bureaucratic and organic control (Schermerhorn, 2007; Ivancevich, 1994). Managers can take direct action to control the behaviour of others. This is a strategy of *bureaucratic control* that occurs through personal supervision and the use of formal administrative systems. Bureaucratic control attempts to control through formal arrangements. It attempts to gain employee conformance through strict administration of rigid, straightforward policies and procedures, and its reward system focuses on individual employee compliance with a formal, written code of behavioural standards. As such, bureaucratic control allows limited employee input into organisational activities.

Alternatively, managers can rely on people to exercise self-control over their own behaviour. This strategy of *organic control* allows motivated individuals and groups to exercise self-discipline in fulfilling jobs. Organic control attempts to regulate employee's functioning by relying on informal structural arrangements. It tries to stimulate strong employee commitment by vigorously encouraging employee input and group participation rather than setting strict behavioural standards as in bureaucratic control systems. Organic control relies on self-control and informal group norms to effectively create a relaxed, but sharply focused, working environment.

Organisations with effective control typically use both strategies to good effect. However, the trend today is to increase the emphasis on internal or self-control. This is consistent with the renewed emphasis on participation, empowerment and involvement in the work place.

It should be kept in mind that large organisations consist of a number of departments, which can differ widely in their emphasis on bureaucratic or organic controls. For example, if the production department faces a relatively stable environment and the marketing department faces a changing one, managers of the two departments are likely to choose different ways to divide and co-ordinate the work.

The contrast between bureaucratic and organic control also exists in quality management in the traditional and quality-based models of control, the latter being based on the principles of the organic model (Ivancevich, 1994).

Figure 5.19 shows the contrast between both models. The traditional approach does not include training for workers; management inspects the results of production, and failures to meet production specification results in worker reprimands. In contrast, the quality-

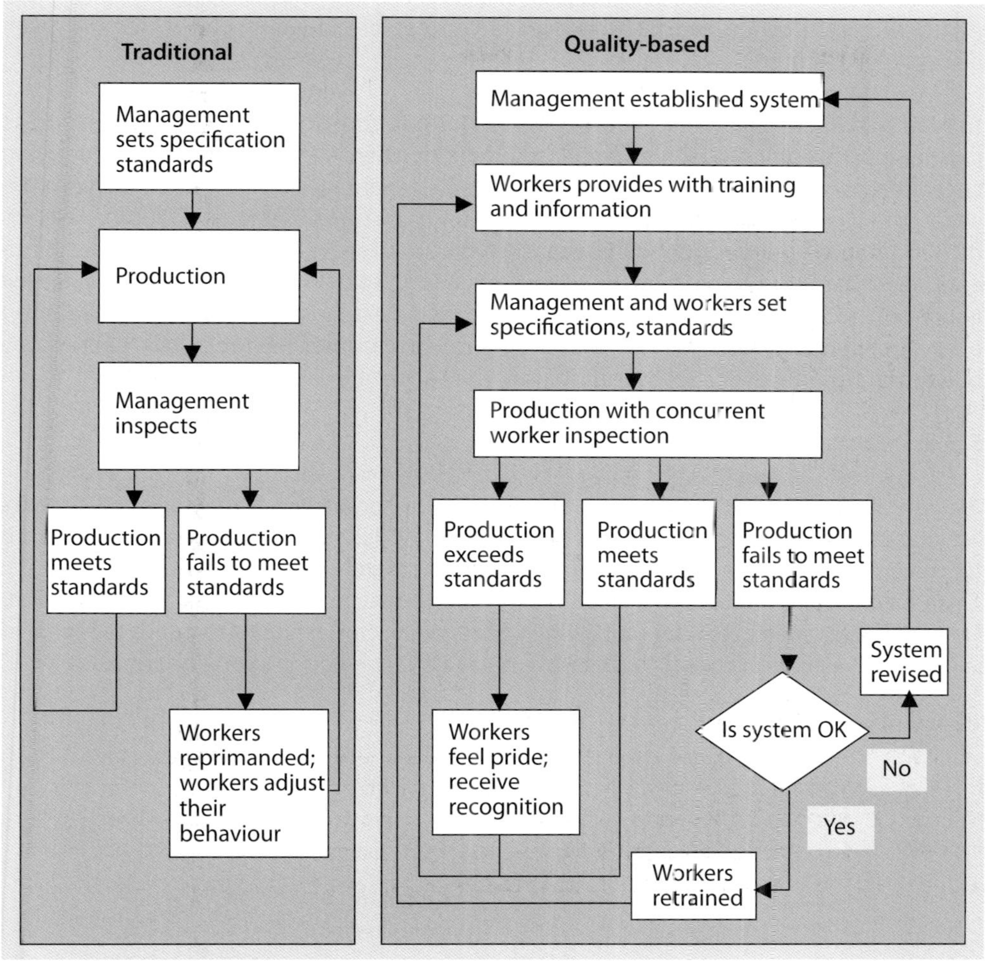

Figure 5.19. Models of traditional and quality-based control.

based control model includes worker training; workers inspect the results of production during the production process, and failures to meet production specification results in revision of the system. In quality management *employee involvement* can be considered as an essential ingredient (Heizer and Render, 2006). Employee involvement means including the employee in every step of the process from product design to final packaging. Building communication networks that include employees encourages employee involvement. This can be achieved through:

- open and supportive supervisors;
- moving quality responsibility from the quality control department and inspectors to production employees;
- building high-morale organisations;
- using formal techniques such as quality circles (see Chapter 7).

All these measures are consistent with the concept of empowerment, which has been discussed in Chapter 3.

In food manufacturing there is a restriction on decentralisation, because specialised tests must be carried out in the lab, and not at the production line. The central issue in the decision concerning on-site or lab inspection is whether the advantages of specialised lab tests are worth the time and disruption needed to obtain the results (Stevenson, 2006). Reasons favouring on-site inspection include quicker decisions and avoidance of introduction of extraneous factors (e.g. damage or alteration of samples). On the other hand, specialised equipment and a more conditioned test environment (less noise and confusion, reliability) offer strong arguments for using a lab. Many food companies rely on self-inspections by operators. This places the responsibility for errors at their origin.

5.5.3 Costs and benefits of control

As with all organisational activities, controls should be pursued if the expected benefits are greater than the costs of performing them. Figure 5.20 shows a cost-benefit model for judgement of the effectiveness of an organisation's control system (Hellriegel *et al.*, 2005). The horizontal axis indicates the amount of organisational control, ranging from low to high. The vertical axis indicates the costs and benefits of control, ranging from zero to high. For simplicity, the curve of cost of control is shown as a direct function of organisational control.

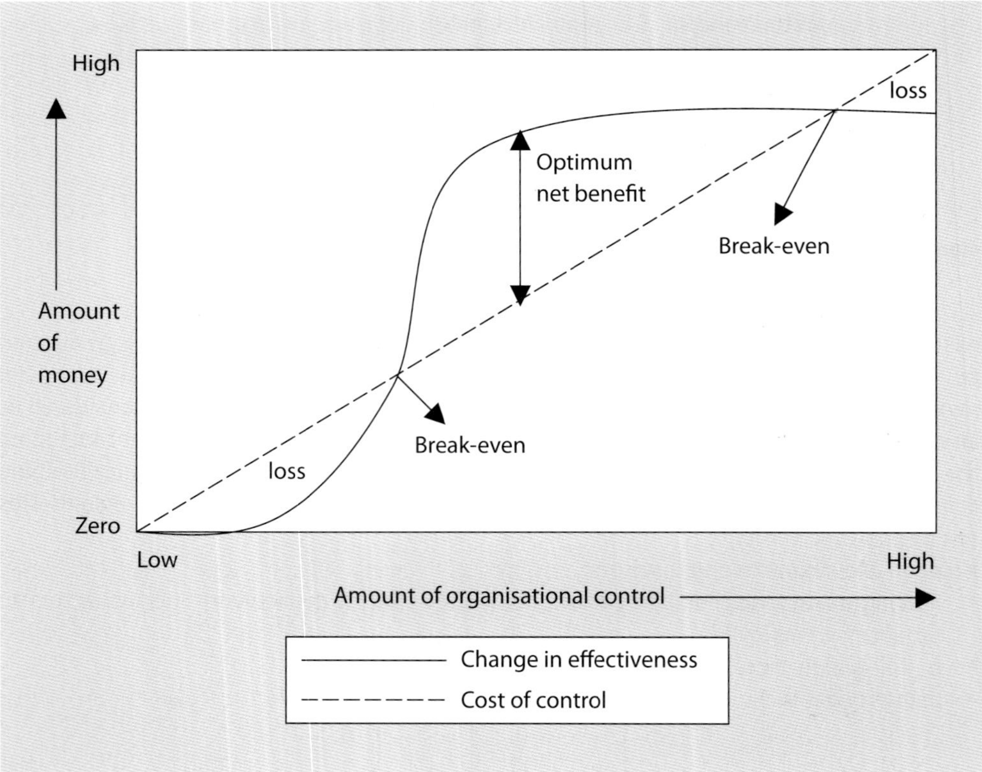

Figure 5.20. Cost-benefit model of organisational control.

Managers have to consider trade-offs when choosing the amount of organisational control to use. With too little control, costs exceed benefits and the organisation controls are ineffective. As the amount of control increases, effectiveness also increases up to a certain point. Beyond this point, further increases in the amount of control result in decreased effectiveness. For example, an organisation might benefit from intensifying inspection of finished goods, reducing the number of defects in shipped goods. However, if all batches containing defective products are detected by good sampling, more inspection only causes losses because the inspected items have to be eliminated. Figure 5.20 shows two break-even points. They indicate where the amount of organisational control moves from a loss to a net benefit and then returns to a net loss. The optimal point is difficult to calculate. Effective managers probably come closer to achieving it than ineffective managers do.

5.6 Quality control in practice

A typical problem complicating control in the food industry is the location of specialised engineering knowledge. This problem originates from the typical process character of food industry, which can be typified by repetitive flow processes. Control in such situations is mainly focussed on process parameters, e.g. temperature, and pressure during operation. Moreover, many control systems are automatic. Most knowledge about the production process is actually located in a specialised department, like the engineering department. Often the knowledge is not even available in the company itself but is held in the companies supplying the equipment. This results in fewer possibilities to influence and control the production process. This problem becomes even more pronounced when, due to higher demands on quality and flexibility, more influence and control on processes is required.

The functional organisation form is very popular in the food industry, and forms a basis for the type of problems described above. In functional organisations, specialisation, and expertise are clustered in functional departments (e.g. purchase, engineering and production department). Specialised departments like R&D and quality assurance often feel responsible for quality. They provide other departments with quality services and knowledge and so make them dependent on the quality and/or R&D department. A possible consequence of this situation is that control is focused on controlling each other instead of controlling the process to comply with consumer demands.

In addition to these organisational hurdles, control as such is rather difficult. Difficulty in control refers to the large variation in product and process conditions on the one hand, and the high number of product attributes that are commonly required (e.g. colour, taste, nutritional value, but also low costs) on the other. Based on the complexity of control one would expect to see statistically supported control systems in the food industry. In practice, however, the use of statistics (e.g. statistical process control approach) is very limited in the food industry and sometimes control frequency depends only on the number of available inspectors.

In this type of situation the techno-managerial approach would favour management activities that are focused on major points in the process that can be technologically or technically influenced. The basic idea behind it is that only a restricted number of control points determine the quality as requested by consumers. In other words, control measures should focus on those points that are relevant for influencing the final product quality. An advantage of this approach is that control activities can be restricted and attention of employees and operators is then focused only on those points in the process relevant for quality (e.g. sensory properties, safety, etc.). It is important therefore to develop proper measuring and information systems. Quick measuring units, which can be used by operators themselves, can be very supportive.

Concepts such as HACCP (Chapter 7), FMEA (Chapter 4) and the Taguchi principles (Chapter 4) enable companies to perform an in-depth analysis of the process and assess point, which are critical with respect to safety and quality. The use of these 'tools' should be combined with management actions like decentralisation and training, as described in the model of quality-based control (Figure 5.19). This combination creates opportunities to control and take corrective actions when required.

Chapter 6

Quality improvement

In food and agribusiness quality improvement of product and processes is of great importance. Competition demands innovation and improvement of products and processes and reduction of costs (Scott *et al.*, 2009; Antony *et al.*, 2005; Terziovski and Sohal, 2000). Juran has distinguished quality control and quality improvement as two different activities, alongside quality planning. They are all part of Juran's trilogy: planning, control and improvement (Juran, 1990). Quality planning is the continuous process of the development of products or services, which meet customer demands. Quality control includes activities undertaken to ensure that actions lead to achievement of objectives. It involves evaluating actual quality performance, comparing this with quality objectives and acting on any discrepancies. Quality improvement actually embodies the need for change, breaking through the status quo. It implies that the firm has to 'learn'. It requires facilitating structures, such as means of communication, procedures and reward systems, focused on the identification of changes in internal and external business environment and on avoiding routine and rigid structures in the firm. Typical goals of quality improvement include increasing customer satisfaction, achieving higher quality levels, reducing cost, increasing productivity, and accelerating the process.

In this chapter the improvement process will first be analysed (section 6.2), followed by a summary of the quality guru's ideas about improvement (section 6.3). Then, typical improvement tools and methods are described (section 6.4). In the last section attention is focused on managing the improvement process, with topics such as training, working with improvement teams, and -due to the importance of management and employee involvement- on strategies of organisational change (section 6.5). In the last section the typical situation in the food industry with respect to quality improvement is briefly described and considered from the techno-managerial perspective (section 6.6).

6.1 Introduction

There is a widely held belief that an organisation would experience few, if any, problems if workers would only do their jobs correctly (Edge, 1990). In fact, the potential to eliminate mistakes and errors lies mostly in improving the systems through which the work is done, not in changing the workers. Juran and Deming have proposed since the early 1950s that at least 85% of an organisation's failures is due to faults in management-controlled systems. Workers control fewer than 15% of the problems. This observation has evolved into the rule of thumb that at least 85% of problems can only be corrected by changing the systems (which are very largely determined by management). For example, a worker in the production line cannot do a top quality job when working with defective tools or parts. Even when it does appear that an individual is doing something wrong, the trouble often lies in how that worker was trained, which is a system problem (Crosby, 1979; Ishikawa, 1985; Deming, 1986, 1993; Juran, 1990).

So, an important base for quality improvement is an approach with both management and employee involvement. Improvement will be only accepted if there is sufficient commitment.

Management should initiate the driving force for this process, because improving means changing, getting away from routine structures. In many cases within firms, routine structures exist that cause rigidity and hamper the improvement process.

In Figure 6.1 the quality improvement function is described in its main elements. It includes a decision-making process in which quality improvement objectives are translated into higher quality within lower tolerances by specifying quality problems, analysing relevant influencing factors, and finding solutions for those problems.

As in the quality control function, this process is dependent on the ability to measure product properties and processes (both technological processes and people processes). But in quality improvement measuring is not meant to measure the actual state of the product or process as in quality control; it is intended to measure the behavioural patterns of products, processes, and people. When there is information about these patterns, they can be analysed on typical deviations from the standards and on the structural causes of such problems.

Special tools, using basic statistics, have been developed for such analysis, as well as typical methods for supporting the improvement process. In addition, typical organisational

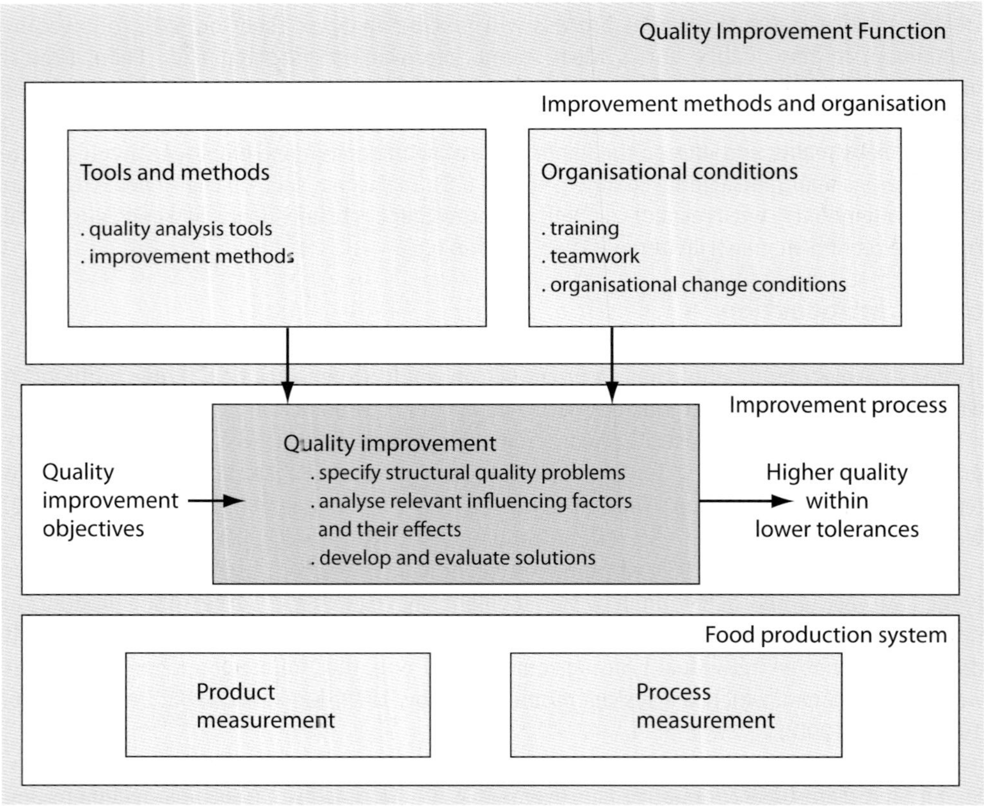

Figure 6.1. Quality improvement function.

measures can support the effectiveness of the improvement process, like teamwork and project management. These factors will be described in the following sections.

6.2 Quality improvement process

Quality improvement is a basic activity in food quality management and involves a systematic approach to improving a system by paying attention to structural causes and solutions. It is aimed at changing people, processes, and resources in order to bring them to a higher level of quality performance and into 'new zones of control', which means working with lower tolerances (Juran, 1990).

> *The core mechanism of quality improvement is realising a higher quality level with lower tolerances by analysing behavioural patterns of product properties and processes, structural causes of insufficiencies, and by modifying the system for better functioning on a continuous basis in a way that final quality will meet or exceed customer/consumer needs and legal requirements.*

In this section the steps in the quality improvement process, as well as the decisions involved, are described.

6.2.1 Quality improvement activities

Quality improvement is a systematic process. It involves documentation, measurement and analysis for the purpose of improving the functioning of a system. Quality improvement can be perceived as a double-looped cybernetic process, in which the control circle is provided with an additional circle aimed at structural improvements (Figure 6.2).

The first loop is identical to normal control processes and deals with incidental problems, bringing the process back between its control limits. The second loop arises when problems have a structural nature, for instance quality problems always occurring at the end of the week during production peak, or quality problems that only arise when materials from one specific supplier come in. This kind of problem must first be *analysed* before corrective action can prevent these problems in the future. After analysis, the next step in the loop is modifying the system to a higher level of quality performance.

Stevenson (2006) provided three crucial steps in the improvement of processes within a system:

1. *Map the process.*
   - Collect information about the process; identify each step in the process; for each step, determine inputs and outputs, people involved and decisions that are made.
   - Document measures as time, cost, space used, waste, employee morale and employee turnover, accidents and/or safety hazards, working conditions, revenues and/or profits, quality and customer satisfaction, as appropriate.
   - Prepare a flow chart that accurately depicts the process; note that too little detail will not allow for meaningful analysis, and too much detail will overwhelm analysts and will be counter-productive. Make sure that key activities and decisions are represented.

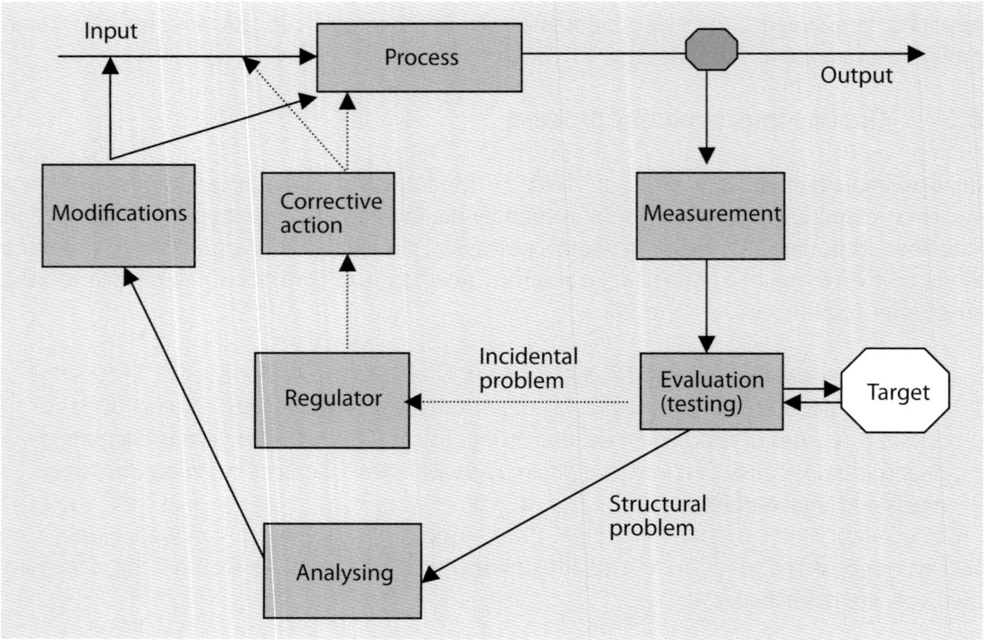

Figure 6.2. Quality improvement process and activities.

2. *Analyse the process.*
   - Ask the following questions about the process: is the flow logical, are any steps or activities missing, are there any duplications?
   - Ask the following questions about each step: is the step necessary, could it be eliminated, does the step add value, does any waste occur at this step?
   - Analyse for improvement by asking: what causes arise for known problems, could they be eliminated, can the process be improved by shortening time or by reducing costs, can the process be improved by providing better conditions for quality?
3. *Redesign the process.*
   Use results of analyses for redesign of the process. Document improvements; potential measures include reductions in time, cost, space, waste, employee turnover, accidents, safety hazards, and increases in employee morale working conditions, revenues/profits, quality, and customer satisfaction.

Quality improvement is in a way contrary to the belief in optimisation that was very strong in the 1960s and the 1970s. Optimisation had to be done within a fixed framework. Improvement, however, is partly a matter of ridding ourselves of this framework and starting to think along entirely new lines (Bergman and Klefsjö, 1994). To a great extent, successful companies are using special programmes for quality improvement. The first applications were started with workers in their own work environment, dealing with problems that arise in their daily work. These programmes worked on the basis of improvement teams or quality control circles (QC circles, see section 6.5), small groups of employees who discuss different problems and come up with suggestions for improvement. QC circle activities were not just confined to the manufacturing industry, however they were initiated there. There are, for example, QC circles in sales organisations, department stores, hotels, banks and the restaurant trade.

Quality improvement processes always include an analysis step. Analysing past performance from structured product and process data is a learning process that builds up people's experience, and changes their mental models. So, improvement processes not only contribute to better quality, they also strengthen people's competencies. But food products and food production processes are complex and not easy to understand, which requires adequate execution of the improvement process. Learning and complexity will be discussed in the following sections.

Continuous improvement

Tenner and De Toro (1992) developed a model to implement total quality management (TQM, see Chapter 8 for description of this concept) based on the combination of various quality gurus' teaching. They defined *continuous* improvement as the main objective of the TQM system and described three other TQM principles from the continuous improvement perspective:

Customer focus: collection and analysis of customer requirements; when these requirements are understood and accepted, they must be met.

Process improvement: the concept of continuous improvement is built on the premise that work is the result of a series of interrelated steps and activities that result in an input. Continuous attention to each of these steps in the work process is necessary to reduce the variability of the output and improve the reliability of the process. The first goal is to achieve reliable processes, this means that they should produce the desired output each time. If the variability has been overcome and the results are still unacceptable, the second goal is to redesign the process to be better able to meet the consumer's requirements.

Total involvement: this element includes the active leadership of senior management and the use of the talent of all employees. Employees at all levels are empowered to improve their outputs by coming together in new and flexible work structures to solve problems, improve processes, and satisfy consumers.

The analysis of Tenner and De Toro shows the importance of improvement as a cornerstone of quality management. However, they mentioned some important requirements for senior management:

- understand the principles of total quality management;
- provide the information the employees need on the mission, vision, direction, and strategy of the organisation as well as the skills they need to secure quality improvement and resolve problems;
- bring about the change necessary to implement a quality strategy and to understand the concepts of quality;
- communicate to all employees a sincere commitment to change;
- reward teams and individuals who successfully apply the quality process; and

- measure external customer satisfaction to determine the extent to which customers perceive that their needs are being meet. The collection of customer data provides an objective, realistic assessment of performance and is useful in motivating everyone to address real problems.

6.2.2 Quality improvement decisions

The food quality decisions grid (Figure 3.17) shows which decisions can be seen as the result of quality improvement activities, i.e. decisions on:
- structural out-of-tolerance of product properties and necessary modifications (influencing food dynamics), e.g. incoming material problems result in changing specifications, or consumer complaint analysis results in reducing acceptable variation in product composition;
- structural deviation from requirements on process parameters, equipment and buildings and necessary modifications (influencing technological conditions), e.g. production data show a regular out-of-control of temperature which results in modification of regulating equipment;
- structural out-of-tolerance of people's actions and necessary modifications (influencing human dynamics), e.g. data showing lack of hygienic behaviour result in a more supportive leadership style;
- structural deviation from requirements on people, procedures, and information systems and necessary modifications (influencing administrative conditions), e.g. when procedures can never be complied with, they are modified, or final product inspection data show structural quality defects from one specific shift and work practices in this shift are reorganised.

These different types of decision show the different opportunities for quality improvement. All decisions are the outcome of a well-organised process with the involvement of all employees who influence the problem situation.

6.2.3 Complexity of food production

In Chapter 5 a special definition of quality is mentioned (Montgomery, 2008): 'Quality is inversely proportional to variability'. This definition implies that if variability in the important characteristics of a product decreases, the quality of the product increases. Thus, quality is inversely proportional to variability. According to this definition of quality, quality improvement can be defined as follows (Montgomery, 2008): 'Quality improvement is the reduction of variability in processes and products'.

The previous definitions of both quality and quality improvement are in accordance with Deming's philosophy. His philosophy focuses on bringing about improvements in product and service quality by reducing uncertainty and variability in the design and manufacturing process. In Deming's view, variation is the chief culprit of poor quality. As Deming stated: 'If I had to reduce my message to management to just a few words, I'd say it all has to do with reducing variation'. Another quality guru, Crosby (1979), described quality as 'complying with clear specifications, whereby the management is responsible for establishment of univocal specifications'.

Food products are variable and significantly different from non-perishable products. But, besides variation, there are other factors that make food products complex. Generally, the term complexity is defined intuitively as the number of basic elements in a system. Nevertheless, numeric size in itself is not sufficient to fully define complexity. The concept of complexity also refers to diversity of forms, to emergence of coherent patterns out of randomness and to the ability of frequent switching among such patterns. Complexity is the combination of three basic factors: numeric size, variety, and relation. Additionally, the degree of product novelty is mentioned as a complicating factor, because when a product involves a new architecture or new technologies, there is not a stable, well-understood set of interactions between components (Novak and Eppinger, 2001).

Because food products consist of a wide variety of chemical and biological compounds, these components can affect each other and thus also the final intrinsic quality attributes of food products. Food products and food production are also relatively complicated, in terms of complex chemical, enzymatic, physical and physiological reactions (Luning and Marcelis, 2006). So, food products and food production processes are really complex and therefore difficult to understand. The complexity can be typified in:

- *Product complexity:* diversity of the product and the dynamic character that is inherent to food products, that is often not well known or understood.
- *Process complexity:* diversity of the process given by the number of steps in the process line, and the controllability of the process, and the often rather unpredictable influence of technological conditions chosen for certain production and storage processes.

More specifically, the dynamic character of the product is a result of the variability in food properties and the typical characteristics of food processes (see Figure 2.9). Variability of properties involves variation in the composition of food products (including raw materials, ingredients, manufactured or fresh produce) like content of sugars, flavour concentration, initial contamination of bacteria, and variation in behaviour profiles of product-inherent microbial, chemical, physical, physiological and biochemical processes. Examples of these processes are growth or inactivation of pathogens, enzymatic reactions, moisture diffusion, or respiration of fruits; these processes change food properties and related quality attributes (e.g. colour, taste, safety, etc.) over time.

The controllability of the production process depends on the degree of exceptions that emerge during the process and the degree to which the technology used can be analysed (Perrow, 1967) The number of steps in the process depends largely on the type of production (unique products, batch production, or mass production).

Technological conditions are measures to achieve and/or maintain the required desirable product properties within acceptable tolerances (see Chapter 2 section 2.4.2). Examples of these conditions include harvesting and handling circumstances, hygienic design situation, processing conditions (e.g. time temperature, pressure, etc.), packaging concepts or storage conditions. The separate effects on the quality of food properties and processes as related to technological conditions are mainly understood but the effects of their interactions on quality are not (Luning and Marcelis, 2006).

Table 6.1 shows an example of food complexity. For fruits, the important quality attributes, and the components and or factors that influence these attributes are listed (Salunkhe *et al.*, 1991). Separate attributes cannot be singled out to be the cause of 'good quality fruit'. Final quality judgement is judging combinations of different attributive values.

Theoretically, it is clear that technological complexity has a relationship with quality improvement. On one hand it complicates the analysis process, on the other hand understanding the nature of the products and processes results in better predictability of them and finally in better control. Analysis and learning are important for food quality improvement due to a food product's complexity.

Table 6.1. Complexity of quality attributes (Salunkhe *et al.*, 1991).

Factors	Components
Appearance	Size: dimensions, weight, volume Shape and form: diameter/depth ratio, smoothness, compactness, uniformity Colour: uniformity, intensity Gloss: nature of surface wax Defects: external and internal, morphological, physical and mechanical, physiological, pathological, entomological
Texture	Firmness, hardness, softness Crispness Succulence, juiciness Mealiness, grittiness Toughness, fibrousness
Flavour (taste, smell)	Sweetness Sourness (acidity) Astringency Bitterness Aroma (volatile compounds) Off flavours and off odours
Nutritive value	Carbohydrates Proteins Lipids Vitamins Minerals
Safety	Naturally occurring toxicants Contaminants (chemical residues, heavy metals) Mycotoxins Microbial contamination

6.2.4 Learning

Quality improvement processes always include an analysis step. Analysing past performance from structured product and process data is a learning process that builds up people's experience, and changes their mental models. So, improvement processes not only contribute to better quality, they also strengthen people's competencies. But food products and food production processes are complex and not easy to understand, which requires adequate execution of the improvement process.

Thus, improvement requires a commitment to learning (Bessant and Francis, 1999), and is essential for building an organisation that is able to learn. A learning organisation is an organisation skilled at creating, acquiring and transferring knowledge, and at modifying its behaviour to reflect new knowledge and insights (Garvin, 1993). Senge (1990) identified five skills that a learning organisation should have; system thinking, personal mastery, mental models, shared vision and team learning.

Garvin (1993) took Senge's work and defined these skills as five activities:

- *Systematic problem solving.* This recommends using different tools for problem solving, e.g. the Deming cycle, to use data rather than assumptions for decision making, and to use statistical tools.
- *Experimentation.* This activity involves systematic searching for and testing new knowledge. The difference with problem-solving methods is that this is not used to solve a current situation but to test a new idea.
- *Learning from the past.* Companies must review their successes and failures, assess them systematically, and record the lessons in a form that employees understand. 'Those who cannot remember the past are condemned to repeat it'. In this respect, analysis of data over the past is necessary.
- *Learning from others.* Besides learning from self-analysis, learning is also about looking outside one's immediate environment to gain a new perspective. That enlightens people and helps them create new ideas and take useful knowledge from outside. The most well-known tool here is benchmarking.
- *Transferring knowledge.* For learning, knowledge must be spread quickly and efficiently throughout organisations. A variety of mechanisms spur on this process, including written, oral, and visual reports, site visits and tours, personnel rotation programmes, education and training programmes.

Paying attention to these skills makes learning significant as a basis for continuous improvement processes in the whole organisation.

Learning process

Learning is the process of acquiring knowledge, skills, attitudes, or values, through study, experience, or teaching, that causes a change in behaviour or allows an individual to formulate new conceptual knowledge, such as attitudes or values. It is a process that depends on experience and leads to long-term changes in the possible behaviour of an individual in a given situation in order to achieve a goal.

Kolb (1984) gave the definition of learning as the process whereby knowledge is created through the transformation of experience. His model of experiential learning can be found in many discussions of the theory and practice of adult education, informal education and lifelong learning. Kolb (1984) has described this learning process as a four-phase cycle in which the learner: (1) does something concrete or has a specific experience which provides a basis for (2) the learner's observation and reflection on the experience and their own response to it. These observations are then (3) assimilated into a conceptual framework or related to other concepts in the learner's past experience and knowledge from which implications for action can be derived and (4) tested and applied in different situation (Figure 6.3).

Argyris (1982) divided learning into two categories both of which illuminate the relationship between what humans believe and how they actually behave.

Single-loop learning is learning within the confines of our daily actions; people behave according to their reasoning and beliefs. *Double loop-learning* recognises what happens when there is incompatibility between what we say we believe and the way we behave. This means that people do not necessarily behave on the basis of their values and beliefs. People often challenge beliefs, culture and values on the basis of their experiences.

6.3 Quality gurus on improvement

Deming (1986) urged everyone to learn a common method of describing and tackling problems. This commonality is an absolute requirement if personnel from different parts

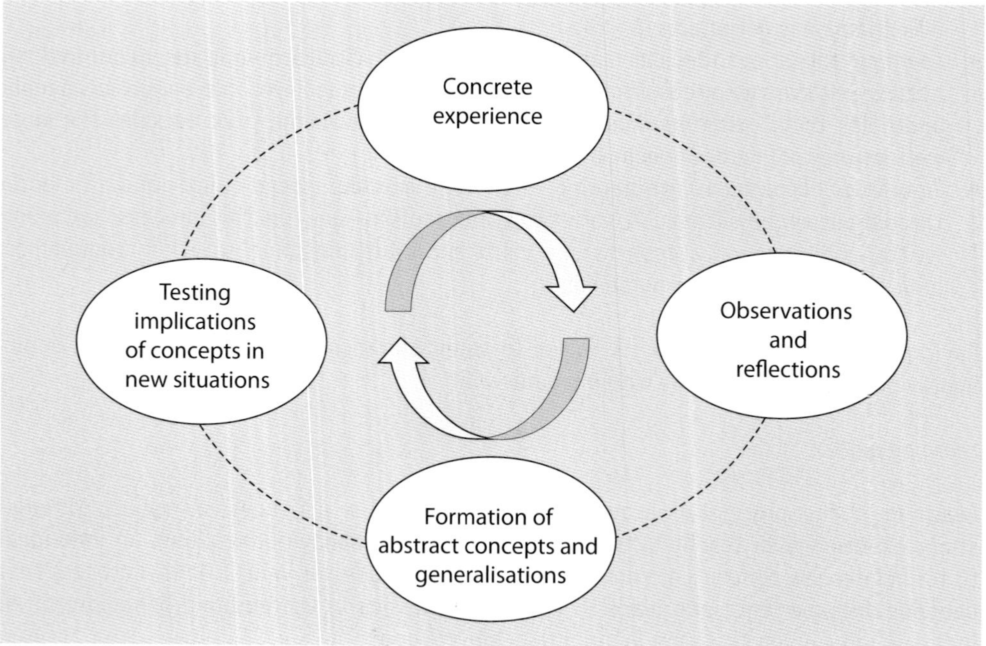

Figure 6.3. Experiential learning.

of the company have to work together on company-wide quality improvement (Edge, 1990). Deming therefore introduced a framework; the Deming cycle, where all parties can discuss problems and suggest improvements (Figure 6.4).

The plan-do-study-act (PDSA) cycle, also referred to as either the Stewart cycle or Deming wheel, is the conceptual basis for continuous improvement activities (Stevenson, 2006; Melnyk and Denzler, 1996). Representing the process with a cycle underscores its continuing nature. Going upwards suggests higher levels of quality to be reached. There are four basic steps in the cycle:

1. *Plan*. Begin by studying the current process and document that process. Then collect data to identify problems. Next, survey data and develop a plan for improvement. Specify measures for evaluating the plan.
2. *Do*. Implement the plan, on a small scale if possible. Document changes made during this phase. Collect data systematically for evaluation.
3. *Study (check)*. Evaluate the data collected during the 'do' phase. Check how closely the results match original goals of the plan phase.
4. *Act*. If the results are successful, standardise the new method and communicate the new method to all people associated with the process. Implement training for the new method. If results are unsuccessful, revise the plan and repeat the process or cease the project.

In replicating successful results elsewhere in the organisation, the cycle is repeated. Similarly, if the plan was unsuccessful and you still wish to make further modifications, repeat the cycle. Employing this sequence of steps provides, according to Deming, a systematic approach to continuous improvement.

According to Juran (1990) quality planning is necessary to establish processes that are capable of meeting quality standards. Quality control is necessary in order to know when corrective action is needed and quality improvement will help to find better ways of performance (Juran's trilogy). A key element of Juran's philosophy is the commitment of management to continuous improvement. His approach urged companies to 'break the chain of the past' and reach improved quality levels on a continuous basis (Figure 6.5).

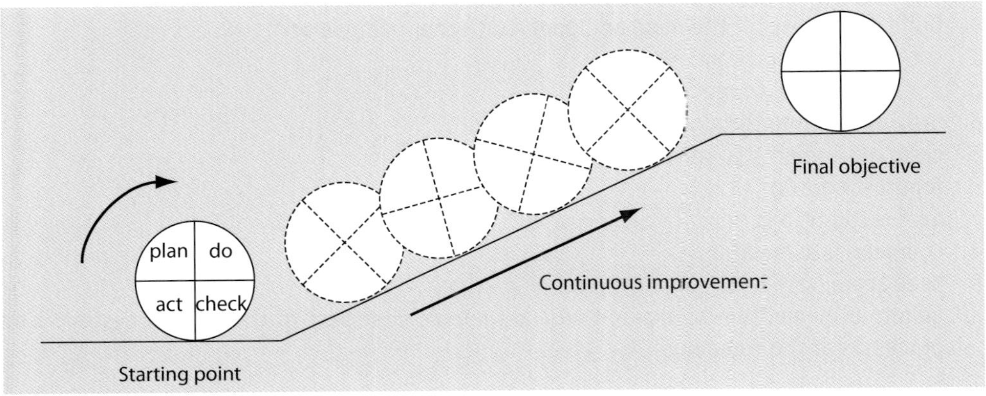

Figure 6.4. The Deming cycle.

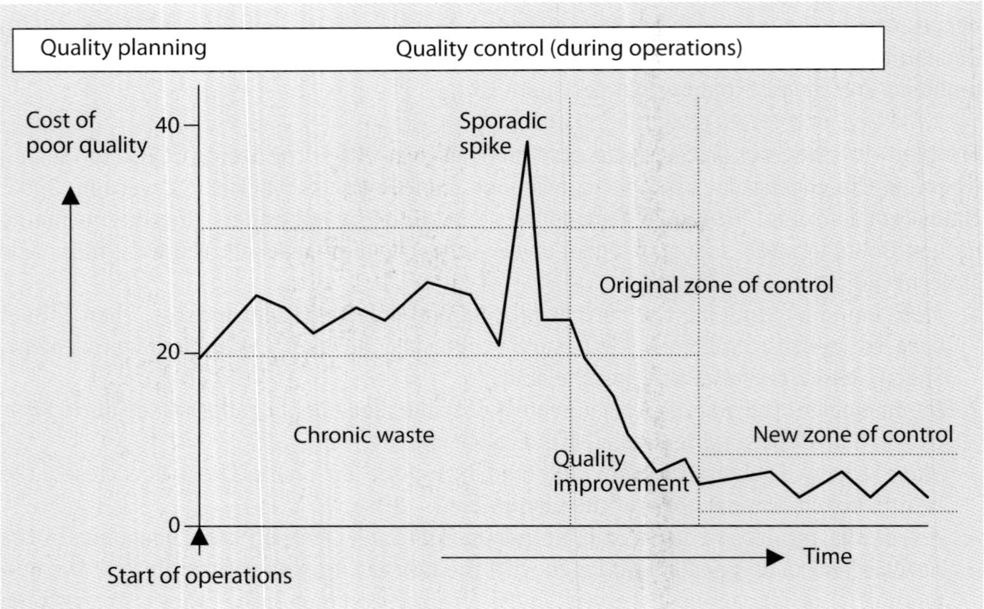

Figure 6.5. Juran on improvement.

Juran described improvement as a *'managerial breakthrough'* being the organised creation of beneficial change (Edge, 1990). It is this organised, step-by-step approach to creating management breakthrough that is the hallmark of the Juran quality improvement process. Juran provoked that quality improvement is not a one-off process. In order to tackle quality problems successfully, organisations must develop the attitude of creating annual improvements year after year. It is an attitude that refuses to accept today's quality levels as good enough.

The quality improvement effort continues forever and must become a formalised part of the company's planning and operational practices bringing them into 'new zones of control' on a higher quality level. In order to achieve this, Juran proposed ten steps for quality improvement (Juran, 1951):

1. build awareness for the need and opportunity for improvement;
2. set goals for improvement;
3. organise people to reach the goals;
4. provide training throughout the organisation;
5. carry out projects to solve problems;
6. report progress;
7. give recognition;
8. communicate results;
9. keep score;
10. maintain momentum by making annual improvement part of the regular systems and processes of the company.

Ishikawa was strongly influenced by Deming and Juran, although he made significant contributions of his own to quality management (Stevenson, 2006). Among his key contributions were the development of the cause-and-effect diagram (also known as a fishbone diagram) for problem solving and implementing *quality circles* that involve workers in quality improvement (see section 6.7 for quality circles). He was the first expert who paid attention to the *internal customer,* i.e. the next person in the process. He was a strong supporter of the need for companies to have a shared vision in order to join everyone in the organisation in a common goal. In addition, he is widely recognised for his efforts to make quality control 'user friendly' for workers.

His contribution can be placed in the context of the features of quality work in Japan (Ishikawa, 1985; Edge, 1990):

1. A company-wide control programme. Every department and all levels and types of personnel within the company are engaged in systematic work, guided by written quality policies, that are endorsed by higher management.
2. Top management is subject to quality audits. A quality executive team visits each company department to identify, isolate and help solve any obstacles to production of quality products or services.
3. Industrial education and training. Since company-wide quality requires participation by everyone involved, education and training in quality must be given to everybody in all departments at each level.
4. Quality circle (QC) activities. A quality circle is 'a small group which meets voluntarily to perform quality improvement within the workshop to which they belong'. The quality circle provides a forum to discuss the department's problems.
5. Application of statistical techniques. Statistical techniques enabled the Japanese to quantify quality and make it understandable, giving it at the same time a high degree of 'scientific' objectivity.

Imai (1986) introduced the elements of Kaizen. Kaizen is a Japanese term for a formal system to promote continuous improvement. It begins with the notion that an organisation can assure its long-term survival success only when every member of the organisation actively pursues opportunities to identify and implement improvements every day.

Kaizen sets no conditions for the size of improvements. In fact, it often favours small, incremental improvements over large innovations. Pursuit of small improvements keeps people thinking about the process and its current operation. They identify potential improvements by understanding the functions of the current system and its weaknesses or relative inefficiencies. Furthermore, small improvements gain returns for the firm without the need for large investments.

Kaizen refines its emphasis on daily incremental improvement to develop three guiding principles:

1. Process view of the system; the basic unit for analysis is the process.
2. Success comes from people; any successful Kaizen programme relies heavily on people's knowledge and their insights and intuition.

3. Constant sense of urgency; a successful Kaizen programme depends on awareness of the need for change; everyone must feel that they can improve their performance (Melnyk and Denzler, 1996).

Like Deming and Juran, *Crosby* (1979) viewed the march toward quality as a never-ending journey (Melnyk and Denzler, 1996). Before beginning this journey, managers must first assess their firm's current quality culture. The management maturity grid (see Chapter 8) can support this self-assessment by positioning the firm in one of five states of quality awareness and quality culture. Once managers have determined their position, they can begin improving quality by implementing Crosby's 14 steps for quality improvement.

This programme emphasises prevention and elimination of process faults rather than inspection. The programme seeks to improve product quality by changing the corporate culture. Detailed statistical tools, while important, remain secondary to this general change in culture. The programme accepts no standard short of zero defects – designing and building products free of any quality problem. The 14 steps are:

1. create management commitment;
2. establish a quality improvement team with clear access to top management;
3. establish clear measures of quality stated in meaningful terms that relate specifically to individual activities;
4. identify the price of non-conformance and the price of conformance by collecting data on cost components;
5. promote awareness of quality;
6. take corrective action;
7. zero-defects planning: how to move the company from correcting problems to totally eliminating them;
8. train managers and employees at all levels, because education is the primary catalyst for change;
9. schedule zero-defects day to signal that the company is moving to its goal of zero defects;
10. set goals that represent progress towards this target;
11. encourage employees to remove causes of error;
12. give recognition to motivate and to set examples;
13. establish a quality council for information and co-ordination; and
14. do it all over again!

6.4 Tools and methods for quality improvement

Improvement abilities include problem-solving skills and active participation. But first of all the people involved should understand the problem situation. The complexity of food products and production processes could restrict the effect of improvement processes, due for instance to low education levels. On the other hand, many problems in food production are caused by a lack of understanding of food and food production systems (Luning and Marcelis, 2006). So, better understanding of learning processes could bring unexpected improvements. To support this, the use of statistical tools and research methods in the improvement process is essential.

6.4.1 Tools for improvement

As a basis for improvement work, data are required together with an analysis of these data (Bergman and Klefsjö, 1994). In Japan it was quickly realised that everyone in a company had to participate in the improvement work. This meant that statistical tools, which have to be used, had to be simple yet efficient. Seven methods, or 'tools', were put together by Ishikawa, amongst others. These methods are called 'the seven QC-tools' where QC means quality control. Since the beginning of the 1960s these tools have been taught to workers and foremen in Japanese industry who use them systematically for problem solving. The tools aid in data collection and interpretation and provide a basis for decision making. Figures 6.6 and 6.7 give an overview of tools that can be used to support quality improvement processes.

A *check sheet* is a simple tool frequently used for problem identification. It provides a format that enables users to record and organise data in a way that facilitates collection and analysis. Check sheets are designed on the basis of what users are attempting to learn by collecting data.

A *flow chart* is a visual representation of a process. As a problem-solving tool, a flowchart helps investigators in identifying possible points in a process where problems occur. A flowchart consists of procedures and decision points.

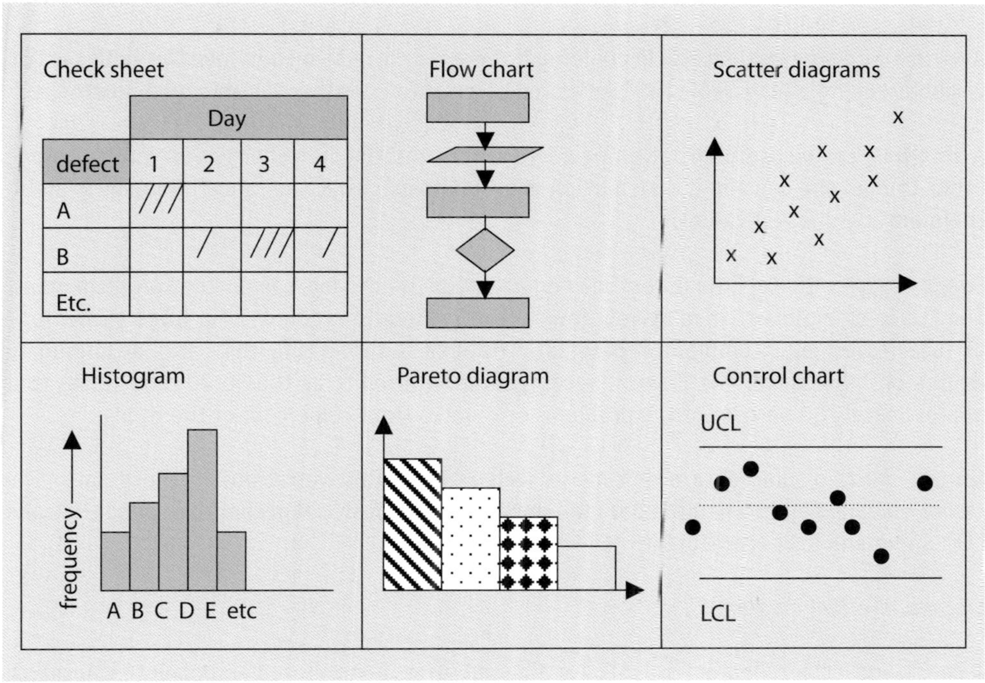

Figure 6.6. Common techniques for process improvement.

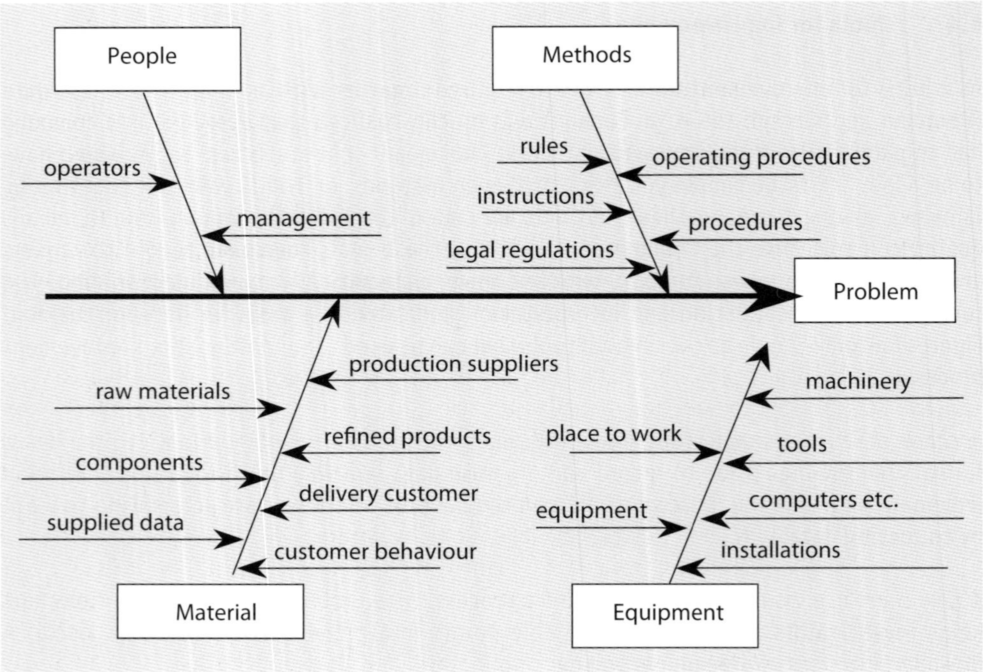

Figure 6.7. Cause-effect diagram of Ishikawa.

A *scatter diagram* can be useful in deciding if there is a correlation between the values of two variables, using one variable to make a prediction about another variable or characteristic.

A *histogram* can be useful in obtaining a sense of the distribution of observed values. Among other things, one can see if distribution is symmetrical, what the range of values is and if there are any unusual values.

Pareto analysis is a technique for focusing attention on the most important problem areas. The concept implies that relatively few factors generally account for a large percentage of total cases (e.g. complaints, defects). The idea is to classify the cases according to degree of importance. The Pareto concept, often referred to as the 80-20 rule, states that approximately 80 percent of the problems originates from 20 percent of the items.

Control charts display data over time as well as computed variations in those data. The charts visually depict when the data fall outside a previously set acceptable range (UCL and LCC: upper and lower control limit).

Cause-and-effect diagram

A cause-and-effect diagram or Ishikawa diagram offers a structured approach to the search for the possible cause(s) of a problem (Figure 6.7). This tool helps to organise problem-solving efforts by identifying categories of factors that might be causing problems. This tool is often used after brainstorming sessions, to organise the ideas generated (Stevenson, 2006).

6.4.2 Methods for improvement

Over the years several methods have been successfully introduced to execute improvement processes in a structured way. They all rely on the basic principles of the improvement process, but they differ in scope and implementation.

Dashboard metrics use visual aids such as graphs and charts (see tools mentioned in section 6.4.1) to summarise and communicate performance to employees and management. By monitoring dashboard metrics, organisations can identify opportunities for continuous improvement (Scott *et al.*, 2009; Evans and Lindsay, 2005).

Total productive maintenance (TPM) is designed to maximise equipment effectiveness (improving overall efficiency and reduce downtime to zero) by establishing a comprehensive productive-maintenance system covering the entire life of the equipment, spanning all equipment-related fields (planning, use, maintenance, etc.) and, with the participation of all employees from top management down to shop-floor workers, to promote productive maintenance through motivation management or voluntary small-group activities (Tsuchiya, 1992). TPM is typically a resource-oriented method. In the seventies the concept of TPM came up in the manufacturing industry. TPM encompassed all production equipment. Equipment may break down, potentially disrupting a firm's entire operation (Melnyk and Denzler, 1996). Typical losses are then downtime, speed losses and product defects. TPM is a method designed to eliminate these losses by identifying and tackling all causes of equipment breakdowns and system downtime. Instead of accepting maintenance as necessary, TPM tries to achieve the ambitious goal of zero breakdowns. Five principles guide TPM programmes:

1. maximise equipment effectiveness (reduce downtime to zero);
2. establish a thorough system of preventive maintenance for the entire life span of equipment, from design and acquisition to disposition;
3. implement TPM in all organisational areas;
4. involve every single member of the organisation from top manager to worker on the shop floor;
5. assign responsibility for the programme to small autonomous groups of employees rather than managers.

Based on these five principles TPM tried to encompass all kind of activities that influence equipment uptime: keeping a well-organised shop floor, proper operating procedures, improving maintenance plans, improving weaknesses in equipment design and improving operation and maintenance skills. Figure 6.8 presents an overview of TPM goals, participants and activities. Through these activities TPM tackled variance and waste by preventing problems and eliminating the causes of defects (Shirose, 1992).

McKone *et al.* (2001) found out that TPM has a positive and significant relationship to low cost (as measured by higher inventory turns), high levels of quality (as measured by higher levels of conformance to specifications), and strong delivery performance (as measured by higher percentage of on-time deliveries and faster delivery times).

Overall workshop improvement, developing optimum human-machine conditions

	Individual improvements to raise equipment efficiency: eliminating the six big losses (losses due to breakdowns, setup and adjustments, idling and minor stoppages, reduced speed, quality defects and rework, and startup)	Establishment of autonomous maintenance system	Establishment of planned maintenance system	Establishment of skills development system for operators and maintenance workers	Establishment of MP design and early equipment management system
Goals	• To achieve the 'zero' target (zero breakdowns, defects, etc.) • To get equipment operating at optimum availability (operating rate)	• To train operators in equipment-related skills • To enable operators to take care of their own equipment	• To raise the efficiency of the maintenance department to prevent the six big losses	• To raise operator skill levels	• To design equipment that is less likely to have breakdowns or defects, and to carry out early equipment management that quickly gets the equipment running reliably to reduce and stabilize startup times
Participants	• Staff • Line leaders	• Operators • Line leaders	• Maintenance staff, leaders, and workers	• Operators • Maintenance workers	• Production engineers • Maintenance staff
Specific activities	• Identifying the six big losses • Calculation of overall efficiency and target-setting • Analysis of phenomena and review of related causes • Implementation of PM analysis • Thorough study of optimum equipment conditions	• Implementation of the 7 steps for autonomous maintenance • Initial cleaning • Eliminating sources of contamination and inaccessible areas • Cleaning and lubrication standards • General inspection • Autonomous inspection • Workplace organisation and housekeeping • Thoroughly implemented autonomous maintenance program	• Daily maintenance and checking measures • Periodic maintenance • Predictive maintenance • Improvements to increase equipment life • Spare parts control • Breakdown analysis and prevention • Lubrication control	• Maintenance fundamentals • Nuts and bolts • Keys • Bearings • Power transmissions • Sealing (leak prevention)	• Establish design goals • Autonomous maintainability • Maintainability • Operability • Reliability (Incorporate into MP) • Life cycling costing • Identifying problems at each stage (design, drawing, fabrication, installation, etc.) • Debugging

Figure 6.8. TPM goals, participants and activities (Shirose, 1992).

Lean manufacturing is a methodology used to transform complex processes to a smooth continuous production flow, which delivers customer value more rapidly, improves workflow, standardises processes and eliminates waste. Waste, in the context of lean manufacturing, is defined as non-value added processes which consume resources (Scott *et al.*, 2009).

A rather extreme improvement concept was introduced in 1993 by Hammer and Champy, called *business reengineering*. Reengineering (also known as business process redesign) is a type of improvement with the potential to dramatically improve quality and speed of work and to reduce its cost by fundamentally changing the processes by which the work is executed (Dean and Evans, 1994).

Reengineering goes beyond the process level by including the entire value chain. It also goes beyond reorganisation by including fundamental redesign and starting from scratch. It is often used when the improvements needed are so great that incremental changes to operations will not be sufficient. For instance, improvements from 50% to 1000%.

The general principles of business process redesign (BPR) include:
- Reduce handoffs: every time a process is handed from one person or group to another, errors can occur.
- Eliminate steps: the best way to save time on a step is to avoid it when possible. If a step does not add customer value, stop doing it.
- Perform steps in parallel rather than in sequence: unless one operation can not be done until another is finished, do both at once.
- Involve key people early in order to avoid doing things over when key people do not give their input until the process is under way. Doing it is consistent with the total quality management principle: 'do it right the first time'.

Six sigma programmes have originated from Motorola (Evans and Lindsay, 2005). This company developed programmes and techniques to realise a quality level of six sigma. Sigma refers to the number of standard deviations from the mean in any statistically measured process. In a normally distributed process 97.73 percent lie within 3 sigma, and 99.9999998 percent lie within 6 sigma. Many companies meeting the 3 sigma level feel rather satisfied, but in areas like food safety much higher requirements are set. To compare, when 3 sigma is 1000 km, 6 sigma is one footstep.

The term 'six sigma process' comes from the notion that if one has six standard deviations between the mean of a process and the nearest specification limits, there will be practically no items that fails to meet the specifications. The six sigma programme seeks to continuously improve quality performance towards this more or less 'zero defects' level by well-organised processes and by the advanced training of people involved.

Six sigma programmes use two key methods, DMAIC and DMADV, both inspired by Deming's Plan-Do-Check-Act cycle.

DMAIC is used for improving existing business processes, and includes the following five steps:

- *Define* the process improvement goals that are consistent with customer demands and enterprise strategy.
- *Measure* the current process and collect relevant data for future comparison.
- *Analyse* to verify relationship and causality of factors. Determine what the relationship is, and attempt to ensure that all factors have been considered.
- *Improve* or optimise the process based upon the analysis using techniques like Design of Experiments.
- *Control* to ensure that any variances are corrected before they result in defects. Set up pilot runs to establish process capability, transition to production and thereafter continuously measure the process and institute control mechanisms.

DMADV is used to create new product or process designs for predictable, defect-free performance, and includes the following five steps:

- *Define* the goals of the design activity that are consistent with customer demands and enterprise strategy.
- *Measure* and identify CTQ's (critical to qualities), product capabilities, production process capability, and risk assessments.
- *Analyse* to develop and design alternatives, create high-level design and evaluate design capability to select the best design.
- *Design* details, optimise the design, and plan for design verification. This phase may require simulations.
- *Verify* the design, set up pilot runs, implement production process and handover to process owners.

Benchmarking is the process through which a company learns how to become the best in some area by carefully analysing practices of other companies that already do extremely well in that area (best-practices companies). Benchmarking helps a company to learn its strengths and weaknesses, and those of other industrial leaders, and incorporate the best practices into its own operations (Dean and Evans, 1994).

Two major types of benchmarking are *competitive* and *generic*. Competitive benchmarking usually focuses on the products and manufacturing of a company's competitors. Generic benchmarking evaluates processes or business functions against the best companies, regardless of their industry.

The benchmarking process can be described as follows:

- Determine which functions to benchmark. These should have a significant impact on business performance and should be key dimensions of competitiveness. If fast response is an important dimension of competitive advantage, then processes that might be benchmarked would include order processing, purchasing, production planning, and product distribution. There should also be an indication that the potential for improvement exists.
- Identify key performance indicators to measure. These should have a direct link to customer needs and expectations. Typical performance indicators are quality, performance, and delivery.
- Identify the best-in-class companies. For specific business functions, benchmarking might be limited to the same industry. For generic business functions, it is best to look

outside one's own industry. Selecting companies requires knowledge of which firms are superior performers in the key areas. Such information can be obtained from published reports and articles, industry experts, trade magazines, professional associations, former employees, or customers and suppliers.
- Measure the performance of the best-in-class companies and compare the results to your own performance. Such information might be found in published sources or might require site visits and in-depth interviews.
- Define and take actions to meet or exceed the best performance. This usually requires changing organisational systems. Simply to imitate the best is like shooting at moving target – their processes will continually improve. Therefore, attempts should be made to exceed the performance of the best (Deans and Evans, 1994).

Despite the fact that they are not typically improvement methods, HACCP (Chapter 7) and TQM (Chapter 8) are also mentioned as improvement methods (Scott *et al.*, 2009).

6.5 Managing the quality improvement process

The most effective means of improvement is to use the people who do the job to identify and implement appropriate changes (Imai, 1986). Quality improvement processes sometimes come up spontaneously, but normally they should be initiated by top management and properly organised in order to support the continuous character of the improvement processes.

Bessant and Francis (1999) defined improvement as an organisation-wide process of focused and sustained incremental innovation. It contributes to the 'dynamic capability' of the organisation, since it offers a mechanism whereby a high proportion of the organisation becomes involved in its innovation and learning process. In other words, improvement is an activity that has to be done by certain routines towards 'doing what we do better' and requires involvement based on the premise that people are capable of creative problem-solving.

This section describes the basic conditions for quality improvement and discusses different forms of teamwork with their impact for improvement. It finishes with some remarks about project management, and a deeper explanation of organisational change strategies.

6.5.1 Basic conditions for quality improvement

Improvement requires a commitment to learning and most of the failures in improvement have been due to the lack of attention to this premise. Therefore, implementing a philosophy of continuous improvement in an organisation may be a lengthy process (Krajewski *et al.*, 2007), and several steps are essential to its eventual success:
- Train employees in the methods of statistical process control (SPC) and other tools for improving quality and performance.
- Make SPC methods a normal aspect of daily operations.
- Build work teams and employee involvement.
- Utilise problem-solving tools within the work teams.
- Develop a sense of operator ownership in the process.

Note that employee involvement is central to the philosophy of continuous improvement. However, the last two steps are crucial if the philosophy is to become part of the every-day organisation. Problem solving addresses the aspects of operations that need improvement and evaluates alternatives for achieving improvements. The sense of operator ownership appears when employees feel as if they own the processes and methods they use and take pride in the quality of the product or service they produce. This sense can derive from participation in work teams and in problem-solving activities, which give employees a feeling that they have some control over their workplace and tasks.

A potential barrier to improvement is the form of organisation (Edge, 1990). Traditional organisation structures, based on a series of departments each with their own goals, specialists and command structures, have a natural tendency to focus on their own priorities, such that overall business and company-wide objectives are often neglected.

The formal authority structures present within each department create a complex communication network, which makes it more difficult to keep people informed. Only top management can impose the need for change in such circumstances. Therefore, there is a need to create simpler, flatter structures with added flexibility, which ensure involvement from the total workforce. The result of this is the creation of shorter links to customers and suppliers, such that changing circumstances can be quickly perceived and dealt with. Figure 6.9 illustrates how relationships include the possibility of joint improvement based on customer requirements, both internal and external (Kehoe, 1995).

A second potential barrier to improvement is lack of information. People must have ready access to reliable and accurate information and have the organisational scope to act on it. This information must embrace manpower, machines, methods and materials and the company environment.

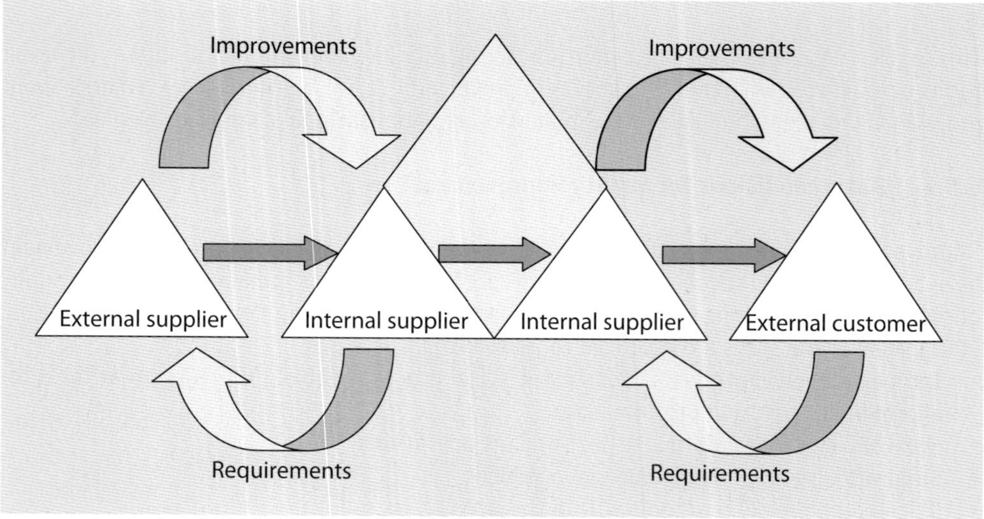

Figure 6.9. The internal and external customer-supplier relationships.

Additionally, Wilkinson *et al.* (1998) reported about a research project concerning difficulties faced by organisations in the improvement of quality. Their findings are summarised in Table 6.2. In the table the difficulties are included, with the percentage of respondents labelling it as a major or minor factor.

Quality management seems to suffer from limitations, with 'lack of resources' and 'cost constraints' emerging as two of the most important difficulties in implementation. Quality management may also be constrained by short-term attitudes, with over 60 percent of organisations seeing the 'emphasis on short-term goals' as a difficulty.

Problems in measuring quality, in communications and with a lack of training also emerged as significant difficulties.

The quality of employees was seen as a 'major difficulty' by only 7 percent of the sample. Interestingly, this compares with 17 percent who saw the quality of management as a major difficulty. In total, over half of the respondents felt that the quality of management presented at least minor difficulties.

6.5.2 Training

If is often said that quality begins with education and ends with education. Thus, the first requirement is to make certain that the supervisors, and all the employees are trained as to what quality is, why control of quality is necessary, where to be concerned about quality, when to make changes and who is responsible (Gould and Gould, 2001).

Training is a process that facilitates learning (Sofo, 1999). It is a planned process to modify attitude, knowledge, or skills. Its purpose in the work situation is to develop the abilities

Table 6.2. Difficulties in implementing quality improvement.

	Major difficulty (% of respondents)	Minor difficulty (% of respondents)
Lack of resources	42	37
Cost constraints	38	36
Emphasis on short-term goals	34	29
Measuring quality	27	38
Communication	25	41
Lack of training	23	41
Clash with other initiatives	22	32
Recession	21	26
Quality of management	17	38
Lack of UK quality infrastructure	12	26
Seen as product/operations concern only	11	24
Quality of employees	7	40
Other	4	1

of the individual and satisfy the current and future manpower needs of the organisation (Armstrong, 1996). Training, as human resource development, is also very important for the management staff of the organisation for three main reasons (Analoui, 1999):

- Training can increase the effectiveness of managers by reducing learning time to achieve specified performance levels.
- Training can improve job performance levels and ultimately increases job satisfaction
- Training ensures future succession of the needs of manpower.

Armstrong (1996) identified four places where training can take place - in company, on-the-job, off-the-job and external locations.

Training systems are meant to motivate the employees to learn, create knowledge and share that knowledge with others in the company (Swanson and Falkman, 1997). However, the organisation should make sure that there is mutual understanding between it and the employees with regard to training issues. This is some of the prior important knowledge if training is to be effective (Walter, 2002):

- Employees should understand why training is needed.
- There should be adequate verification of the employees' prerequisite knowledge, skills and experience.
- Employees should be made aware of the training to be covered.
- Employees should know their performance evaluation criteria.
- Employees should know the trainee evaluation process.
- Employees should be briefed on record keeping.
- Employees' responsibilities should be explained.
- Overall safety and quality issues encountered on the job should be discussed with the employees.

Like any other business, training can become wasteful if not carefully planned and supervised. Systematic training is planned and specifically designed to meet defined needs, which include (Armstrong, 1996) identifying and defining training needs, defining objectives of training, planning and implementing training programmes, and evaluating the training.

More specifically, with respect to food safety, industries have an obligation to ensure that their managers as well as their food handlers receive adequate education and training. Industries should consider food safety as one of their highest priorities and should clearly communicate this to their employees (Käferstein, 2004). Crowther *et al.* (1993) have summarised what supervisors and food handlers need to know regarding food safety education:

- the major sources of microorganisms in the product for which they are responsible;
- the role of microbes in disease and food spoilage;
- why good personal hygiene is required;
- the importance of reporting diarrhoea, vomiting, other illness, lesions and cuts to supervisory personnel;
- the nature of the control required at their job in the process;
- the proper procedures and frequency for cleaning equipment for which they are responsible;
- the procedures necessary to report deviations from control specifications;

- the characteristics of normal and abnormal products at their given step in the process (e.g. colour, texture, package integrity, odour);
- the importance of maintaining proper records;
- how to monitor critical control points (CCP's) of operations for which they are responsible.

There is, however, uncertainty regarding the efficacy of current food hygiene training. Howes *et al.* (1996) have indicated that although training may bring about an increased *knowledge* of food safety this does not always result in a positive change in food handling *behaviour*. It has been suggested that this disparity between knowledge and practice occurs because much of the existing training, particularly formal certificated training, is designed using the KAP model (Rennie, 1995). This approach assumes that an individual's behaviour or practice (P) is dependent on their knowledge (K) and suggests that the mere provision of information will lead directly to a change in attitude (A) and consequently a change in behaviour. It has been suggested that this model is flawed in its assumption that knowledge is the main precursor to behavioural change. Rennie (1995) argues that too little emphasis is placed on changing individuals' beliefs and attitudes and that the model fails to take into account cultural, social and environmental influences. This fact is demonstrated in research conducted by Clayton *et al.*(2002). Despite 95% of respondents receiving food hygiene training, 63% admitted to sometimes not demonstrating food safety behaviour.

6.5.3 Teamwork

Years ago, the decision of companies such as Volvo, Toyota and General Foods to introduce teams into their production processes made news because no one else was doing it (Robbins, 2000). Today, it is just the opposite. It is the organisation that does not use teams that has become newsworthy. How can the current popularity of teams be explained? Evidence suggested that teams typically do better than individuals when tasks require multiple skills, judgement and experience. As organisations have restructured themselves to compete more effectively and efficiently, they have turned to teams as a way to better utilise employee talents. Management observed that teams are more flexible and responsive to changing events than are traditional departments or other forms of permanent groupings. Teams have the capability to quickly assemble, deploy, refocus and disband. Furthermore, teams facilitate employee participation in decision-making and in that way they increase the motivation of employees.

Teams are especially important in total quality management (see for TQM concept Chapter 8). Teamwork enables various parts of the organisation to work together in meeting customer needs that can seldom be fulfilled by employees limited to one speciality. The TQM philosophy recognises the interdependence of various parts of the organisation and uses teams as a way of co-ordinating work (Dean and Evans, 1994).

The formal definition of a *team* is a small group of people with complementary skills, who work together to achieve a shared purpose, and hold themselves mutually accountable for its accomplishment (Schermerhorn, 2007).

The difference with a workgroup is that a workgroup is a group that interacts primarily to share information and to make decisions to help each other to perform within each member's area of responsibility (Robbins, 2000). Workgroups have no need or opportunity to engage in collective work that requires joint effort. So their performance is merely the summation of all group members' individual contributions: there is no positive synergy – the creation of a whole that is greater than the sum of its parts. The presence of synergy means that a team uses its membership resources to the fullest and through collective performance achieves far more than could otherwise be achieved (Schermerhorn, 2007). The usefulness of teams includes the following contributions:

- increasing resources for problem solving;
- fostering creativity and innovation;
- improving the quality of decision making;
- enhancing members' commitment to tasks;
- raising motivation through collective action;
- helping control and discipline members;
- satisfying individual needs as organisations grow in size.

However, teams have their disadvantages as well. Problems that are commonly encountered in teams include:

- Personality conflicts: individual differences in personality and work style can disrupt the team.
- Task ambiguity: unclear agendas and/or ill-defined problems can cause teams to work too long on the wrong things.
- Poor readiness to work: time can be wasted when meetings lack purpose and structure.
- Poor teamwork: failures in communication, conflict and decision making may limit performance and/or hurt morale.

The trend toward greater empowerment is associated with several developments in the utilisation of teams. In Chapter 4, cross-functional teams in product development have already been described. Managers seek to expand opportunities for broad-based participation in workplace matters. Forms of teamwork can be distinguished according to the level of empowerment (Schermerhorn, 2007; Robbins, 2000; Robbins and Coulter, 2007):

1. *Committees and task forces.* Committees and task forces bring people together outside of their daily job assignments to work in small teams for a specific purpose. They typically operate with task agendas and are led by a designated head or chairperson who, in turn, is held accountable for the results. A committee usually operates with an ongoing purpose and its membership may change over time. An example is the quality steering committee responsible for developing quality policy and quality systems.
   A task force usually operates on a more temporary basis. Its official tasks are very specific and time defined. Once the stated purpose has been accomplished the taskforce may disband.
2. *Cross-functional teams.* Cross-functional teams are made up of employees at about the same hierarchical level, but from different work areas, who come together to accomplish a task. They are expected to exchange information, develop new ideas and solve problems and co-ordinate complex projects, and not be limited in performance by purely functional concerns and demands.
3. *Quality circles.* A quality circle (QC) is a group of workers that meets regularly to discuss and plan specific ways to improve work quality. Usually it consists of 6 to 12 members from a work area. After receiving special training in problem solving, team processes and

quality issues, members of the quality circle try to come up with suggestions that can be implemented to raise productivity through quality improvement. Juran (1990) pointed out the differences between quality (control) circles and project teams (task forces and cross-functional teams) – see Figure 6.10. Problems with quality circles raised where basic conditions like training and top management commitment were not fulfilled. This is the main reason for quality circles not to have continuing success.

4. *Self-managing teams.* Self-managing teams are generally composed of people who take over the responsibilities of their former supervisors. Typically, these responsibilities include collective control over the pace of work, determination of work assignments, organisation of breaks and collective choice of inspection procedures. Fully self-managed teams select their own members and the members evaluate each other's performance. As a result, supervisory positions become less important and may even be eliminated. Self-managed teams operate with participating decision-making and multi-tasking, in which team members each have the skills to perform several different jobs (Figure 6.11). Typical characteristics of self-managing teams are as follows:
   - Members are held collectively accountable for performance results.
   - Members have discretion in distributing tasks and scheduling work within the team.
   - Members are able to perform more than one job.
   - Members train one another to develop multiple job skills.
   - Members evaluate one another's performance contributions.
   - Members are responsible for the total quality of team products.

The overall research on the effectiveness of self-managing teams is not uniformly positive. For example, although individuals in these teams tend to report higher levels of job satisfaction, employees in self-managing teams seem to have higher absenteeism and turnover rates than employees in traditional work structures (Robbins, 2000).

The forms of teams just described are examples of formal groups in the organisation. In contrast, there are informal groups, which are also important in every organisation.

Feature	Quality circles	Project teams
•Primary mission	•To improve human relations	•To improve quality
•Secondary mission	•To improve quality	•To improve participation
•Scope of project	•Within single department	•Multi-departmental
•Size of project	•One of the useful many	•One of the vital few
•Membership	•From a single department	•From multiple departments
•Basis of membership	•Voluntary	•Mandatory
•Hierarchical status of members	•Typically in the work force	•Typically managerial or professional
•Continuity	•Circle remains intact after each project	•Team is ad hoc, disbands after project is completed

Figure 6.10. Contrast between quality circles (QC) and project teams (PT).

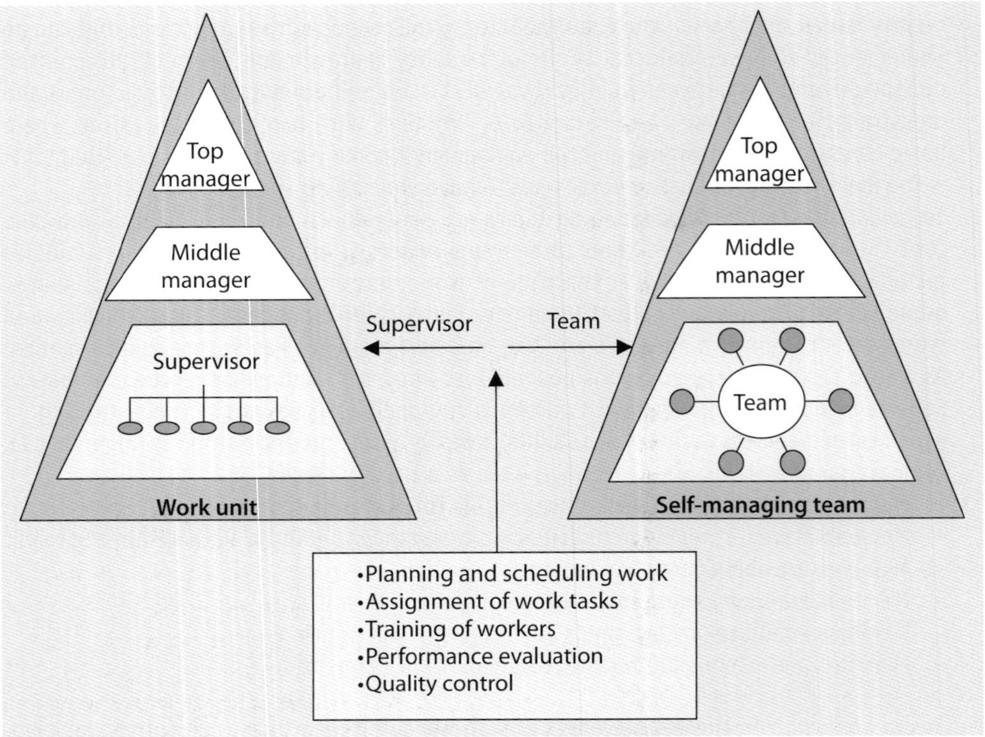

Figure 6.11. Difference between work units and self-managing teams.

These informal groups are not recognised on organisational charts and are not officially created. They emerge as part of the informal structure and from natural and spontaneous relationships among people. Informal groups can have a positive impact on work performance. In particular, the relationships and connections made possible by informal groups may actually help speed the workflow or allow people to 'get things done' in ways not possible within the formal structure. Furthermore, members of informal groups often find that the groups offer social satisfaction, security, support and a sense of belonging (Schermerhorn, 2007).

6.5.4 Project approach

The quality improvement effort must go on forever, according to Juran, and must become a formal part of the company's planning and operational practices, just like the corporate budget and annual marketing plan (Edge, 1990). To attain this objective, Juran called for the establishment of quality improvement project teams starting at the top of the company with a senior management quality team. All members of the senior management team, without exception, should accept the concept of annual quality improvement as a way of life for the future. Key members of the senior management should be part, along with other participants, of a quality steering council or committee. The steering council determines the broad policy for quality improvement programmes, it decides how appropriate training must be implemented and encourages management project teams to initiate the improvement process.

These *project teams* pass through a series of phases ranging from start-up to project planning, implementation and project evaluation. A more detailed description of the project management system has already been discussed in Chapter 4, where project management is analysed in the context of quality design processes.

A typical project team meets weekly for about three months, and should arrive at a quality improvement solution. Once a team is finished, it is dismissed and new teams are formed to define and tackle another quality improvement project. In large companies it is not uncommon to have dozens of quality improvement teams working simultaneously.

6.5.5 Organisational change strategies

Organisational change is very important for quality management, especially when human capacities are to be used to their full extent. This will be explained firstly in this section, and then an overview will be given of organisational change strategies and their advantages for quality management.

The basis of continuous improvement philosophy is the belief that virtually any aspect of an operation can be improved and that people most closely associated with an operation are in the best position to identify changes that should be made (Krajewski *et al.*, 2007).

However, in studies on the implementation of advanced manufacturing technologies it was observed that the most difficult aspects in the implementation of new technology and systems were changing the organisation and people (Edge, 1990). So organisational change is fundamental to quality management in which improvement processes continuously bring about changes in technologies and systems. As Dean and Evans (1994) stated 'change is a way of life'. They also stressed the importance of cultural change, i.e. the change in beliefs and values shared by people in the organisation.

Choosing between *top-down* change and *bottom-up* change, the latter is usually preferable in quality management. In top-down change, strategic changes are initiated which have a comprehensive impact on the organisation and its performance capabilities. However, the change that is driven from the top runs the risk of being perceived as insensitive to the needs of lower level personnel. It can easily fail if implementation suffers from excessive resistance and insufficient commitment to change. The success of top-down change is usually determined by the willingness of middle and lower level workers to actively support top management initiatives.

In bottom-up change the initiatives for change come from persons throughout the organisation and are supported by the efforts of middle and lower level managers. Bottom-up change is essential to organisational innovation and is very useful in terms of adapting operations and technologies in order to change requirements of work. Empowerment, involvement and participation enable this bottom-up change (Schermerhorn, 2007).

When change occurs, there will normally be resistance. Some of the major reasons for resistance to change include uncertainty and insecurity, a reaction to the way change is presented, threats to vested interests, cynicism and a lack of trust, and a lack of understanding. Resistance to change may also result in a reaction to being controlled or getting less autonomy or power.

Dealing with change objectives and resistance to change requires a change strategy. Elements of change strategies are education, communication, and participation. Different strategies favour different elements depending on the specific situation. Behavioural scientist Kurt Lewin (1952) provided one of the earliest models of change. Lewin viewed the change process as a modification of forces that keep a system's behaviour stable. His model described change as consisting of three phases in which driving forces, or pushers for change, compete with restraining forces, which strive to maintain the status quo. These three phases are (see Figure 6.12):

- *Unfreezing*, which involves disruptive forces that maintain the existing state or level of behaviour (state A). This might be done by introducing new information to show discrepancies between state A and one desired by the organisation – i.e. a performance gap.
- *Moving*, which entails a transition period (state B) during which behaviours of the organisation or department are shifted to a new level (desired state C). It involves developing new behaviours, values and attitudes through changes in structure, technology, strategy and human processes.
- *Re-freezing*, which stabilises the organisation in a new state of behavioural equilibrium (state C). This is accomplished by using support mechanisms to reinforce the new organisational state, such as culture, norms, policies and structures.

Lewin's model thus suggests that managers should find ways to unfreeze existing equilibrium before any change will occur.

Change agents (person or group responsible for change) use various approaches when trying to get others to adapt to change. Figure 6.13 summarises three common change strategies, each of them acting different on the unfreezing, moving, and re-freezing phases (Schermerhorn, 2007):

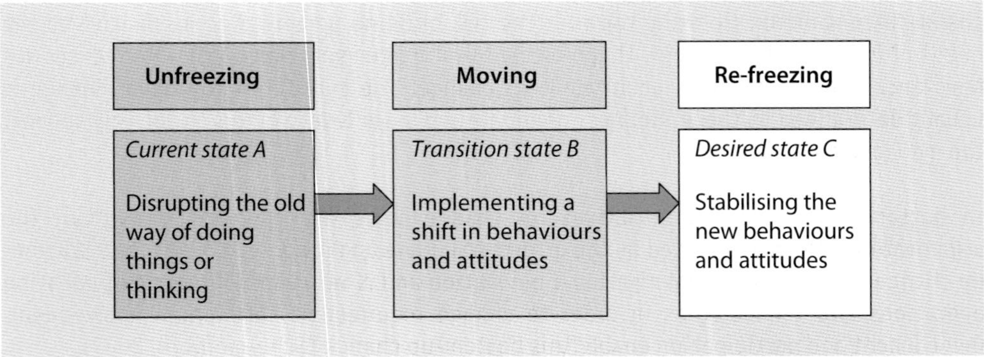

Figure 6.12. Lewin's model for implementing change.

Force-coercion strategies

A force-coercion strategy uses the power bases of legitimacy, rewards and punishments as primary inducements to change. As Figure 6.13 shows, likely outcomes of force-coercion are immediate compliance but little commitment. Force-coercion can be pursued in at least two ways, both of which can be commonly observed in organisations.

In a *direct forcing* strategy, the change agent takes direct and unilateral action to 'command' that change takes place. This involves the use of formal authority or legitimate power, offering special rewards, and/or threatening punishment.

In *political manoeuvring,* the change agent works indirectly to gain special advantage over other persons and thereby make them change. This involves bargaining, obtaining control of important resources, or granting small favours.

In both versions, the force-coercion strategy produces limited results. Although it can be implemented rather quickly, most people respond to this strategy out of fear of punishment or hope for a reward. This usually only results in temporary compliance with the change agent's desires. The new behaviour continues only as long as the opportunity for rewards and punishments is present. For this reason, force-coercion is most useful as an unfreezing device that helps people break old patterns of behaviour and gain initial motivation to try new ones.

Change strategy	Power basis	Managerial behaviour	Likely results	
Force-coercion Using position power to create change by decree and formal authority	Legitimacy Rewards Punishments	Direct forcing and unilateral action Political manoeuvring and indirect action	Fast	Temporary compliance
Rational persuasion Creating change through rational persuasions and empirical argument	Expertise	Informational efforts using credible knowledge, demonstrated facts and logical argument		
Shared power Developing support for change through personal values and commitments	Reference	Participative efforts to share power and involve others in planning and implementing change	Slow	Longer term internalisation

Figure 6.13. Alternative change strategies and their managerial implications.

Rational persuasion strategies

Change agents using a rational persuasion strategy attempt to bring about change through persuasion backed by special knowledge, empirical data, and rational arguments. The likely outcome is eventual compliance with reasonable commitment. This is an informational strategy which assumes that facts, reason, and self-interest will guide rational people when deciding whether or not to support a change.

A manager using rational persuasion must convince others that the cost-benefit value of a planned change is high and that it will leave them better off than before. Accomplishing this depends to a large extent on the presence of expert power. This can come directly from the change agent if he or she has personal credibility as an 'expert'. If not, it can be obtained in the form of consultants and other outside experts or from credible demonstration projects. When successful, a rational persuasion strategy unfreezes and re-freezes a change situation. Although slower than force-coercion, it tends to result in longer lasting and internalised change.

Shared power strategies

A shared power strategy engages people in a collaborative process of identifying values, assumptions, and goals from which support for change will naturally emerge. The process is slow, but it is likely to yield commitment. Sometimes called a *normative-re-education strategy,* this approach is based on empowerment and is highly participative in nature. It relies on involving others, in examining personal needs and values, it relies on group norms, and operating goals. Power is shared by the change agent and other persons as they work together to develop a new consensus to support needed change.

Managers using shared power as an approach to planned change need reference power and skills to work effectively with other people in group situations. They must be comfortable allowing others to participate in making decisions that affect the planned change and the way it is implemented. Because it entails a high level of involvement, a normative-re-education strategy is often quite time-consuming, but it is likely to result in a longer lasting and internalised change.

Among consulting professionals *organisation development*, or OD for short, is known as a shared power strategy that fits well together with the concept of total quality management and employee involvement. Organisation development is supposed to help organisations cope with environmental and other pressures for change while also improving their internal problem-solving capabilities. OD, in this sense, translates the need for continuous improvement into a planned change process.

In organisation development two goals are pursued simultaneously. The *outcome goals of OD* focus on task accomplishments, while the *process goals of OD* focus on the way people work together. It is this second goal that strongly differentiates OD from more general attempts at planned change in organisation. You may think of OD as a form of 'planned change plus', with the 'plus' meaning that change is accomplished in such a way that organisation members develop a capacity for continued self-renewal. OD tries to achieve change while helping organisation members become more active and self-reliant in their ability to continue changing in the future. What also makes OD unique is its commitment to strong humanistic values and established principles of behavioural science. OD is committed to improving organisations through freedom of choice, shared power and self-reliance, and by taking the best advantage of what we know about human behaviour in organisations (Schermerhorn, 2007).

6.6 Quality improvement in practice

Quality improvement is still not a success story in the food industry. In food companies there are sometimes success stories about e.g. good functioning of total productive maintenance (TPM), while experiments with e.g. quality circles failed.

Considering the typical situation in the food industry, the following aspects may explain why improvement activities are not yet very common in this sector:

- Low level of education, sometimes combined with language problems, makes it more difficult to involve operators or employees in problem solving activities.
- The production circumstances are often obstructive, e.g. noise, bad smell, high humidity etc. These conditions are dissatisfying and are not motivating if they are not reduced to acceptable levels.
- As a consequence of the functional organisation structure, most knowledge is centralised in the specialised departments. This implies that the responsibility for improvements is assigned to those specialists, or the heads of departments. On the other hand operators frequently have much experience and knowledge but are not often involved in improvement activities.
- Equipment is often not very accessible for improvement. The knowledge is present at the supplier of equipment, who also carries out modifications. The user of the equipment is often unable to submit any requests or needs and is restricted to the range of equipment offered by (a handful of) suppliers.
- Poor feedback of information about quality performance results. The lowest level of operators in the organisation are particularly inadequately informed or only indirectly about quality results, which is not very stimulating.
- Improvement activities usually go unrewarded. Acknowledgement and support from top management are commonly missing.
- Soft organisation methods are not very common in the food sector. In general, the culture is more solution-oriented and pro-active, whereas less time is spent on discussion and analysis.

From a techno-managerial perspective, quality improvement processes can be encouraged by focusing on those issues that are strategically relevant for the organisation from top to bottom, i.e. the core competences of the company. In the food industry, core competences are usually of a technical or technological nature. Therefore, the techno-managerial approach would suggest focusing on improvement processes on those core competences, which at the same time should be supported by appropriate management activities. Below are some illustrations:

- A managerial condition for improvement is appropriate measurements and information systems. This information should include the critical quality and safety points that are

related to the technological core competences. Furthermore, information must be easy to access and applicable for statistical tools for analysis (e.g. Pareto analysis, standard deviation, etc.), which enables not only those involved but also management to take appropriate decisions.
- Another essential management activity with respect to improvement is training. Training of operators and employees must be aimed at developing relevant technological and managerial knowledge for the specific quality control points they are responsible for. Moreover, training on simple statistical techniques is required to be able to organise and process data effectively, in order to use it as basis for analysis and argumentation.
- Team building is another important management task but cannot be forced and requires good leadership capacities. Managers should not just install teams; they should facilitate team building in such a way that teams can grow spontaneously.

In the food industry a limited number of companies currently work with self-managing teams and use statistical tools to support improvement activities. However, to get to this stage, these companies have consciously worked for years on organisational change processes. In fact, the development of a culture, in which improvement processes are common, requires a lot of time.

Chapter 7

Quality assurance

This chapter describes the general principles of acknowledged quality assurance guidelines and standards commonly used in agribusiness and the food industry. First, the quality assurance process is described, briefly explaining the major QA activities and typical QA decisions (section 7.2). The section on 'primary quality assurance guidelines and standards' describes the principles and characteristics of respectively the basic codes and guidelines (like GMP, PRP), Hazard Analysis and Critical Control Points (HACCP), and ISO 9001:2000 (section 7.3). In the next section typical tools and methods to support quality assurance activities are explained (section 7.4). Then the management of assurance systems is discussed and attention is paid to the principles of auditing and certification (section 7.5). Section 7.6 describes the principles of various quality assurance standards and guidelines with a specific chain focus (e.g. ISO 22000, SQF), and ones that are specifically aimed at certain parts of the agri-food chain. In the last section the typical situation in agribusiness and the food industry with respect to quality assurance is briefly described and considered from the techno-managerial perspective (section 7.7).

7.1 Introduction

The objective of quality assurance is to *guarantee* that quality requirements, such as product safety, quality, reliability, service, etc., are realised by the quality management system. Quality assurance should provide confidence to government, customers, consumers, and other stakeholders that quality requirements will be met. Quality assurance is specifically important for the food industry and agribusiness since the typical characteristics of food production (e.g. hidden safety risks, restricted shelf life, seasonal harvesting, heterogeneity of raw materials, complexity of supply chains) put high demands on assuring ultimate product quality, specifically on the intrinsic attribute of safety (Van der Spiegel *et al.*, 2003, Luning *et al.*, 2008, 2009).

In Figure 7.1 the quality assurance function is described in its main elements. Quality assurance objectives are translated into actual assurance on product quality by ensuring that production processes function well, by putting requirements on the internal quality management system, validating its effectiveness, and verifying its actual performance. By internal and external auditing, systematic documentation and record-keeping, and by communicating the outcome of this process, stakeholders can be shown evidence that the system functions well.

The assurance process depends on the ability to determine critical points in products, processes, and the production environment. Critical points predict more than other points that quality is really under control. Assurance processes are focused on these points because they provide a high level of prediction, while on the other hand assurance costs will be kept as low as possible. For instance, a small company with simple low risk products may have only a few critical points. Large companies with high risks products and processes probably have more critical points. However, having too many critical points (tens to hundreds) may

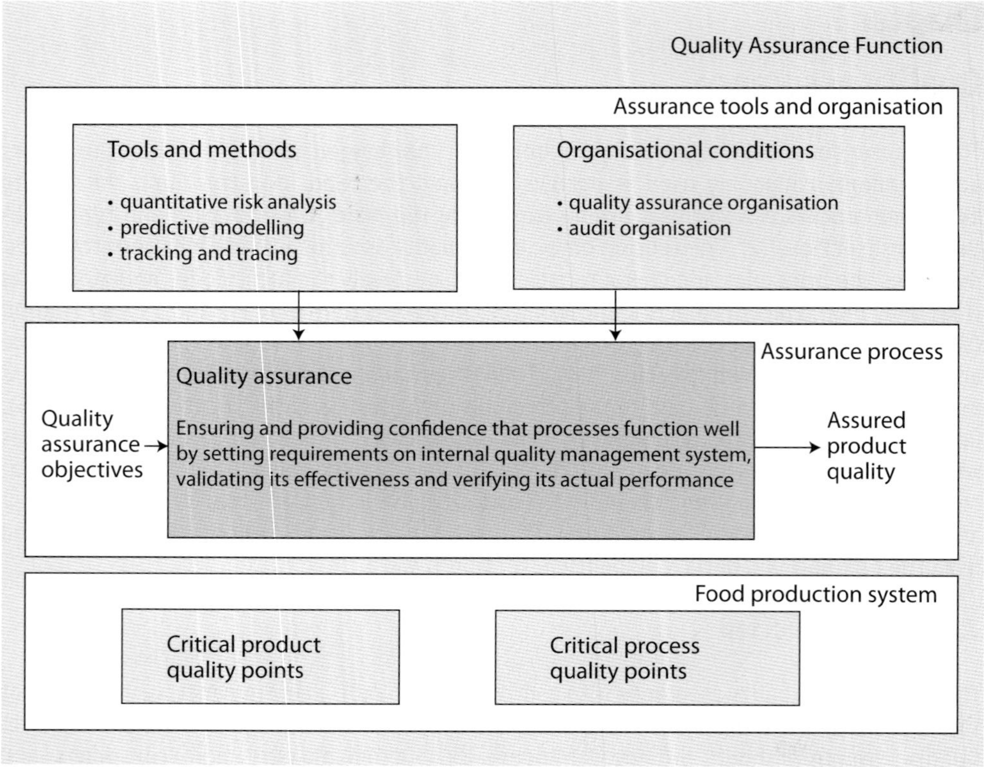

Figure 7.1. Quality assurance function.

show that the production process is not under control, due to poor product design and or inadequate production processes and equipment.

Tools and methods commonly used in quality assurance include quantitative risk analysis, predictive modelling, and tracking and tracing. The latter is important from the chain perspective. Typical organisational conditions for quality assurance are the organisational embedding of quality assurance and especially the role of auditing. Important questions are for instance, which department is responsible and how independent is the auditing process. These factors will be discussed in the following sections.

7.2 Quality assurance process

Quality management refers to all activities that organisations use to direct, control, and co-ordinate quality, including formulating a quality policy, setting quality objectives, quality planning, control, assurance, and improvement (ISO, 2005). Quality assurance specifically deals with setting requirements on the control (but also design and improvement) activities in the quality management system, evaluating its performance, and organising necessary changes (Luning and Marcelis, 2007). Figure 7.2 schematically depicts the quality assurance process.

> *The core mechanism of quality assurance is the translation of assurance requirements of stakeholders into the company specific set up of the quality management system, implementation of the system and verifying if the system is complied to in a way that stakeholders feel assured about delivered product quality.*

Organisations use a quality management system to direct and control implementation of quality policies and achievement of quality objectives (ISO, 2005). A QMS includes the organisational structure, responsibilities, processes, procedures, and resources that facilitate the achievement of quality management (Luning *et al.*, 2002). This definition refers to the company specific quality management system. Stakeholders (e.g. government, retailers) commonly require a specific set-up of the company's quality management system. These requirements are described well in the established quality assurance guidelines and standards, like GMP, HACCP, ISO 9001:2000, ISO 22000, BRC, etc. (Jacxsens *et al.*, 2009; Luning *et al.*, 2006; Escriche *et al.*, 2006). They impose via requirements and guidelines demands on the organisation of the company's system (Figure 7.2).

In this section the activities and decisions of quality assurance will be described in more detail.

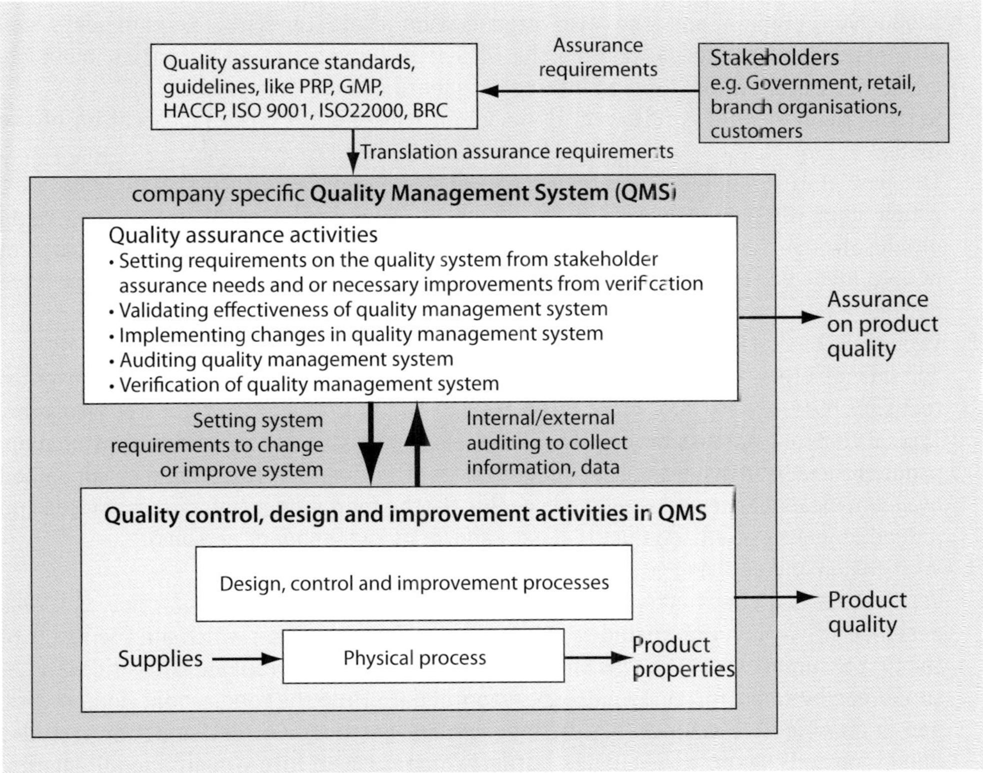

Figure 7.2. Quality assurance process and activities.

7.2.1 Quality assurance activities

Quality assurance can be seen partly as a planning activity, in which external requirements are translated into quality management system specifications, and partly as a control activity (Figure 7.2). By measurement of the physical process and the design, control, and improvement processes, information is collected on how quality is realised and to what extent the assurance needs of the stakeholders are met. The arrows in the figure show a control loop which consists of three basic steps: measurement (auditing), evaluation (verifying) and corrective action (change and improvements).

Figure 7.2 describes the core activities in the quality assurance process (Luning *et al.*, 2009). They are:

1. Setting requirements on the quality system, derived from stakeholder assurance needs and/or necessary improvements, derived from verification.
   Stakeholders make assurance demands on a company's specific QMS by requiring the implementation of various established international (and or national) QA standards and guidelines (e.g. HACCP, BRC, SFQ, etc.). These QA guidelines and standards usually consist of a broad range of technological requirements and organisational procedures. These requirements are translated into specific requirements on the internal quality management system by customising them to the specific circumstances in the internal production process and the own organisation characteristics. Requirements can also originate from feedback from the QMS (e.g. by verification activities, auditing, complaints, etc.), which form an internal trigger to change the system.
   Setting requirements is followed up by actual implementation or modification of the QMS.
   The new and/or modified requirements have to be translated into specific measures, which must be implemented in the company. The implementation process is an organisational change process, which requires careful preparation and participation of people involved. In Chapter 6 organisational change processes are discussed in more detail.
2. Validating to check effectiveness of quality management system
   Validating refers to checking *in advance* the effectiveness of the activities aimed at realising food quality, e.g. objectively testing the effectiveness of control measures. Typical control measures in the planned system, e.g. requiring a certain sanitation plan, should be analysed in advance and judged on its effectiveness. This should be supported by scientific evidence and should be specific for the food production circumstances and judged objectively (e.g. against real data and/or by independent persons).
3. Verification and auditing to check if QMS works as designed
   Verification is a check afterwards on whether the system works in practice as it was designed to. Verification includes a comparison with standards, which are the result of the first assurance activity. This checking should be done in a reliable and valid way, i.e. supported by scientific evidence, experimental data from the food production process, and in an objective manner (e.g. by independent persons). Corrective action includes improving and changing the quality system in order to meet future quality requirements.
   Auditing is aimed at measuring the quality management system. An audit is a means by which management can determine whether people in the organisation are carrying

out their duties and whether the organisation is operating effectively according to the prescribed system. Audits are described in more detail in section 7.5.2

4. Documentation and record-keeping to support above activities
   Documentation and record-keeping support the assurance activities, whereby documentation is aimed at storing knowledge and information and record keeping is aimed at collecting data. Procedures, manuals, research reports, complaints, statistical analyses, etc., are typical sources of information and knowledge for documentation, whereas process and product data are stored in a record-keeping system.

Quality assurance is primarily aimed at giving guarantees to both management and stakeholders about quality and the quality management system. So quality assurance has two faces: to the company it initiates corrective actions and to external stakeholders it provides confidence and ensures credibility.

7.2.2 Quality assurance decisions

Quality assurance is a process of decision-making on the quality system and its necessary changes. The food quality decisions grid (Figure 3.17) shows which decisions can be seen as the result of quality assurance activities, i.e. decisions on:

- Critical properties, and critical points and parameters (influencing food dynamics), e.g. identifying, measuring and evaluating which compounds or agents, and where in the process, are critical for quality.
- Capability of processes, equipment and buildings at critical points and related monitoring systems (influencing technological conditions), e.g. assessing and validating the capability of equipment and processes related to critical points, selection and validation of process control parameters, and choice of monitoring systems.
- Critical decision and action points of people (influencing human dynamics) e.g. setting requirements on the composition of the quality system project team and compliance behaviour towards rules and procedures related to critical quality points.
- Responsibilities and procedures in quality systems and auditing (influencing administrative conditions), e.g. the organisational position of the quality department, implementation of information systems for auditing purposes, and development of rules and procedures.

These decisions show that assurance refers to both the technological activities and conditions and the managerial ones. For instance, physical separation of store rooms in order to prevent cross-contamination and procedures for using the store rooms to ensure that materials are indeed stored separately.

7.3 Primary quality assurance guidelines and standards

Nowadays a wide range of internationally and nationally recognised quality and safety assurance standards, codes, schemes, plans guidelines, and principles are available in practice (e.g. GMP, PRP, HACCP, ISO 9001:2000, ISO 22000, BRC, SQF, etc.). However, it should be recognised that some of them are standards (like ISO 9001:2000 and BRC), which means that a company can be certified against the specific standards. Others just provide guidelines or principles (like HACCP) to support a company in setting up its own QMS. In this chapter we use the term 'quality assurance standards and guidelines' to cover quality assurance standards, codes, schemes, plans, guidelines, principles, etc.

The quality assurance standards and guidelines differ on various aspects, such as focus (e.g. on food safety, food quality, quality management), scope (e.g. food company, primary production, chain perspective), status (e.g. compulsory by law, required by customers, etc.), acknowledgement (national, worldwide), performance requirements on QMS set up (very generic, guidelines, detailed descriptions), requirements on QMS performance assessment (e.g. very strict requirements on auditing), and the possibility to certify the system (J. Kussaga, personal communication 2009).

Food companies and agribusinesses have to translate the different quality assurance standards and guidelines into their own company specific *quality management system (QMS)*. In this section we describe the principles of three major quality assurance standards and guidelines important for companies in the agri-food chain, i.e. the basic codes and guidelines (Codex Hygiene codes, GMP, PRP) (section 7.3.1), Hazard Analysis and Critical Control Point analysis principles (HACCP) (section 7.3.2), and ISO 9001:2000 standard (section 7.3.3). The first types are primarily aimed at creating basic conditions for production and manufacturing of safe and wholesome food; the second one is aimed at assuring that food production systems produce safe products; and the last one is a generic (not food specific) standard which is focused on management and assurance of quality of products and services (Figure 7.3). Other acknowledged quality assurance standards and guidelines like ISO 22000, BRC (British Retail Consortium - Global Standard for Food Safety), IFS (International Food Standard), GFSI (Global Food Safety Initiative), SQF 2000 (Standard Quality Food), etc., combine various elements of GMP, HACCP, and ISO 9001:2000, and will be described in more detail in section 7.6.

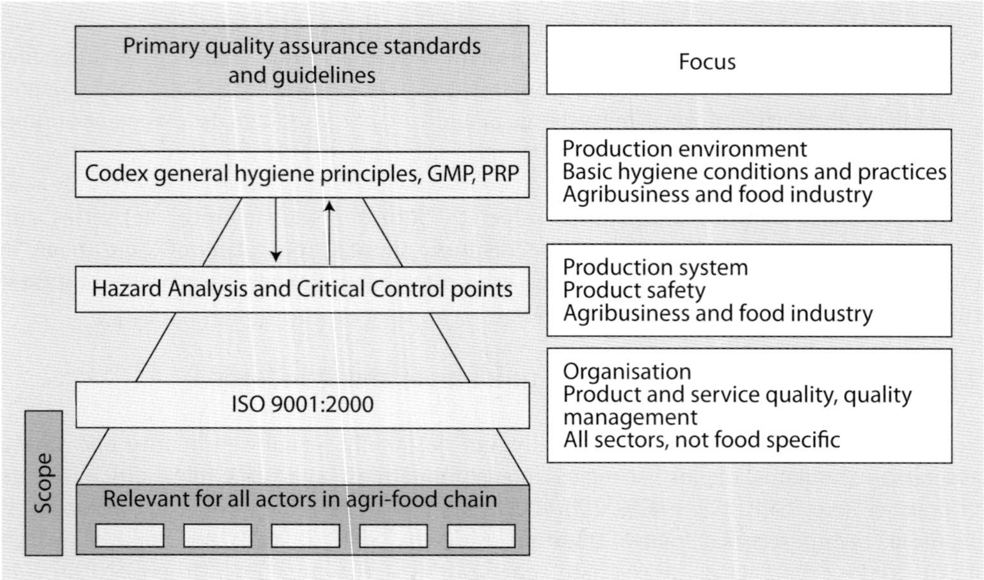

Figure 7.3. Primary quality assurance standards and guidelines.

7.3.1 Principles of basic codes and guidelines

The Codex Alimentarius (2003) advises that, prior to application of HACCP principles, any sector of the food chain should operate according to Codex 'General principles of food hygiene', appropriate codes of practices (GMP), and appropriate food safety legislation. These basic systems are aimed at creating minimum acceptable standards and conditions for processing and storage of food products, whereas HACCP is specifically focused on production processes. Typical conditions that are considered in these basic systems are related to equipment, facilities, personal hygiene, sanitation, raw materials, and waste.

There are various codes, plans, guidelines, etc. available to create basic conditions for hygienic production of safe food, such as good manufacturing codes, and prerequisite programmes. One should recognise that there are different GMP concepts (e.g. GMP codes developed by IFST Institute of Food Science & Technology and GMP regulations in USA). Although they cover similar items they have a different legislative status. In addition, prerequisite programmes (PRP) cover the good hygiene practices (GHP), GMP, and legislation in alignment with the Codex General Principles of Food Hygiene. They differ from GMP in that this latter is more than a working instruction or plan or regulation. PRP's are general control measures that need to be verified for their effectiveness on a regular basis. The principles of these basic codes and guidelines are explained in more detail below.

Codex general principles of food hygiene

The 'General principles of food hygiene' of the Codex Alimentarius (see also Chapter 1) is an elementary document for the production of safe food in agri-food production chains. The principles lay a firm foundation for ensuring food hygiene and should be used in combination with appropriate specific codes of hygienic practice and the guidelines on microbiological criteria. The document follows the food chain from primary production through to final consumption, highlighting the key hygiene controls at each stage, and recommends a HACCP-based approach wherever possible to enhance food safety.

The General principles are commended to governments, industry (including individual primary producers, manufacturers, processors, food service operators and retailers) and consumers. Governments can consider the contents of this document and decide how to encourage the implementation of these general principles to:

- protect consumers adequately from illness or injury caused by food; policies need to consider the vulnerability of the population, or of different groups within the population;
- provide assurance that food is suitable for human consumption;
- maintain confidence in internationally traded food; and
- provide health education programmes which effectively communicate the principles of food hygiene to industry and consumers.

Industry should apply the hygienic practices set out in this document to:

- provide food which is safe and suitable for consumption;
- ensure that consumers have clear and easily-understood information, by way of labelling and other appropriate means, to enable them to protect their food from contamination

and growth/survival of foodborne pathogens by storing, handling and preparing it correctly; and
- maintain confidence in internationally traded food.

Consumers should recognise their role by following the relevant instructions and applying appropriate food hygiene measures (FAO/WHO Food Standards Programme, 2001).

In addition to the general practical principles of food hygiene (as summarised in Appendix 7.1), the Codex 'General principles of food hygiene' document contains guidelines for the application of the Hazard Analysis and Critical Control Point (HACCP) analysis (see also 7.3.2), a chapter about the principles for establishment and application of microbiological criteria for foods, and a chapter about principles and guidelines to conduct a microbiological risk assessment (see also 7.4.1). Although this document has no legislative status it is a fundamental document for each company involved in the production of food.

For more details about the Codex 'General principles of food hygiene', the reader is referred to http://www.fao.org/docrep/005/Y1579E/y1579e02.htm.

Prerequisite programmes (PRP)

Prerequisite programmes involve specific documented activities or facilities corresponding to the Codex General principles of Food Hygiene, the GMPs, and legislation. The programmes consist of general control measures that need to be verified on their effectiveness on a regular basis. The basic idea behind PRP's is that they will prevent various hazards (like microbial contamination by people, cross-contamination due to poor equipment design, etc.) that are not controllable as critical control points (CCP), when they are well elaborated, implemented and regularly verified. PRP's are considered as the necessary foundation for a functional HACCP based food safety management system. Each prerequisite programme should be considered as an independent system, which is not linked to a specific production process but which is applicable to the whole company.

Each PRP must contain a detailed description of all tasks to be performed by the company to achieve the PRP objectives.

Escriche and co-authors (2006) described a structured way to present this information to verify the activities and control their efficiency. Verification provides evidence that tasks in the programme are being carried out. Control of its efficiency will prove whether the results obtained are as expected. Questions that need to be answered to obtain the required information are:
- 'what' tasks need to be controlled?
- 'how' should the control be conducted? For example, visually with checklists, or analytically by measuring microbial counts, etc.
- 'when' will the control be carried out (frequency and moment)?
- 'who' is responsible for carrying out the control (person in company, laboratory personnel)?
- 'where' will the control be carried out (specify the place)?

Each plan should include appropriate corrective actions in case control results show non-compliance or deviations. All activities need to be properly documented.

The PRP's have been divided into 14 groups as shown below and briefly described in Appendix 7.2:

PRP1 *Cleaning & disinfection*: are crucial for prevention of cross-contamination via machines, equipment, tools, cutting boards, conveyor belts, utensils, etc.

PRP2 *Pest control:* presence of insects, rodents and other animals constitutes a serious threat to food safety.

PRP3 *Water & air quality:* water from unsafe sources have been the cause of enteric infections. Water for food use should therefore comply with microbiological and chemical. *Air flow* can provide a route for micro-organisms and can contribute to the dissemination through the company.

PRP4 *Temperature control & registration*: temperature and moisture are critical factors influencing the growth of micro-organisms (including psychotropic ones that can grow close to 0 °C).

PRP5 *Personnel*: food handlers form a potential risk for transferring pathogens to food.

PRP6 *Structure & infrastructure*: the circumstances for food production are very important for safety.

PRP7 *Technical maintenance and calibration:* in order for food equipment to be used safely it needs to be inspected and maintained on a regular basis.

PRP8 *Waste management*: waste, in particular animal waste, often contains a large number of spoilage micro-organisms and possibly pathogens.

PRP9 *Control of raw material*: poor quality raw materials can lead to a serious risk of contamination.

PRP10 *Traceability, recall, returned goods, rejections/no conform products:* taking poor quality (unsafe) products off the market is also important for food safety (and quality).

PRP11 *Allergens:* since an increasingly large group of people appear to be allergic to various compounds, a general allergy policy should also be outlined.

PRP12 *Physical and chemical contamination*: this is a description of how contamination with physical and chemical hazards is controlled in the company.

PRP13 *Management of product information*: methods of preparation are company specific but are important for food safety and quality.

PRP14 *Work methodology (procedures)*: hygienic working is crucial for food safety.

Depending on the risk level of products and their production processes, the need for strict and detailed PRP's can vary. For example, in situations where air contamination forms a major risk for the end product (e.g. production of gluten free products) much attention needs to be paid to air control. In other circumstances (e.g. in production of fresh meat) control of incoming raw materials is an essential preventive measure to keep microbial levels below acceptable levels.

Principles of good manufacturing practice (GMP) codes

Good practice (GP) codes are not a static concept but an evolutionary mechanism by which overall improvements can be made and maintained. The development of good practice codes was initiated in the 60's in anticipation of a worldwide need for clarity about safe and hygienic manufacturing procedures. The most common code is 'good manufacturing practice' (GMP). For pharmaceutical companies, GMP guidelines have been developed in anticipation of the resolution adopted by the World Health Organisation (WHO), concerning correct preparation and quality control of pharmaceuticals. These GMP codes include the following four topics:

- *Means*, which includes prescriptions on buildings, equipment and utilities.
- *Materials*, which refers to requirements for raw materials and supporting materials.
- *Methods*, which includes standard operating procedures (SOP) and work instructions.
- *People*, which refers to knowledge and experience of employees (e.g. job description, training programmes).

GMP codes have also been developed for food and drinks. These GMP codes involve guidelines that are aimed at assuring minimum acceptable standards and conditions for the processing and storage of food products. GMP codes vary from general guidelines that can be applied to all foods, to specific guidelines that can be applied to a horizontal product branch or to a vertical production chain. Specific aspects of food production can also be prescribed in good practice codes. For example, good hygiene practice (GHP) codes specifically focus on all hygienic aspects of food production. GP codes can also refer to other topics than food production, e.g. good laboratory practices (GLP), good clinical practices (GCL) or good agricultural practices (GAP).

GMP codes have been drafted, and are still being drafted, by governments, international bodies, and organised interest groups like consortia from the food industry and control authorities. GMP codes are commonly *rules of self-discipline* and have no legislative status. However, the codes must be applied in view of current (inter)national regulations, which means that legislative norms and requirements must be taken into account. Moreover, GMP codes are sometimes required within a branch (e.g. GMP for animal feed) or have even been enacted in national legislation (e.g. GMP regulation food hygiene).

A well-known GMP code was developed by the Institute of Food Science & Technology (IFST) over 20 years ago. The IFST is the Europe-based professional qualifying body for food scientists and technologists. Due to the expertise of its members and its total independence from government, industry and lobby/special interest groups, IFST is uniquely placed to develop guidance on good practice. The aim of these GMP codes is to combine procedures for manufacturing and quality control in such a way that products are manufactured consistently to comply with their intended use. Recently, IFTS has brought out the 5th edition. *GMP 5* sets out to ensure that food manufacturing processes deliver products that are uniform in quality, free from defects and contamination, and are as safe as it is humanly possible to make them. The fifth edition builds on previous editions whilst reflecting important new developments in the food industry such as the BRC Global Standard for Food Safety, the Global Food Safety Initiative, and the introduction of ISO 22000:2005 Food Safety Management Systems. It includes new and updated chapters on manufacturing activities, cleaning, sanitising, personnel and training, infestation control, calibration, and foreign body controls (www.ifst.org).

In the USA the Food and Drug Administration (FDA) under the authority of the Federal Food, Drug, and Cosmetic Act (FDCA) developed the *GMP regulations*. These regulations, which have the force of law, require that manufacturers, processors, and packagers of drugs, medical devices, some food, and blood take proactive steps to ensure that their products are safe, pure, and effective. GMP regulations require a quality approach to manufacturing, enabling companies to minimise or eliminate instances of contamination, mix-ups and errors. This in turn, protects the consumer from purchasing a product which is not effective or may even be dangerous. The failure of firms to comply with GMP regulations can result in very serious consequences including recall, seizure, fines, and jail sentences.

The FDCA provides the regulatory basis for today's food GMPs. Current food GMPs are the result of an extended rulemaking process that spans decades and are published in Title 21 of the Code of Federal Regulations, Part 110 (21 CFR 110). The GMPs describe the methods, equipment, facilities, and controls for producing processed food, and the minimum sanitary and processing requirements for producing safe and wholesome food.

Topics that are described in the food GMP regulation include requirements on:
- *Current good manufacturing* practices: criteria for determining adulteration, food covered by specific GMPs is also covered by umbrella GMPs.
- *Personnel:* disease control, education & training, supervision, etc.
- *Exclusions:* excluded operations (raw agricultural commodities), FDA can issue special regulations to cover excluded operations.
- *Plant and grounds:* description of adequate maintenance, plant construction and design to facilitate sanitary operations and maintenance.
- *Sanitary operations:* cleaning/sanitising of physical facilities, utensils, and equipment, storage of cleaning and sanitising substances, pest control, sanitation of food contact surfaces, and storage and handling of cleaned portable equipment and utensils.
- *Sanitary facilities and controls:* water supply, plumbing, sewage disposal, toilet facilities, hand-washing facilities, rubbish and offal disposal.
- *Equipment and utensils*: the design, construction, and maintenance of equipment and utensils.
- *Processes and controls*: processes and controls for raw materials and other ingredients, manufacturing operations.
- *Warehousing and distribution:* storage and transportation of food must protect against contamination and deterioration of the food and its container.
- *Defect action levels:* FDA has established maximum defect action levels (DAL's) for some natural or unavoidable defects; compliance with DAL's does not excuse violation of 402 (a)(4), and food containing defects above DAL's may not be mixed with other foods.

The food GMP regulation is an important part of the regulatory control of safety of food supply, and it serves as one basis for FDA inspections. For details about the food GMP regulation the reader is referred to www.cfsan.fda.gov.

The Codex Alimentarius hygiene codes, PRP's, GMP5, and food GMP regulations have various topics in common. They all aim at creating basic conditions for safe and wholesome food production. However, they differ in legislative status, formulations (as guidelines or as plans), the exact topics, and the way in which they are structured.

7.3.2 Principles Hazard Analysis and Critical Control Point (HACCP)

Food safety is still a major concern worldwide for all food companies and agribusinesses (Ropkins and Beck, 2000; Luning *et al.*, 2008). Many efforts are (still being) made to implement quality assurance standards and principles into the company-specific quality management system and maintenance of their system. A food safety management system (FSMS) is that part of a company's QMS that is specifically aimed at controlling and ensuring that food safety requirements are met. HACCP is an internationally recognised quality assurance guideline for the development of an effective FSMS. The HACCP principles should support companies in setting up their own specific FSMS. The principles of HACCP and how to develop and maintain a system based on the principles is explained in more detail below.

Historical background

Hazard Analysis Critical Control Points (HACCP) is a systematic approach to the identification, evaluation and control of those steps in food manufacturing, which are critical to product safety. It is an analytical tool that enables management to introduce and maintain a cost-effective, ongoing food safety management system. The basic objective of the HACCP principles is assuring the production of safe food products by prevention instead of by the inspection of end products. The basis for HACCP originated from the need for safe food supply for manned space flights by NASA (National Aeronautics and Space Administration) in 1959. They required food with a quality level as close to 100% assurance as possible, because any illness or injury in space might end in an aborted or catastrophic mission. Quality control at that time could not give that high level of assurance. Therefore, a system was developed in collaboration with Pillsbury Company that participated in the Space programme. This system controlled raw materials, the process, the environment, personnel, storage and distribution, beginning as early in the system as possible. This method of control combined with record keeping had to ensure that the final packaged product did not require any testing other than for monitoring purposes.

Although introduced in 1959, the HACCP principles were first formally presented to the general public in 1971. It was not until 1985 that HACCP was seriously considered for broad application in the food industry. In 1993 HACCP principles were included in the Codex Alimentarius and in 1997 and 1999 some modifications were made. In 2003 some additional changes were included for small and/or less developed businesses. In 1993, the European Union (EU) also officially recognised the HACCP approach as a guideline for food manufacturers to implement and maintain a production control system. The recognition was established in 'The Hygiene of Foodstuffs Directive 93/43/EEC' in Europe. From that moment, European Member States had 30 months to implement this Directive in their laws. As a consequence, all food-manufacturing facilities within the EU were forced to work in accordance with the HACCP principles from the 1st of January 1996.

HACCP is basically designed for application in all parts of the agri-food production, ranging from growing, harvesting, processing, manufacturing, distribution and merchandising up to preparing food for consumption. Application of the HACCP principles is obligatory for each company that prepares, and/or processes, and/or handles, and/or packages, and/or transports and/or trades food. Besides the food industry, also small and medium enterprises, traditional manufacturing enterprises and the hotel and catering industry are now obliged to integrate the HACCP principles into their own FSMS.

There are different ways for companies to comply with the HACCP principles (as required by legislation), i.e. application of branch hygiene codes or development of their own FSMS according to the HACCP principles. Branch organisations are allowed to develop hygiene-codes based on the HACCP principles. Companies with processes similar to the ones described in the branch hygiene code, can confine themselves to application of the rules described in that code. If a company's processes deviate from the branch hygiene code, then the additional processes must be evaluated according to the HACCP principles. Branch organisations can also develop framework codes, which are more global than the hygiene codes. The framework codes can be helpful in the development of an internal FSMS system.

Development of an FSMS based on HACCP principles

Prior to the implementation of HACCP principles into a company-specific FSMS, each company should create basic conditions for safe food production by ensuring that factors external to the process, like equipment, personal hygiene, food handlers facilities, etc., are controlled. The objective is to create a solid basis for an effective FSMS. In the words of Codex Alimentarius: 'Prior to application to any sector of the food chain, that sector should be operating according to the Codex General Principles of Food Hygiene, the appropriate Codes of Practice (GMP), and appropriate food safety legislation'.

Besides these basic *technological requirements* on the food production circumstances, basic *organisational conditions* are also important to enable people to take the right food safety decisions when designing and working in the FSMS. Two crucial conditions that should be considered are people commitment and competences (Chapter 3). Various studies have stressed that management commitment is of crucial importance for an effective quality (food safety) management system. Commitment should be reflected in a clear quality and or safety policy statement and allocation of resources (e.g. QA budget, people, time, investment in equipment and facilities). A company should also create conditions to hire and retain people with the right competences to do their food safety tasks adequately. Typical issues related to these organisational conditions are acquiring personnel with appropriate safety knowledge, regular and tailored training and instructions, involvement of personnel in designing and modification of QMS/FSMS.

To translate the HACCP principles into their specific food safety management system (FSMS) companies can follow a stepwise procedure as described in Table 7.1. It shows which activities should be undertaken to achieve the specified objectives. The seven principles, which are part of this stepwise procedure, form the core of an HACCP-based FSMS and are explained in more detail below.

Principle 1: hazard analysis

Table 7.1 shows that the hazard analysis (principle 1) is step seven in the implementation of the HACCP standard. The primary objective is to identify potential hazards and to consider control measures. It results in a *list of potential hazards* and an estimation of their

Table 7.1. Overview of steps involved in developing an HACCP plan (compiled from Codex Alimentarius 2003, Jacxsens *et al.*, 2009, Escriche *et al.*, 2006).

step	Objective	Activities
0	To define a policy statement on food safety	Defining *what* company wants to achieve (safe products, for which groups, etc.), *how* to achieve it (via GMP, compliance to legislation, HACCP plan, supplier selection, T&T system, etc.), *who* is responsible (QA team, QA manager, etc.), *which* means are available (personnel, training, information sources, finances, etc.), *when* objectives should be reached (targets for QA manual, auditing, etc.), and *what* will receive extra attention (e.g. shelf life tests, purchase spec, etc.)
1	To assemble an HACCP team	*Selection* of people with *specified knowledge and experience* with product, production process, packaging and related hazards, and work methodology, and additional knowledge (legislation, transport, laboratory analyses, etc.). Moreover, a representative from management should be involved (because of power of decision-making)
2	To define the scope of study	*Defining different products groups* and corresponding production processes (process lines, storage, etc.), and defining hazards (commonly microbial, chemical and physical), sometimes quality aspects (e.g. related to shelf life) are also targeted
3	To describe a product	*Collecting information* about *raw materials* (e.g. type of ingredients, shipping method, packaging, percentage in final product, official microbial and chemical criteria, storage conditions, etc.) and *finished product* (composition, physical and chemical properties, process information, packaging, storage and distribution requirements, etc.)
4	To identify intended use and consumer	*Identification* of *target customers* (retailers, catering, etc.) and *consumers* (do they belong to high-risk groups?) and describing handling and preparation procedures
5	To construct a flow diagram	*Defining* and *describing* different *steps* of full production process (including information about raw materials and packaging, sequence of process steps, process conditions, storage and distribution conditions), and design flow chart (using standard patterns as indicated by ISO 10628)
6	To confirm flow diagram	*Checking* flow diagram by performing an on-site review of operation to verify accuracy and completeness of flow chart
7	To identify potential hazards and consider control measures	Making an *inventory* of *potential hazards* (biological, chemical, physical, allergens, see also Chapter 2) that can occur in food production process under study; making an *inventory* of *control measures* for each potential hazards; and *selecting hazards* by *analysing* the probability of them occurring and their effect on public health (= *risk*)
8	To identify critical control points (CCP)	*Identifying critical control points*, which are points, process steps or activities, where *loss of control* of hazards results in an *unacceptable risk* for public health and where concrete measures can ensure *efficient and quick control*

Table 7.1. Continued.

step	Objective	Activities
9	To establish critical limits for each CCP	*Identifying* the *critical conditions* associated with the CCP (i.e. which product/process conditions can lead to loss of control, e.g. wrong pH, poor initial temperature, viscosity, etc.). *Accurately defining* the critical *limits* for each CCP, limits are linked to critical product and process parameters
10	To establish a monitoring system for each CCP	*Choosing* a *monitoring system* which is able to quickly detect a loss of control at a CCP (i.e. a CCP that exceeds critical limits); systems should preferably be *calibrated, on-line,* and *continuously* monitoring to detect deviations and adapt/modify product and or process conditions to avoid loss of control at a CCP
11	To establish corrective actions	*Defining* of *corrective actions* that should be undertaken in case of out-of-control situations at a CCP, i.e. *how* to *adapt/modify* the process to attain the required product and process parameters within limits; what to do with *non conformance* products/batches, analysis of *causes of deviation*
12	To establish verification procedures	*Checking in advance* if selected control measures are effective (validation of HACCP system design), and checking *afterwards* if performance of control strategies is in compliance with the design. These activities are crucial for testing the effectiveness, and maintaining and improving the system. Full verification of the HACCP system should be done at least once a year
13	To establish documentation and record-keeping	*Deciding* on how to *manage* (i.e. collect, store, maintain) *documents* such as quality manual, procedures, specifications, instructions, & registrations forms, and how to *keep records* (of product and process parameters). Preferably all documents and records should be *kept up to date, online,* and *integrated* through the whole company
14	To keep HACCP system updated	*Review* HACCP system when important changes occur between two verifications

risk. It helps to establish whether these hazards should be addressed by the prerequisite programmes (PRP), which are control measures that are not specifically linked to a certain production process (step), or by critical control points.

The hazard analysis consists of the following activities:
1. making an *inventory* of all hazards that could potentially occur in the companies' food production system or under food preparation/handling circumstances;
2. considering what *control measures* can be applied to each hazard, if any; and
3. *analysing* the risk of each identified hazard.

The QA manager or team starts with the *inventory* of hazards and makes a list of all potential hazards that are reasonably likely to cause any injury or illness if not properly controlled. Within this scope, the word hazard is limited to safety. Hazards can be divided into biological (i.e. micro-organisms and their metabolites), chemical (e.g. residues of agricultural chemicals, packaging materials, additive or environmental residue), and physical (e.g. pieces of glass, metal, wood, etc.). Nowadays, allergens are often considered as a separate class of hazard as well (see also Chapter 2).

For conducting the hazard inventory, the QA manager/team systematically analyses hazards that may derive from the product, process, and environment. Potential hazards may occur due to:

- *Raw materials, ingredients* and *semi-finished products* as a source of e.g. residues of pesticides, initial content of micro-organisms.
- *Conditions* of the *food* which favour microbial growth, e.g. composition, pH, a_w.
- *Packaging,* e.g. is packaging resistant to damage, tamper-proof, does it contain proper labelling and instructions?
- *Process equipment/facilities design,* e.g. are appropriate time-temperature conditions obtained, can equipment/facilities be properly cleaned and sufficiently controlled?
- *Survival* of important process steps, e.g. heat-stable toxins.
- *Deviations* in managing the *process,* e.g. process delays, technical trouble.
- *Contamination* by equipment, personnel or environment (air, water)?
- *Lay-out design production area,* e.g. is an adequate separation of raw and processed materials guaranteed?

Next, the QA manager/team must consider which *control measures,* if any, could be applied for *each* hazard. Control measures are measures that are necessary to guarantee the safety and suitability of final products with regard to the most probable use. Control measures are not necessarily of a technological nature (e.g. measuring of physical parameters such as temperature, pH, etc.); they can also include education and training to improve knowledge and experience of personnel. There are roughly three types of control measure (Jacxsens *et al.,* 2009, Escriche *et al.,* 2006):

1. *Once-only measures to eliminate the hazard,* which commonly concern the infrastructure. This type of measure requires little inspection, but needs to be evaluated periodically with regard to the actual situation. For example, appropriate physical separation of raw and processed products.
2. *Measures to limit risks by controlling the hazards,* which typically concern the general measures that are not product and process related but which are aimed at preventing, avoiding or reducing contamination. These measures are covered by prerequisite programmes (PRP).
3. *Measures to limit risk by eliminating or reducing hazards to an acceptable level,* which are measures that are specifically directed at the process. Examples of these specific controls are heat treatment, cooling, application of preservatives, and metal detection.

The last activity in this phase is the *analysis* of hazards. The QA manager/team has to systematically evaluate and estimate whether hazards are such that their elimination or reduction to an acceptable level is *essential* to the production of safe food. In conducting the hazard analysis, the following should be included wherever possible (Codex, 2003; Jacxsens *et al.,* 2009):

- *Severity* of adverse effect on health of the potential hazard, i.e. magnitude and duration of illness or injury.

- Likely *occurrence* of hazard, which is usually based on a combination of experience, epidemiological data, and information in scientific technical literature.
- Qualitative and/or quantitative *evaluation* of the presence of hazards.
- *Production* or *persistence* in foods of toxins, chemical or physical agents.
- *Survival* or *multiplication* of micro-organisms of concern.
- Conditions leading to the above.
- Number of people potentially exposed to the hazard.
- Age/vulnerability of those exposed.

This procedure provides a *qualitative* analysis of risks of identified hazards and enables a company to *focus* on control of those hazards that may, if not properly controlled, result in a high risk on negative health effects for consumers.

In practice, hazard analysis appears to be one of the most difficult steps in the HACCP procedure, because appropriate identification of potential hazards and assessment of their risk is rather complex and requires much technological knowledge and information. There are often differences of opinion, even among experts. Because different assessors of risks need to perform the process in a uniform way, the Scientific Committee for Foods (SCF, 1996) and Codex Alimentarius (1998) proposed guidelines, with the essential elements of a (semi) *quantitative* risk assessment (see section 7.4.1 for details). The hazards analysis step is often missing or insufficiently established in companies. In these cases, there is no additional step between the inventory of hazards and identification of critical control points (Escriche *et al.*, 2006), which may lead to an undesirable high number of critical control points.

Principle 2: identification of critical control points (CCP)

The primary objective of this step is to determine where the selected hazard(s) with a high risk, as determined in the previous step, needs to be controlled in the production process. A critical control point (CCP) is a step (i.e. point, procedure, operation, or stage in the food production system) at which control *can* be applied, and where control is *essential* to prevent or eliminate a food safety hazard or to reduce it to an acceptable level. CCP's are unique for each process at each facility. There is no such thing as a standard CCP in every food production process. Moreover, there is no limit on the number of CCP's that may be identified in the flow diagram, however it is important to strive for a lean system, and prevent an excessive number of CCP's.

CCP's are determined by applying the *CCP decision tree*, at each step of the process, for each hazard that was selected in the hazard analysis (Figure 7.4). The CCP decision tree is a sequence of questions to determine whether a step is a CCP or not. The decision tree consists of four questions (Q1-Q4) which can all be answered with a yes or no. The answers should lead to a subsequent question or mark the process step as CCP or No CCP (Stop). More specifically, for each hazard as identified in the previous principles, the following questions should be asked using the flow diagram with the different production steps (Jacxsens *et al.*, 2009):

- *Question 1 asks if any preventive measures exist for the hazard.* If the answer is YES proceed to question 2. If NO, is control necessary at this step for product safety? If control is NOT necessary, then this step is NOT a CCP, and the decision tree for the following identified

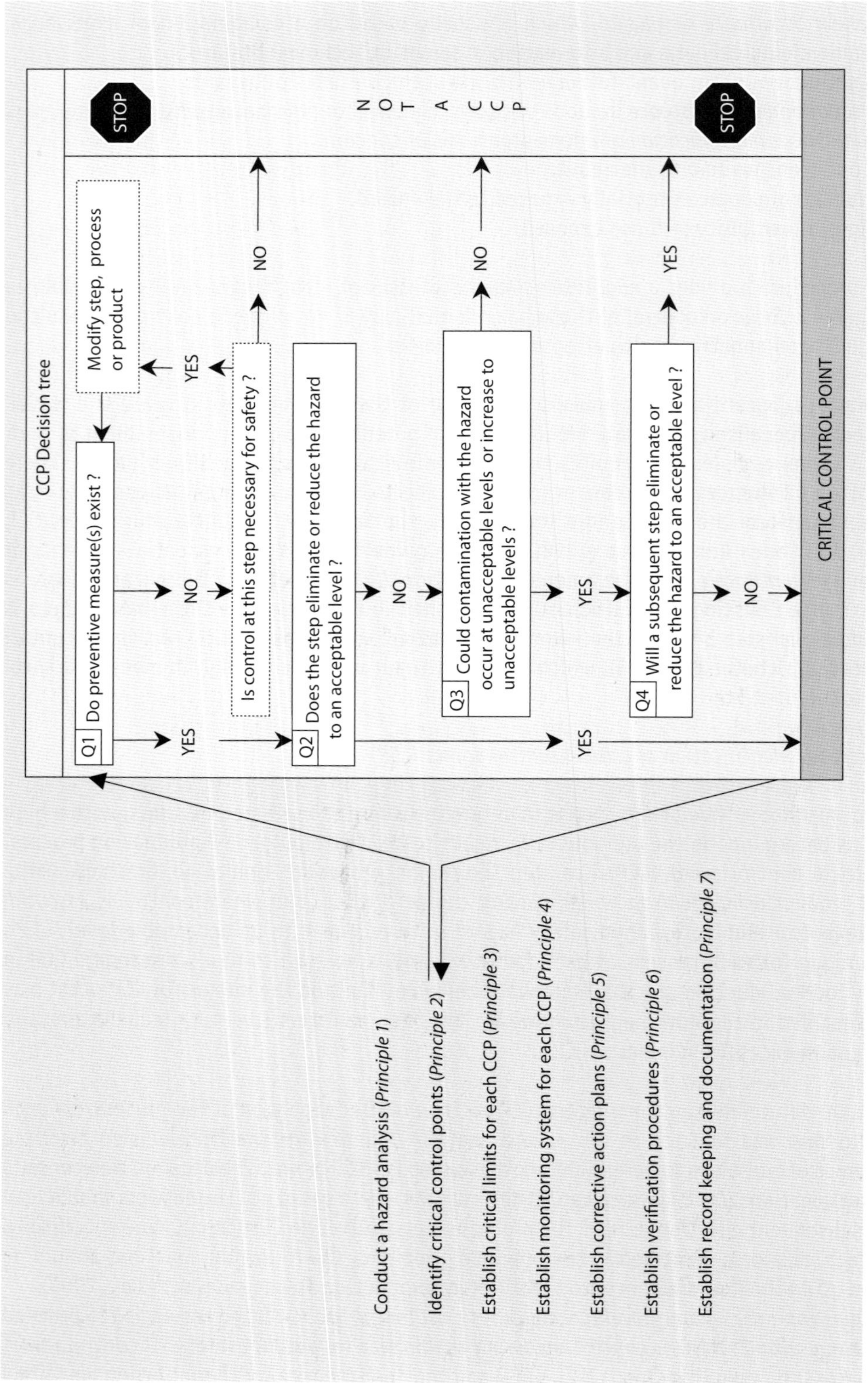

Figure 7.4. CCP decision tree to support the 2nd principle of HACCP.

hazards should be applied. If control is necessary, then the step, process or product should be modified in such a way that the hazard can be controlled.

- *Question 2 asks if the step eliminates or reduces the likely occurrence of the hazard to an acceptable level?* For each step in the flow diagram it should be asked if it is designed to eliminate or reduce the hazard to an acceptable level. For microbial hazards, product properties (e.g. pH, a_w, preservatives) and process conditions (T-t, pressure, etc.) can be specifically designed to reduce the microbial hazard. If the answer to question 2 is YES, then the step under consideration is a CCP. The QA manager/team must define what is exactly critical (e.g. an ingredient, process step, location, certain handling associated with the process step, etc.). If the answer is No, then question 3 must be considered for the *same* process step.
- *Question 3: could contamination with the identified hazard occur at that step at an unacceptable level or could it rise to an unacceptable level?* The QA manager/team should determine whether the immediate processing environment might be a source of the hazard at the process step under consideration. With respect to unacceptable increase the QA manager/team must also consider the possibility of increase at the step under study and the possibility of accumulation to unacceptable levels due to previous and subsequent steps. If the answer is NO then it is NOT a CCP; if it is YES, then the QA manager/team should ask question 4 for the same process step.
- *Question 4: does a subsequent step eliminate or reduce the hazard to an acceptable level?* If the answer to question 3 is yes, then the QA manager/team should proceed with the remaining process steps in the flow diagram to determine whether any subsequent step will eliminate the hazard to an acceptable level. If the answer is NO, then a CCP has been identified, and the QA manager/team should again specify exactly what is critical. If the answer is YES, the decision tree should be applied for the next process step.

Although the decision tree seems rather straightforward, in practice QA managers and teams still face difficulties in establishing the CCP's because it requires profound insight into the behaviour of hazards under the applied food production circumstances and insight into the potential risk of poor control.

It should be noted that raw materials can be an important source of contamination and unexpected food safety problems. Therefore, it is important that companies with high-risk raw materials that could cause cross contamination, which is not controlled in the production process, consider receipt of raw materials as a CCP as well. There are various ways to control raw materials, such as systematic and regular analysis of incoming raw materials, inspection of specifications at receipt, selection of preferred suppliers (supplying products with reliable safety levels), or auditing the QA system of suppliers. Considering raw materials as a CCP is especially important when one deals with high-risk raw materials that do not undergo subsequent inactivation or reduction steps (e.g. fresh packed meat products).

Principle 3: establishment of critical limits

In order to control a hazard at a critical control point one needs to have standards and tolerances (limits) against which the actual product and process conditions can be judged (Chapter 4). Firstly, the QA manager/team needs to specify which factors are associated with the CCP. In other words, which circumstances of a CCP are critical for proper control of the hazard at the CCP. For example, the pasteurisation step of milk is allocated as a CCP for certain microbial hazards. Critical factors associated with the pasteurisation are, for example, functioning of the valves (supplying heated water), timer (influencing time of heat exposure) and batch volume (affects necessary heat transfer time). For each factor influencing the appropriate control of the CCP, the desired standard/norm and tolerances (limits) should be established. The limit depends on the point or level at which the influencing factors become critical for public health. In other words, the conditions under which the hazards are no longer effectively controlled and may lead to a risk for public health.

Establishment of the critical limits need to be well motivated and, where available, legislative norms should be used. Limits can also be derived from guidelines, literature, experiments, tests in company food production systems, and expert input. It should be recognised that limits as controlled in a CCP are linked to the specific product and process parameters and not to the final product, whereas legal limits are commonly set for the end product. This means that these limits should be taken as starting point for assessing the critical limits in the production process.

More specifically, one can differentiate between microbial, chemical and physical limits. Microbial and chemical limits are commonly set for final products. Moreover, microbial limits are not directly used to judge the situation at a CCP, because microbial analyses take at least 48 hours which is much too slow in terms of feedback for a production system. Therefore, indirect measures are used for microbiological hazards, e.g. time-temperature regimes. The microbial tests are then used for validation of the effectiveness of the applied t-T regimes in advance and for verification purposes, i.e. to check afterwards if equipment is still functioning properly in practice as it was designed to (Luning *et al.,* 2009).

When limits are associated with processing equipment, one should also determine process capability in order to judge whether the process equipment is capable of producing within the set limits (Chapter 4).

Principle 4: establishment of a monitoring system

Monitoring is the scheduled measurement or observation of a CCP relative to its critical limits. It is necessary to assess whether the CCP is under control and to provide written documentation for verification purposes. Important aspects of a monitoring system that should be considered are (Jacxsens *et al.*, 2009; Luning *et al.,* 2008):

1. *Method of monitoring*. CCP can be monitored by observation or by measurements. Observations can be visual monitoring of colour, appearance as indicators for e.g. poor time/temperature history of products. Visual monitoring at CCP's must make use of checklists with defined standards and limits. Measurements are commonly chemical (e.g. pH, moisture level) and physical (e.g. temperature, pressure, time), which can be in-line, at-line, on-line, or off-line (Chapter 4).
2. *Measuring plan*. It is very important to measure product and process parameters at the right *location* with the correct *frequency*. Preferably, the process should be continuously monitored but that is often not possible. When the QA manager/team is determining the frequency then the following issues should be considered:

   - stability of the production process; if there are large variations between production parameters then more monitoring is required;
   - the production rate and the time between storage and delivery, to be sure that any corrective action could take place before the product goes to market.
3. *Adequacy of measuring equipment*. It is crucial for the performance of the monitoring unit that the measuring equipment provides reliable data. Aspects like accuracy, preciseness, reliability, and sensitivity are therefore important considerations (see also Chapter 4).
4. *Responsibility for monitoring*. Last but not least, to be sure that monitoring is done in the right way with the desired frequency, it is essential that it is clear who is responsible for the task. This entails assigning a person with relevant knowledge, who evaluates and signs the monitoring data.

So basically each monitoring unit should include instructions describing:
- How monitoring should be executed.
- Which measuring equipment is to be used and how to operate it.
- The frequency of measuring.
- Who is responsible for monitoring.
- What data needs to be registered.
- Who can take corrective actions and what kind of corrective actions (5th principle).

Note that for monitoring purposes physical and chemical tests and visual inspection are usually applied. Microbiological testing is seldom an effective means of monitoring because of the time required to obtain results.

Principle 5: establishment of a corrective action plan

If monitoring data reveal that the product and/or process parameters deviate from critical limits, then corrective action must be taken. Corrective actions must ensure that the CCP will be brought under control again. A comprehensive corrective action plan considers the following issues:
- Analysis of the cause(s) of non-compliance or non-conformance; which technological or people factors caused the deviation (e.g. ingredients with excessive initial contamination levels, improper food handling, or equipment deviations due to maintenance problems)?
- Characterisation of the non-compliant product / process; what are the exact product and/or process parameters (e.g. too high pH, too low heating temperature)?
- Adjusting the process; how to adjust the process depending on the extent of the deviation (e.g. decrease temperature, full restart of the process; instruction of food handler)?
- Handling of non-compliance products; what to do with the non-compliance products (e.g. store for further analysis, rework, throw away in waste containers)?
- Recording of corrective actions taken; which activities must be undertaken and which data must be recorded, and by whom?
- Structural analysis of deviations; when same deviations occur frequently, then it might be necessary to analyse the structural causes of deviation in more detail, in order to improve the production conditions (Chapter 6).

Some studies have provided evidence that corrective actions are not always well established in food companies (including food service establishments), which might induce a risk for human health.

Principle 6: establishment of verification procedures

Before describing what this principle includes, we will firstly explain the terms verification and validation, because they sometimes create confusion in practice (Panisello and Quantick, 2001; Scott, 2005). The Codex Alimentarius (2003) defined validation as obtaining evidence that the elements of the HACCP plan are effective, but it is considered as part of the 6th principle 'establishment of verification procedures'. Verification concerns the application of methods, procedures, tests and other evaluations, in addition to monitoring to determine compliance with the HACCP plan. In other words, validation activities involve checking *in advance* the effectiveness of designed control measures, whereas verification activities concern checking *afterwards* whether control activities are operating in practice as they were designed to. Validation and verification thus have a distinct function and are typical assurance activities (Jacxsens *et al.*, 2009; Luning *et al.*, 2009). The function of quality assurance is to define the quality system and provide evidence and confidence to stakeholders that assurance requirements are being met (Luning and Marcelis, 2007).

More specifically, *validation* activities are applied to determine whether the HACCP plan has been properly developed (appropriate hazard analysis, well-selected preventive measures, intervention systems, and monitoring systems). In other words, to check whether the selected control measures are effective under the typical production circumstances (or handling, preparation) of the food company. For this purpose, the QA team/manager should use external support (for an independent opinion), whereby scientific evidence and actual (trials) experiments should be applied to prove the effectiveness of selected control measures.

Verification activities should include verification of preventive measures (as established in the PRP's), intervention processes (all treatments to inactivate or reduce hazards to an acceptable level), and monitoring systems (to check whether they really accurately, precisely and reliably monitor deviations).

- Typical equipment related performance must be verified, for instance analysing equipment records, calibration of measuring/analytical equipment and tools, actual reduction levels of hazards due to intervention equipment (e.g. heat treatment, metal detector) or intervention methods (e.g. lowering pH, use of preservatives).
- People-related performance (e.g. actual compliance to procedures, actual competence level, etc.) should also be verified.

Sources of information for verification are record-keeping systems (using records related to core control activities), observations of people behaviour, physical and microbiological measurements and analyses, documents (quality manual, procedures and instructions), consumer and/or customer complaints.

Verification can be achieved with regular inspections, and internal and external audits. As with validation, it is important to have an independent opinion so as to reassure stakeholders (Jacxsens *et al.*, 2009; Luning *et al.*, 2009).

Principle 7: establish record keeping and documentation

Documentation and record keeping are essential for an effective FSMS and they support typical assurance activities like validation and verification. Documentation and record-keeping each have a distinct objective. Documentation aims at storing knowledge and information, whereas record keeping aims at collecting data. Procedures, manuals, (work) instructions, flow diagrams, research reports, complaints, statistical analyses, etc., are typical sources of information and knowledge for documentation, whereas process and product data, specifications of packaging materials, records of distribution and storage are typical sources for the record-keeping system. It is important that information and data in the documentation and record-keeping system are kept up to date (e.g. provide reliable information) and that it is available at the right location and time (preferably a company integrated system with access for all those involved). Connections to external databases and/or scientific libraries also contribute to a more sophisticated system (Luning *et al.*, 2009).

In practice, it appears to be rather difficult for some types of food company or sector (especially SME's, micro-businesses, food service establishments) to carry out adequate documentation and record keeping, and it is commonly regarded simply as a lot of paperwork. To what extent information and data should be collected and maintained and how to store it systematically are typical questions. Generally speaking, registration forms have to be kept for least the shelf life of the product, whereas document for verification purposes should be kept at least 2 years (Jacxsens *et al.*, 2009).

Although the HACCP principles seem to be rather straightforward, in practice companies (especially SMEs, micro businesses, food service establishments, food companies in developing and emerging countries) still face difficulties in translating them into an effective company specific FSMS/QMS. Note that a company specific FSMS can be designed and maintained based upon HACCP principles, GMP codes, PRP's, and other safety management standards. A quality management system has a broader scope than just safety; the implementation of HACCP principles is just one specific part of the whole QMS.

Sector hygiene codes

Hygiene codes involve descriptions of practical hygiene measures for a specific branch or sector. The objective of such hygiene codes is to obtain insight into the potential hazards and risks inherent to a certain production process, in order to realise better production control. The codes can be developed by representatives of a sector and are primarily aimed at food safety. Such codes are primarily for small and medium enterprises that are not able to develop a HACCP-based food safety management system themselves.

7.3.3 Principles ISO 9001:2000 standard

The International Organisation for Standardisation (ISO) is a worldwide federation of international standard bodies, which was founded in 1947. The major objective of ISO is the promotion of standardisation and related activities throughout the world to prevent technical barriers to trade. The preparation of the International Standards is carried out by ISO technical committees (ISO/TC). ISO standards are voluntary, unless a business sector makes them a market requirement, or unless a government issues regulations making their use obligatory. The main part of the adopted standards is concerned with quality, health, safety or environmental aspects. In this section we focus on the ISO 9001:2000 standard because it is commonly used in the food industry to design a QMS.

Historical background

In 1987, the Technical Committee TC/176 developed the first ISO 9000 series, which provided a framework for quality management and quality assurance. These standards were generic and independent of any specific industry or economic sector. The 9000 series consisted of two major groups, i.e. standards for internal and for external quality assurance.

Internal quality assurance was aimed at improving efficiency as well as the quality of products and/or services within the organisation. For this purpose, the ISO 9004 document was recommended, which provided guidelines for the development and implementation of quality management. It included typical elements of quality systems but also paid attention to economic aspects (i.e. benefits, costs and risks!) of quality assurances. External quality assurance was focused on assuring customers that the products or services met the required specifications. For this purpose three 'certification' standards ISO 9001, ISO 9002 and ISO 9003 were recommended. In these standards specific quality system requirements were described, covering topics ranging from management responsibility to inspection.

In 1990, ISO/TC 176 adopted a two-phase revision process. The first phase allowed limited changes to the standards and was completed in 1994. The second, more thorough revision was completed in 2000, and was developed in anticipation of the needs of users and customers of ISO series. A large survey was carried out in 1997, which revealed that the revised standards:

- should have a common structure based on a process model;
- should be simple to use, easy to understand, and use clear language and terminology;
- should facilitate self-evaluation;
- must be suitable for all sizes of organisations, operating in any economic or industrial sector;
- should include demonstration of continuous improvement;
- should address effectiveness and efficiency;
- should increase compatibility with the ISO 14000 series of Environmental Management System Standards (ISO, 1998).

This serious revision resulted in the current ISO 9000 series.

Principles of ISO 9001:2000 for development of a quality management system (QMS)

The ISO 9000:2000 series include standards for developing a QMS addressing the principles and processes surrounding the design, development and delivery of a general product or service. These standards are not specific for agri-business and the food sector and have a broader scope than safety alone (as compared to e.g. HACCP). Within the 9000 series there are three documents, i.e. ISO 9000:2000, ISO 9001:2000, and ISO 9004:2000. The ISO 9000:2000 document provides guidance on quality principles and on the common language used by quality professionals. The ISO 9001:2000 document provides guidance on the development of a QMS, whereas ISO 9004:2000 provides guidance on improvement methods (http://www.iso.org).

ISO 9001:2000 is structured and organised according to a process model; a process is considered as any activity or operation that receives inputs and converts it to outputs. Based on this basic concept the 'quality management process' model has been developed by ISO/TC 176 in 1999 (ISO, 1999), as shown in Figure 7.5. Characteristics of the model are:

- The four major topics of the quality management system are represented, i.e.:
  1. management responsibility;
  2. resource management;
  3. product and/or service realisation;
  4. measurement, analysis and improvement.

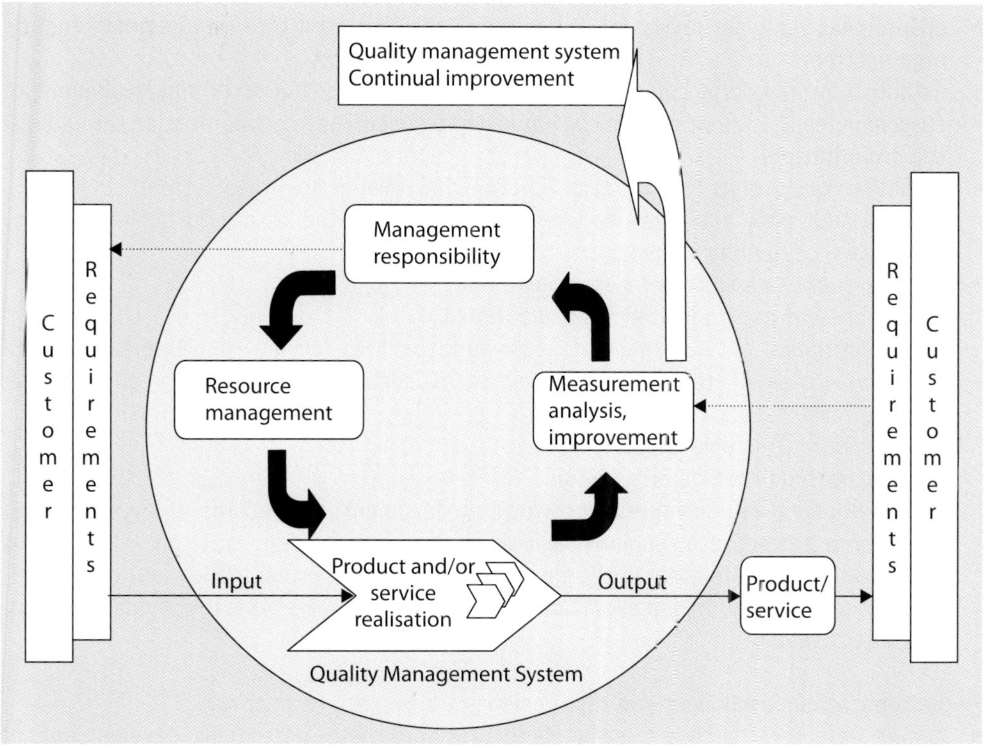

Figure 7.5. Quality management process model (ISO 9001:2000).

- The continuous improvement aspect and measuring of customer satisfaction are clearly expressed.
- Both products and services are explicitly incorporated.
- The model shows interaction between processes, and processes are closed loops. For example, the input for product and/or service realisation is customer-driven. Subsequently, the output is measured by consumer satisfaction measurements. This information is used as feedback to evaluate and validate whether customer requirements are achieved. The management, in its turn, shall ensure that customer requirements are fully understand and met.

ISO 9001:2000 starts with requirements with respect to the quality management system. The organisation must define and manage processes that are necessary to ensure that products and/or services conform to customer requirements. For the implementation and demonstration of such processes, the organisation must develop, document and maintain a quality management system according to the international standard. The typical topics are briefly described in more detail below.

In the topic *management responsibility*, requirements are set on commitment and responsibility of (top) management. Major aspects are that (top) management must:

- *demonstrate commitment* with respect to realisation of customer demands by, amongst other things, creating awareness about the importance of fulfilling customer demands, by ensuring availability of resources, and by establishing quality policy, objectives and planning;
- ensure that *customer needs are determined* and translated into requirements for the organisation;
- establish an *appropriate quality policy* to ensure, amongst other things, fulfilment of customer needs, commitment to continual improvement, and communication throughout the organisation;
- establish *quality objectives* for each function and level in the organisation;
- identify and plan activities and resources which are necessary to achieve quality objectives, i.e. *quality planning;*
- develop *quality management system* including:
  - assignment of responsibilities and authority;
  - appointment of management representatives to ensure implementation and maintenance of the quality management system and awareness of customer requirements;
  - procedures for communication;
  - preparation of the quality manual;
  - development of procedures for controlling documents and records, at a system level.
- establish a procedure for *management review*, i.e. management must provide a status with modifications related to the quality management system, policy and objectives, which is based on audit reports, customer complaints analysis, status of quality policy and objectives.

In the topic *resource management*, requirements are set with respect to:

- *human resources*, such as assignment of appropriate personnel, development of procedures to determine training needs and evaluate its effectiveness;

- *information*, such as procedures for managing information on process, product, suppliers, etc.;
- *infrastructure*, such as workspace, associated facilities, equipment, but also support services;
- *work environment*, such as health and safety conditions, work methods and ethics.

The topic *product and/or service realisation* consists of general requirements and five subtopics. The general requirements refer to determination, planning and implementation of realisation processes (e.g. criteria and methods to control processes, measuring arrangements, monitoring and corrective actions, and maintenance of quality records).

The five subtopics are:

- *Customer-related processes*, which set requirements on identification and review of customer requirements, and customer communication (such as customer complaints, product and/or service information).
- *Design and development* sets requirements on planning and control of design and/or development of products and/or services. The plans must include stages of design/ development processes, review, verification and validation activities and the responsibilities and authorities of personnel involved.
- *Purchasing* puts demands on the purchasing process to ensure that products and/or services comply with organisational requirements. The purchasing documents must contain appropriate information and purchased products and/or services must be verified. The organisation must ensure the adequacy of these documents and verification activities.
- *Product and service operations* refer to planning and control of actual operations. It sets general requirements, for example on the use of suitable production equipment, implementation of suitable monitoring and verification activities, and the availability of clear work instructions. Moreover, it put demands on:
  - identification and traceability (where and how to record);
  - handling, packaging, storage and preservation and delivery;
  - validation of processes, i.e. to detect deficiencies, which may only become apparent after the product is in use or the service is delivered.
- *Control of measuring and monitoring devices* sets requirements on control, calibration and maintenance of measuring and monitoring devices.

In the topic *measurement, analysis and improvement* general requirements refer to planning and implementation of these measuring, analysis and improvement processes to ensure that the quality management system, processes and products and/or services comply with requirements.

- *Measuring and monitoring* requirements refer to the establishment of quality system performance (i.e. customer satisfaction, internal audits), measurement of processes, product and/or services.
- *Control of nonconformity* relates to identification, recording and reviewing of products and/or services, which do not conform to requirements.
- *Analysis of data for improvement* sets demands on data analysis for determining the effectiveness of the quality management system, and for identifying possible improvements.

- *Improvement* relates to the establishment of a system procedure that describes use of quality policy, objectives, internal audit results, analysis of data, corrective and preventive action and management review to facilitate continual improvement.

ISO 9004:2000 is the other part of the consistent pair of quality management systems, the other being ISO 9001:2000. The two standards were designed as a pair, but can be used as stand-alone documents as well. ISO 9004 provides comprehensive guidelines for achieving an excellent organisation through continuous improvement of business performance. The standard is focused on improving internal processes of the organisation. Furthermore, the eight management principles have been integrated in this standard. These quality principles, which have been established by ISO for achieving total quality management, include:

- *customer focused organisation*, i.e. focus the organisation on sustained customer satisfaction while giving benefits to other interested parties as well;
- *leadership*, i.e. provide a visible personal leadership;
- *involvement of people*, i.e. encourage employee efforts;
- *process approach*, i.e. consider activities as processes with an input which is transferred to an output;
- *system approach to management*, i.e. look at the *total* system and not sub-systems;
- *factual approach to decision making*, i.e. make decisions based on facts and not on feelings;
- *continuous improvement*, i.e. develop of a culture for continuous improvement;
- *mutually beneficial supplier relationships*, i.e. create a valued relationships with suppliers.

The structure of topics in the new ISO 9004 is similar to ISO 9001, which facilitates extension of the quality system to total quality management. However, the scope of ISO 9004 is different. In ISO 9004 those elements that are necessary for improvement of performance of the organisation are extensively considered, while ISO 9001 provides quality management system requirements to demonstrate its ability to fulfil customer requirements. The revised ISO 9004 version is thus *not* a guideline to implement the revised ISO 9001:2000.

The structure of the new ISO 9000 family aligns with the structure of ISO 14000 to facilitate compatibility with the series of environmental management system standards. For further information, the reader is referred to Tricker (2006) and the official site of ISO (http://www.iso.org).

7.4 Tools and methods supporting quality assurance

In this section the most important tools and methods supporting quality assurance processes are described. Quantitative risk assessment is especially important for HACCP and other food safety standards (e.g. ISO 22000, BRC), and predictive modelling can help validate the effectiveness of quality assurance system. Tracking and tracing is a sophisticated information tool contributing to assurance in supply chains.

7.4.1 Quantitative risk assessment

Nowadays, quantitative risk assessment is extensively used for determining chemical and microbial risks in food. This concept helps to systematically and scientifically judge whether certain hazardous compounds may reach an unacceptable risk level when food is consumed. Risks assessment helps in the development of effective quality assurance systems, by identifying and characterising those risks that really need to be controlled and assured.

In the last few decades much attention has been paid to the assessment of microbial risks due to their typical differences as compared to chemical risks. Some of these differences are outlined below:

- Microbial risks are mostly due to a single *exposure* (except for microbial toxins), whereas chemical risks are affected by chronic duration of exposure.
- *Response* to infective pathogens is probably more variable as compared to chemical agents, due to different sub-populations and depending on immune status.
- In contrast to chemical agents, micro-organisms are dynamic and can adapt to situations, e.g. antibiotics or heat treatment. So, the risks identified with specific micro-organisms may change over time.

In order to obtain a uniform approach in assessing the risks of potential hazards in food, the European Scientific Committee for Food proposed a framework for risk assessment as part of risk analysis. The United States developed similar principles and guidelines for the Codex Committee on Food Hygiene (Codex Alimentarius, 2003). Risk analysis is a structured and multidisciplinary approach to identifying risks and consists of the following elements (Figure 7.6).

- *Risk assessment*, which is the scientific evaluation of, known or potential, adverse health effects which result from human exposure to foodborne hazards. The purpose of risk assessment is to provide a scientifically based *estimation of risk* of that hazard to a population.

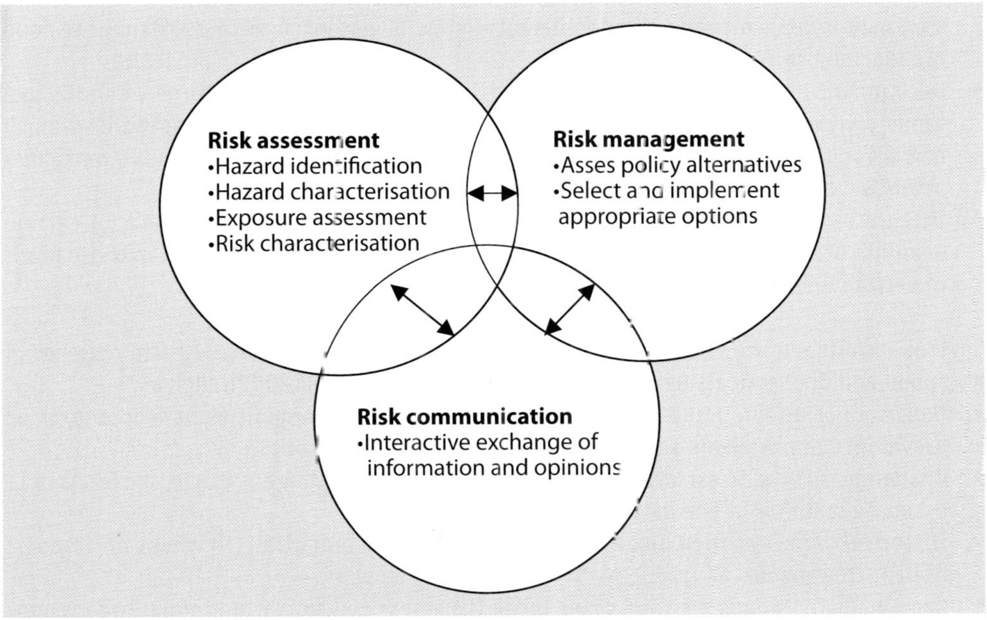

Figure 7.6. Framework of risk analysis.

- *Risk management*, which is the process whereby policy alternatives to accept, minimise or reduce the assessed risk, are balanced and appropriate options are implemented. The aim of risk management is to identify acceptable risk levels and implement control options to comply with public health policy.
- *Risk communication*, which is an interactive process of information exchange and opinion on risk, among risk assessors, risk managers and other interested parties.

Risk assessment is an important tool for supporting the basis on which hazards should be included in the HACCP plan. Risk assessment consists of four elements: hazard identification, exposure assessment, hazard characterisation, and risk characterisation (Figure 7.7).

Hazard identification is the first step in risk assessment and includes the identification of known or potential health effects associated with a particular biological, physical or chemical agent. Hazards can be identified by a systematic evaluation of each step in food manufacturing (up to consumption), whereby information and data must be collected from relevant sources, such as scientific literature and databases from the food industry and government. Typical data can originate from surveillance, clinical studies, laboratory animal, epidemiological and microbiological studies agencies (SSC, 2000a,b; Barlow, *et al.*, 2002; Van den Brandt *et al.*, 2002). To support the identification of hazards the Scientific Steering Committee, advising the European Commission, developed a systematic approach to identify potential hazards (SSC 2000a).

Exposure assessment is the qualitative and/or quantitative estimation of the probable intake of microbial (biological), chemical or chemical agents, via food as well as exposure from other relevant sources. In exposure assessment one should consider aspects like:
- Quantitative determination of the *possible presence* of a hazard in individual foods and diets at the moment of consumption. In this part all factors should be considered that may influence the level and characteristics of potential hazards during the food production chain.
- Determination of *potential intake*, by analysing consumption patterns of individual foods containing the relevant hazard. In this part, it should be evaluated how much of the foods, potentially containing these substances, are consumed, paying particular attention to high-risk groups.
- Integration of the likelihood of consumers eating (occasionally or regularly) certain amounts of contaminated foods and the chance that the hazard is present in these foods at certain levels.

Several techniques and information sources can be used to *assess the possible presence* of microbial and or chemical hazards (Parmar *et al.*, 1997; Tennant, 2002), such as:
- Data collection, on prevalence and distribution of micro-organisms in food and/or on prevalence of chemicals and their reaction mechanisms in food.
- Challenge testing, to estimate likely growth of relevant pathogens in specific foods or to estimate stability of chemicals in specific foods.
- Historical data from food process studies (e.g. Decimal reduction (D) values of survivors of heat treatment).
- Examination of foods involved in outbreaks to assess concentrations of pathogens and/ or examination of foods to assess typical concentrations of chemicals.

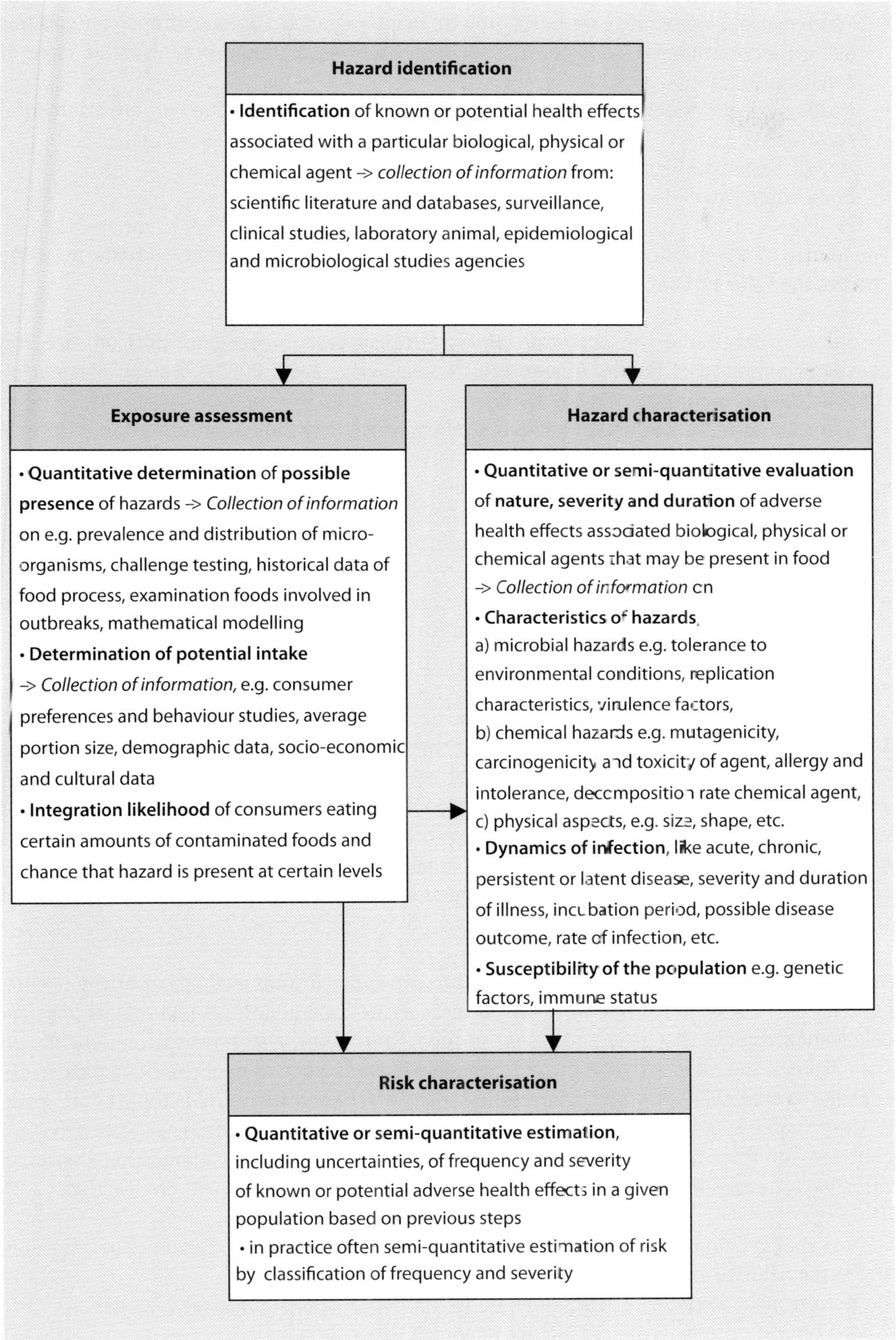

Figure 7.7. Steps and activities in risk assessment.

- Mathematical modelling to predict growth, inactivation or survival of micro-organisms due to environmental conditions, and likely number of micro-organisms at time of consumption.
- Mathematical modelling to predict stability of chemicals due to environmental conditions.
- Market basket surveys.
- Body burden/excretion measurements.

The choice of the method will depend on what information is available and how accurate and detailed the estimate needs to be.

In order to determine the *potential intake*, information is needed on portion size and consumption patterns (Codex Alimentarius Commission, 1998, 2003). Typical information that can be useful for this analysis includes:

- Consumer preferences and behaviour studies, which may provide information on choice and amount of food intake.
- Average portion size as consumed per time.
- Demographic data, e.g. age distribution, susceptible groups, etc.
- Socio-economic and cultural data, which may provide information on consumption patterns.
- Data on population dietary exposure from cohort studies, dietary data from population controls of case-control studies and food consumption surveys.
- Food/component disappearance figures.
- Weights and analyses of foods.

In this phase of risk assessment it is important to (1) critically review facts provided in food composition tables and reasons for differences reported between countries and/or within countries, (2) consider ways foods are categorised and thus made comparable (or not) in food consumption surveys, and (3) explore how one can try to refine assessments as more information becomes available (Kroes *et al.*, 2002). The validity and reliability of any conclusion and advice to risk managers depend on the quality, reliability and relevance of the available exposure data.

Hazard characterisation is the quantitative or semi-quantitative evaluation of the nature, severity and duration of adverse health effects associated biological (microbial) physical or chemical agents that may be present in food. The hazard characterisation depends on the nature of the toxic effect or hazard. For some hazards, such as genotoxic chemicals, it is considered that there may be no threshold for the effect and therefore estimates are made of the possible magnitude of the risk at human exposures (dose-response extrapolation). In contrast, for other hazards, such as non-genotoxic effects, it is considered that there is a threshold of exposure below which no biologically significant effect will be revealed.

Factors that should be considered include characteristics of hazards, dynamics of infection, and susceptibility of the population.

1. With respect to *characteristics* of microbial, chemical and/or physical hazards:
   - Typical *microbial* aspects include, amongst other things:

  - tolerance to environmental conditions can affect the survival chances of micro-organisms;
  - microbial replication characteristics, e.g. self replication, generation time;
  - virulence factors, such as the ability to circumvent host's immune response, the ability to attach to the surface, ability to synthesise various toxins.
- Typical *chemical* aspects include:
  - mutagenicity of the agent;
  - carcinogenicity of the agent;
  - toxicity of the agent;
  - allergy;
  - intolerance;
  - decomposition rate of the chemical agent.
- Typical *physical aspects* include size, sharpness, hardness, and decomposition rate of the physical agent.

2. With respect to *dynamics of infection*, this is only relevant for biological hazards and includes latency, i.e. delayed onset of the clinical infection after exposure by the hazard, and disease pattern, e.g.:
   - acute, chronic, persistent or latent disease;
   - severity and duration of illness;
   - incubation period;
   - possible disease outcome, such as recovery, mortality, chronic sequelae;
   - persistence of the micro-organisms in certain individuals resulting in continued excretion and risk of spread;
   - rate of infection.
3. With respect to *susceptibility of the population*, the extent of infection can differ markedly between different (sub)-populations. Infection can be more severe for groups with reduced resistance. Factors that can influence sensitivity of the (sub)population involve:
   - genetic factors that influence immune response;
   - immune status of (sub)population e.g. pregnant women, children, and the elderly can have a reduced immune status as compared to healthy adults;
   - breakdown of physiological barriers can lead to increased susceptibility (e.g. use of antibiotics, concomitant infections).

The stages in hazard characterisation according to Scientific Steering Committee (SSC, 2000a) are:

- establishment of the dose-response relationship for each critical effect;
- identification of the most sensitive species and strain, etc.;
- characterisation of the mode of action/the mechanisms for the critical effects (including the role of possible active metabolites);
- high to low dose (exposure) extrapolation, interspecies extrapolation;
- evaluation of factors that can influence severity and duration of adverse health effects in order to make an estimate.

Dose-response assessment is the process of obtaining quantitative information on probability of human illness following exposure to a hazard; it is the translation of exposure into harm. For some hazards, dose-response curves have been determined. Dose-response curves show the relationship between exposed dose and probability of a response to a specific hazard. Since validated dose-response relationships are scarce, various other inputs are used to underpin the hazard characterisation phase of risk assessment, such as NOAEL, BMD, minimal effective dose (typical for chemicals).

- NOAEL (no-observed-adverse-effect level) is defined as the highest dose at which no adverse effects are observed in the most susceptible animal species. The NOAEL is used as a basis for setting human safety standards for Acceptable Daily Intakes (ADIs) taking into account uncertainty factors for extrapolation from animals to humans and allowing for inter-individual variability in humans. In some cases no-adverse endpoint can be identified, such as for many naturally-occurring compounds that are widespread in foods. In that case an 'ADI Not Specified' is allocated.
- Benchmark dose (BMD) is an alternative method of defining an intake. The BMD is by definition greater than the threshold. The advantages of BMD are that it rewards good dose-response data because this will be associated with narrower confidence intervals (SSC, 2000a).
- For chemicals, the severity of any adverse effect associated with a chemical used as a food additive is usually directly related to the dose. For most chemicals, hazard characterisation is based on estimations of an intake for humans that would be below the dose necessary to produce adverse effects. For carcinogens, where just one molecule has the theoretical ability to induce a tumour, increasing the dose increases the probability that an individual will contract a tumour (Tennant, 2002; Renwick, 2004).

Exposure assessment and hazard characterisation thus provide an estimate of severity of the negative health effect(s) and the probability that these adverse effects occur after exposure to the specific hazard.

Risk characterisation is the last step in the risk assessment procedure. It is the quantitative or semi-quantitative estimation, including uncertainties, of frequency and severity of known or potential adverse health effects in a given population based on the previous steps (SSC, 2000a). Risk characterisation is the step within risk assessment that integrates information on hazard and exposure to estimate the magnitude of any risk. Comparison of the numerical output of hazard characterisation with the estimated intake will give an indication of whether the estimated intake is a health concern.

The degree of confidence in the final estimation of risk depends on variability, uncertainty, and assumptions identified in all previous steps. The nature of the information available for risk characterisation and the associated uncertainties can vary widely, and no single approach is suitable for all hazard and exposure scenarios.

In practice, it is often impossible to make a quantitative estimation of the risk, and therefore a semi-quantitative estimation can be applied based on the Failure Mode Effect Analysis principles (see also Chapter 4). In the FMEA procedure, severity (S), occurrence/frequency (O), and ability to detect (D) a potential failure (or hazard) are classified using a scale from 1-10. Subsequently, a risk priority number (RPN) is calculated, which is the product of severity (S), occurrence (O), and ability to detect (D), thus $RPN=S\times O\times D$. Likewise, scales to classify the *frequency/probability (P)* and *severity (S)* of known or potential adverse health effects, have been developed to calculate a risk number ($RN=P\times S$). More simple classifications are also used in practice as shown in Table 7.2. In this classification table it is not clearly described what is considered as low or high frequency and great or small severity.

It should be emphasised that classifications into great, moderate, etc., are arbitrary. What does a great severity consist of? Does it mean that lethality is ⩾ or ⩽ to 0.1 or 1% of the exposed population? Likewise, what is considered as high frequency? Is it <1 to 1000, or <1 to 100, or other? This is actually an ethical question about the acceptability of risk. Risk assessment data will serve as a scientific basis for discussions between governments and stakeholders to establish *food safety objectives*. Food safety objectives should specify tolerable levels of risk, which are quantifiable and measurable directly or indirectly. These food safety objectives have not yet been widely established.

For the current situation in practice, it is important that *within* the company (and preferably also for relevant suppliers and purchasers) there is a consensus about the interpretation of categories.

For more information the reader is referred to Brown and Stringer (2002), or Mack and co-authors (2006).

7.4.2 Predictive food microbiology

Information plays a crucial role in the development of a company-specific QMS and/or FSMS. Increasing accurate and reliable information about dynamic food properties and applied technological conditions, and the impact of people will enable better predictions of the food quality and safety outcome. Companies increasingly use various information systems to support their decision making process when designing, improving and executing their QMS/FSMS. Information systems are concerned with the capture, storage, analysis and retrieval of data providing the opportunity for the cumulative gathering of knowledge and the capability for a more informed interpretation of the significance of new data collected to monitor or investigate natural phenomena (McMeekin *et al.*, 2006). In this section we focus on methods/systems that can be very useful in the design, improvement and execution of an effective QMS/FSMS. These methods/systems help in, for example, the identification of potential hazards, in characterising and evaluating hazards (e.g. growth potential of pathogens, inactivation profiles, dose-response relationships, etc.), and in

Table 7.2. Classification of estimated severity and frequency of known or potential adverse health effects in a given population, using a scale of 1-4 (low to high).

Severity of hazard	Frequency		
	Low	Moderate	High
Great	3	4	4
Moderate	2	3	4
Low/small	1	2	3

assigning critical control points. Firstly, we briefly explain the major principles of predictive food microbiology and mention some available application software tools. Secondly, the function of decision support systems is briefly described.

Principles of predictive food microbiology

Predictive food microbiology can be defined as the use of mathematical expressions to describe microbial behaviour in a food product (Whiting and Buchanan, 1994). The basic idea behind it is that the response of a population of micro-organisms to environmental factors (such as pH, a_w, temperature) are reproducible. By characterising environments in terms of those factors that affect microbial growth and survival most, it is possible from past observations to predict the responses of those micro-organisms in other but similar environments. This knowledge can be compiled in mathematical models which can be used to predict quantitatively the behaviour of microbial populations in foods (such as growth rates, inactivation rates, or toxin production) from knowledge of the environmental properties of the food over time (Devlieghere *et al.*, 2006).

Classically, the microbial models are divided into three groups (Devlieghere *et al.*, 2006):
1. *Primary models* describe the bacterial evolution (often expressed as log (cfu/ml) as a function of time (expressed in e.g. days). When the conditions are favourable for the bacteria then it will be a *growth model,* whereas under stress conditions it reflects an *inactivation model.* Typical models that are used in this perspective are 'sigmoidal growth curve', 'Gompertz model', and 'model of Baranyi and Roberts'.
2. *Secondary models* describe the effect of intrinsic (pH, a_w, etc.) and extrinsic (temperature, gas composition) factors on the growth of a micro-organisms within a food matrix. Typical secondary models are Arrhenius models (empirical expression relating growth rate to environmental temperature), Square root (or Belehradek) models, response surface (or polynomial) models (pure black-box models).
3. *Tertiary models* are software packages in which the previously described models are integrated into a ready-to-use application tool.

In addition to the classical distinction between primary, secondary and tertiary models in microbial modelling, it is important to realise that other classifications can be made based on different approaches, assumptions, and type of outcomes. Within the area of predictive modelling, there are *kinetically-based* and *probability-based* models (Devlieghere *et al.*, 2006). Kinetic models can predict the extent and rate of growth of a micro-organism, whereas the probability-based models consider the probability of some event occurring within a certain period of time. Kinetic models are useful for prediction of changes in microbial cell density as a function of time, even if controlling variables that may affect growth are changing. A major drawback is that the development of such models requires a lot of data collection, to be able to model interaction effects between different environmental models. Probability models are commonly used for predicting spore-forming bacteria and pathogens, because the focus here is on predicting the probability that a certain phenomenon occurs. Such models may help in deciding on product formulation, processing and storage conditions.

Another distinction that can be made is between *empirical* and *mechanistic* models. An empirical model, also called a black-box model, is a model where fitting capacities are the only criteria. In other words, how well the data found fit with the model. The aim is to describe observed data in the best possible way, without any knowledge of the underlying biochemical mechanisms. Mechanistic models are based on understanding the underlying biochemical processes and the intercellular and intracellular processes that control cell behaviour. However, the actual microbial knowledge is still insufficient to generate such complex mathematical models (Devlieghere *et al.*, 2006).

To summarise, models that give insight into growth limits are helpful in identifying potential microbial hazards in specific foods, whereas growth and inactivation models provide a quantitative link between measurements used to monitor the process (e.g. time, temperature, pH, adding salt to lower a_w) and potential responses of specific pathogens. This information is useful when limits for critical control points need to be specified or when corrective actions must be determined to achieve compliance with performance criteria. Insight into growth and inactivation of pathogens along the food supply chain is fundamental to microbial assessment to be able to obtain estimations of consumer exposure to pathogens in food at the time of consumption (McMeekin *et al.*, 2006).

For the technical details about the mathematics behind the different groups of models, the reader is referred to Devlieghere and co-authors (2006), Brul and co-authors (2007).

Software package for predictive microbiology

Nowadays various microbial modelling software packages are (freely) available to support companies in estimating microbial hazards and their potential risks. Two well-known software packages are PMP and Growth Predictor. The *Pathogen Modeling Program* (PMP) was developed in the USA, and includes over 35 models of about 11 pathogens. This software package allows the prediction of growth or inactivation of pathogens under different combinations of constant temperature, pH, NaCl/a_w, and in some cases other conditions like organic acid type and concentration, atmosphere, and nitrate. For details the reader is referred to another well-known programme, which was developed in the UK, called Growth Predictor (and includes previous software of the Food MicroModel). It includes over 18 models and allows the prediction of growth or inactivation at constant conditions of temperature, pH, NaCl/a_w, and in some models, additional parameters (McMeekin *et al.*, 2006). Various other software packages are also available but they are often more specific (e.g. for certain product group) and companies may have their own specific software package. Currently, there are also software packages (e.g. Risk Ranger) that assist in predicting micro-organisms as part of risk assessment, which helps in the ranking of risks.

More information can be found on the internet about respectively PMP (http://www.arserrc.gov/mfs/PATHOGEN.HTM), Growth Predictor (http://www.ifr.ac.uk/Safety/GrowthPredictor), and Risk Ranger (http://www.foodsafetycentre.com.au/riskranger.php).

Decision support systems

In addition to the specific tools to support the prediction of microbial growth and/or inactivation and/or production of toxins, other systems can be used for design and improvement of QMS and FSMS. Examples are expert systems or decision support systems.

Expert systems imitate the performance of a human expert and can simulate expert-like reasoning and decision-making within a special domain of experts (McMeekin *et al.*, 2006). They are also called knowledge-based systems or decision support systems. The term decision support is more general and can include both knowledge-based (qualitative) information and quantitative information to support decision-making processes. The strength of such combined decision support systems is the combination of expertise, experiments, literature, quantitative microbiology, and other quantitative data on quality, instead of focusing on one area alone (McMeekin *et al.*, 2006).

It should be noted that each predictive model, expert system or decision support system has its limitations. The user should therefore check the limitations and shortcomings of the systems and tools to prevent prediction of unrealistic scenarios (Ross and McMeekin, 2003).

For more details about modelling food quality and food safety the reader is referred to McMeeking and co-authors (2006), Devlieghere and co-authors (2006), and Van Boekel (2008).

7.4.3 Traceability systems

Traceability is an important element in controlling quality along the whole supply chain from farm to fork. The European Commission's White Paper on Food Safety states: 'A successful food policy demands the traceability of feed and food and their ingredients'. Adequate procedures to facilitate such traceability must be introduced. These include the obligation for feed and food businesses to ensure that adequate procedures are in place and to withdraw food and feed from the market where a risk to the health of the consumer is posed. Operators should also keep records of suppliers of raw materials and ingredients so that the source of a problem can be identified.

For the *purpose of traceability*, information needs to be exchanged between different locations of a single company, but also between different chain participants in the agri-food chain, and with stakeholders in the environment (e.g. government, consumer organisations). The information exchange should therefore enable both horizontal and vertical communication of information within and outside the company. Traceability at company level involves the exchange of information about the location and history of the product (both product and process information). At chain level, product and history information and information about the origin are also important (Trienekens and Van der Vorst, 2006).

The *essence of a traceability system* is to provide accurate and reliable information about a certain product to enable backward (also called tracing) and forward (also called tracking) traceability. By tracking and tracing it is possible to determine (track) where a certain item is located and to trace the history of that item. On the basis of that information, it should be possible to determine the source of any (quality) problem of an item, and it should be possible to find out where other items with the same problem are located in the supply chain.

The narrow purpose of a traceability system is to track and trace products with the objective of redrawing products from the chain because of quality or safety problems. From a broad perspective, traceability systems are not only used for tracking and tracing non-compliant products, but for controlling and optimising processes within a company or a chain level (Trienekens and Van der Vorst, 2006).

There are no univocal guidelines for designing traceability systems. However, typical elements that should be considered when designing such a system are (Mgonja, 2007) (Figure 7.8):

1. Defining *strategy* and *objective* of traceability system;
2. Identifying *unique identity* of products, logistic units and locations;
3. *Capturing and recording* of data;
4. *Data retrieval* and link to management;
5. *Communication* of traceability data.

Core aspects of traceability system	Detailed aspects & technical tools
Defining **strategy** and **objective** of traceability system	**Various strategies**, like compliance-oriented strategy , process improvement-oriented strategy, market-oriented (branding) strategy. **Various objectives**, like comply with regulatory requirements, reduce business risks, facilitate product recall/withdrawal, comply with trading partner specifications, obtain efficient logistics management, support effective quality management, support food safety.
Identifying **unique identity** of products, logistic units and locations	Any trade item and/or location, which need to be traced or tracked, must have **a unique identity**. Precise identification can only be done when a trade item, also called a **Traceable Resource Unit** (TRU), has been well defined.
Capturing and recording of data	Traceability requires pre-defined **data to be captured and recorded** throughout supply chain. For both large and small companies, the recommended enabling technology is **Automated Identification and Data Collection/Capture** (AIDC). For **identification** and **collection** purposes **various technological tools** are available such as barcodes, scanners, DNA recognition, or RFID (Radio Frequency Identification).
Data retrieval and link to management	Ability to **retrieve traceability data** in a fast and accurate manner along a supply chain is **critical. AIDC systems must be integrated** with administrative systems to simplify analysis and exchange of data. **Examples** are Warehouse Management Systems (WMS), Laboratory Information Management Systems (LIMS), and Enterprise Resource Planning (ERP) systems.
Communication of traceability data	**Essential feature** of any traceability system is exchange of information. To enable communication of traceability data **systems for communication** are needed such as (simple) e-mail, Electronic Data Interchange (EDI) software or eXtended Markup Language (XML).

Figure 7.8. Core aspects of a traceability system.

Ad 1. Traceability system strategy and objective

Companies have a variety of motivations to adopt traceability. These include increasing transparency, reducing exposure to liability, improving effectiveness of recalls, enhancing logistics, improving control of livestock epidemics, easier product licensing and price premium, reducing levels of information asymmetry, decreasing transaction costs, increasing trust levels, and thus facilitating contracting. Deciding which traceability *strategy* is the best for a specific company or supply chain depends on the trade-off between costs and benefits and can differ for each specific product group. High levels of traceability require investments in processes, infrastructures, information systems, and people/relationships. When this results in effective recall management, effective risk management, process improvement, and competitive advantages the investments might be worthwhile.

Van der Vorst (2003) distinguished three main strategies of traceability, namely compliance-oriented strategy, process improvement-oriented strategy, and market-oriented strategy.

- In a *compliance-oriented strategy* companies basically comply with rules and regulations by implementing minimum requirements. Companies that adhere to a compliance-oriented strategy focus on the registration of incoming and outgoing materials and leave the process as a black box.
- In a *process improvement-oriented strategy*, companies strive for control of the traceability of products within the own link of the production chain by means of production-integrated measures that achieve both compliance with governmental rules and regulations and a better return. Examples of process-oriented measures are local ICT systems that register all process data and if necessary change process control.
- In a *market-oriented (branding) strategy* companies aim for the establishment of full traceability within the supply chain to achieve competitive advantage (by creating added value in the market place). This requires the redesign of processes to separate small production lots, standardisation of information carriers, adjusted planning and control of production processes, and so on.

The design and performance of a traceability system depends, amongst other things, on its *objective*, typical examples of which are:

- To comply with regulatory requirements (e.g. EU Regulation 178/2002 on food law.
- FDA requirements on establishment and maintenance of records (21 CFR 1) and guidance on recalls (21 CFR 7.59), USDA requirements and guidance on recalls (FSIS Directive 8080.1).
- To reduce business risks above and beyond legal compliance.
- To facilitate product recall and withdrawal (notably to achieve a greater degree of precision, to demonstrate control, increase efficiency and reduce the cost of product recall or withdrawal).
- To comply with trading partner specifications.
- To obtain efficient logistics management.
- To support effective quality management.
- To support product safety.
- To provide information to consumers and trading partners.
- To verify the presence or absence of product attributes.
- To protect the brand.
- To enable product authentication and anti-counterfeit policies.

Ad 2. Identifying unique identity

Any trade item and/or location, which needs to be traced or tracked, must have a unique identity. It is important to note that a precise identification can only be made when a trade item, also called a traceable resource unit (TRU), has been well defined. Defining a TRU may be one of the most difficult steps involved in the design of a traceability system. A product identifier should uniquely identify the unit of batch, be secure (fraud proof), be permanent, retain identity throughout the product life cycle, be simple to read, capture identifying data, and not hinder its host (Loftus, 2005; Thompson *et al.*, 2005; Mgonja, 2007).

Ad 3. Capturing and recording of data

Traceability requires pre-defined data to be captured and recorded throughout the supply chain. It is important to bear in mind that the traceability information has to be shared between partners and/or stored by each trading partner wherever relevant and applicable. For smaller companies with lower trading volumes and less complex business processes, data capture through manual documentation using traditional approaches (archives, folders) can be a viable and functional solution. Nevertheless, for both large and small companies, the recommended enabling technology is *automated identification and data collection/capture* (AIDC).

For identification and collection purposes various technological tools are available such as barcodes, scanners, DNA recognition, or RFID (radio frequency identification). A barcode is a machine readable code consisting of a series of bars and spaces printed in defined ratios. Barcode technology encompasses the symbologies that encode data to be optically read, the printing technologies that produce machine-readable symbols, the scanners and decoders that capture visual images of the symbologies and convert them to computer compatible digital data, and the verifiers that validate symbol quality. DNA identification refers to identification based on DNA analysis which provides unique information. One application in the food sector is the DNA identification of animals. RFID systems cover a range of data-carrying technologies, which transfer data from the identifier to the reader by a radio-frequency link (Trienekens and Van der Vorst, 2006).

Ad 4. Data retrieval and link to management

In the majority of supply chains, products are tracked and traced by their production lot, which has undergone the same transformation (production process) and by their transport/storage path (distribution process). The ability to retrieve traceability data in a fast and accurate manner along a supply chain is critical. This requires the management of successive links between what is received, produced, packed, stored and shipped across the entire supply chain. If one of the partners in the supply chain fails to manage these links, this will result in the disruption of the information chain and in the subsequent loss of traceability. It is impossible to attain full product traceability without correctly identifying products in all their configurations at each different point of the supply chain.

Systems that register tracking and tracing data (AIDC) should preferably be integrated with administrative systems to simplify the analysis and exchange of data. These data can then also be used for planning and evaluation of production. Examples of such systems are warehouse management systems (WMS), laboratory information management systems (LIMS), and enterprise resource planning (ERP) systems.

Ad 5. Data communication

An essential feature of any traceability system is the exchange of information. Traceability requires associating the physical flow of products with the flow of information about them. To ensure the continuity of the information flow, each supply chain actor must communicate pre-defined traceability data to the next one, enabling the latter to apply traceability principles. To enable the communication of traceability, data systems for communication are needed such as (simple) e-mail, electronic data interchange (EDI) software or eXtended Markup Language (XML). For an efficient and effective data exchange in the supply chain, a common and standardised infrastructure is needed to facilitate the exchange of data through the whole supply chain. Some standards have been developed by the European Article Number Association (EAN) to be used in logical systems in close cooperation with the American equivalent (the Uniform Code Council, UCC). They define, for example, global unique codes for industrial producers to facilitate communication (Trienekens and Van der Vorst, 2006).

It should be noted that the required technology investments in traceability systems are very high, which often results in a decision to opt for relatively low levels of traceability. In that case, prevention via high-quality management becomes essential to diminish the risk of product recalls. Efficient Consumer Response Europe (ECR Europe http://www.ecrnet.org) reported that the cost of implementation of traceability systems is likely to vary enormously between business and sectors depending on the type of technology adopted, the amount of information required to be stored and the complexity of the supply chain. Therefore, it is more likely that traceability systems will be introduced rapidly where commensurate benefits exist in logistics and process control or where brand market share would be jeopardised without the introduction of such a system. *These differences lead to a huge spread in the traceability* performances along the supply chains. In many cases companies still focus on their own business instead of the complete supply chain (Van der Vorst, 2003, Trienekens and Van der Vorst, 2006; Mgonja, 2007).

7.5 Managing quality assurance

Quality assurance activities often lead to structural interference in the organisation and/or technology. The type of interference depends on the objective of quality assurance (e.g. to guarantee product safety, supply, service quality, etc.). As a matter of fact, introduction and implementation of a quality management system is not a simple process. It requires insight into relevant managerial processes. Most of these processes have been described in detail in previous chapters, i.e. in Chapter 3 (in specific quality behaviour and decision-making) and in Chapter 6 (improvement processes and organisational change). In this section attention is paid to the managerial aspects specific to the quality assurance process, with a special focus on auditing and certification processes.

7.5.1 Quality assurance in the organisation

People are neither well suited to exact repetition of methods nor to application of consistent standards (Smith and Edge, 1990). It is therefore necessary to provide written standards and procedures in order to make a quality management system effective. Since quality standards and procedures are for day-to-day use they must be concise and readable. In practice they are described as a hierarchy of procedures. The general principles and rules are described in a top level document (quality manual), which refers to subsequent levels of procedures and work instructions that provide the details.

A *quality manual* describes general quality policy and organisation of the company, together with responsibilities for quality. It contains outlines for specific areas for control and it contains lower level documents, which are needed to carry out quality control. *Quality procedures* are listed in the quality manual and provide detailed instructions. The type of information given includes actions to be taken, routing of forms (such as defect sheets, purchase orders, engineering change notes and so on), information to be recorded, responsibilities for each action, etc. *Work instructions* deal with a lower level of activities than quality procedures. Whereas procedures describe who does what, work instructions provide specific information about standards to be achieved in work execution (such as work methods, hygiene instructions, use of equipment).

When using procedures the problem of compliance to procedures arises. Section 3.5 has already discussed how in formalised organisations this compliance can play a role. One should realise that people only follow procedures to a certain degree, dependent on different factors like understanding of procedures, quality attitude, time pressure, and responsibility. Non-compliance is one of the factors contributing to dynamic behaviour of employees.

Quality assurance department

In most companies the quality department is responsible for quality assurance and the quality manual. It keeps knowledge about standards up to date, organises audits, reports on evaluation results, organises corrective actions and communicates with external stakeholders. Normally quality assurance will only be a part of a quality department's activities. However, when quality control has already been delegated to production departments, quality assurance becomes the main task of the quality department. It is then often called a quality assurance (QA) department, and is hierarchically positioned just below top management level. One role of the QA department is to support line activities with advice, recommendations, research, and technical expertise.

The quality assurance department maintains intensive communication with departments in the organisation that have assurance responsibilities for other aspects, for instance for environmental care, work conditions, and personal welfare. Many companies are trying to integrate these assurance tasks into one integrated care/assurance system.

Fit between quality system and administrative concept

Four administrative concepts have been mentioned in Chapter 3 section 3.3.5, i.e. the authority, bureaucratic, business, and flexible concept. They differ in level of centralisation and formalisation. The basic idea of a system that is based on procedures fits with higher levels of formalisation. So, the bureaucratic and business concept will encounter no problems then. But in the authority concept, which is often applied in small companies, a QA system may be seen as overkill, and is consequently not well accepted. In the flexible concept a QA system may become too strict and hinder flexible operations, like bringing new products into the market within a short time period.

When looking to the quality maturity grid of Crosby (see Chapter 8 for details about the grid), it can be concluded that higher stages of quality maturity require less strict systems due to the required flexibility. But less strict or 'lean' systems are in such situations applicable due to the quality attitude and competencies of those involved. The quality maturity of the organisation is so advanced at this stage, that there is a climate of quality consciousness where continuous improvement is a daily challenge for all employees.

Change processes with the real participation of people, who are going to use the system, are very important for obtaining adequate systems design and implementation. Note that external pressure can force change processes when companies feel obliged to meet external standards, such as ISO 9001 or BRC, in a relatively short time period. Choosing an adequate change strategy is also very important in this case in order to get control of these processes. See Chapter 6 section 6.5.4 for a more in-depth discussion of this type of problem.

7.5.2 Quality auditing and certification

Audits are a systematic and independent examination to determine whether quality activities and related results comply with planned arrangements and whether these arrangements are implemented effectively and are suitable to achieve objectives (ISO 9000:2005). In other words it is an assessment of a quality management system to determine whether it is doing what it is meant to do.

Auditing is a way of diagnosing an organisation. An expert collects information, analyses the information, and informs the client about his findings. During an audit a management system is assessed on three aspects:

1. Compliance to the standard: the auditor considers the extent to which the apparently prescribed system conforms to the appropriate standard by examination of documents before execution of the actual audit.
2. Implementation of the prescribed system: examination of what people are doing will reveal the extent to which they are operating as established in their own documents.
3. Effectiveness, each procedure, both written and practical, is there for a specific reason. It must produce some result and there must be a means of showing what has been achieved.

Information will come to the auditors in many ways. The information must be objective evidence otherwise it is of no use for an audit.

For auditing it is essential to establish a procedure and audit criteria. A general procedure for auditing involves collection of information, verification of information, establishing objective evidence, summarising audit findings and preparation of report. Audit criteria are procedures, norms and/or requirements against which the auditor compares objective evidence about the specified process, management system, product and/or service and/or information about these matters. Various quality assurance standards have described detailed procedures for auditing the specific standards. For example, the ISO standards include a special procedure for auditing 'ISO 19011:2002: Guidelines for quality and/or environmental management systems auditing'. The reader is referred to the official websites of the specific standards for the details (see also sections 7.3 and 7.6).

Quality audits provide several advantages, such as:
- giving feedback on the completeness and effectiveness of the quality system;
- when conducted periodically, revealing the seriousness of continuous improvement for the company to everyone;
- indicating those elements of the quality system that are inadequate or need to be improved;
- providing a permanent record of progress in achieving the goals of the quality system (Evans and Lindsay, 2005).

The audit will contribute to a situation of control and continuous improvement. Its characteristics are also described in principle 6 of the HACCP system, verification procedures. The procedures described here review the implementation of a quality system. It can be considered as a system support element (Dillon *et al.*, 2000).

Three types of audits can be distinguished, one of them being an internal and two of them being an external audit (Staples, 1990; Evans and Lindsay, 2005) (Figure 7.9).

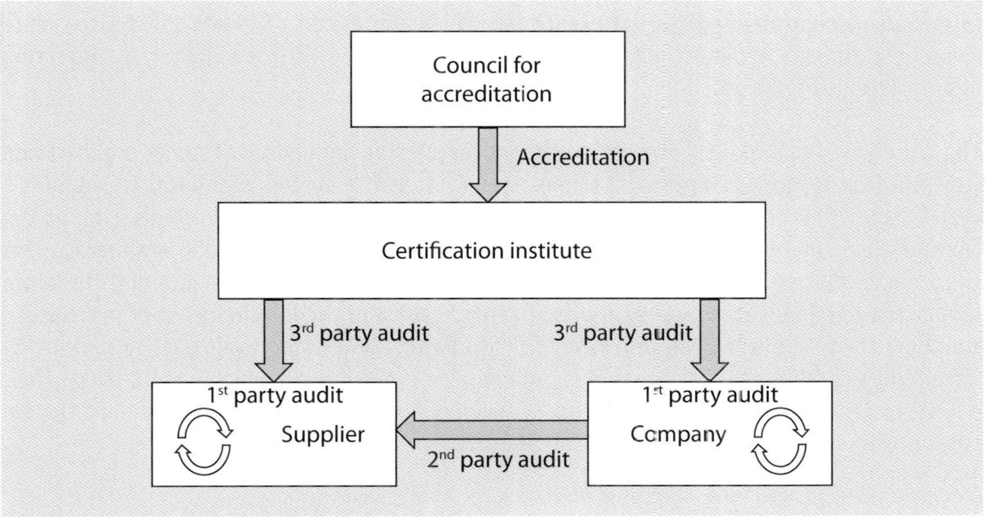

Figure 7.9. Auditing, certification and accreditation.

The *first party audit* is an *internal audit* and is carried out by the company for management purposes such as improvement of the internal quality system.

Audit criteria for internal audits relate to typical issues of the quality system, such as:

- Management involvement and leadership, which refers to the extent to which all levels of management are involved.
- Product and process design, which relates to compliance of products with customer needs.
- Product and process control, which concerns the extent of defect prevention.
- Customer and supplier communication, which includes the extent to which extent customers and suppliers communicate with each other.
- Quality improvement programmes, which refer to the presence of quality programmes and the results of these programmes.
- Employee participation, which relates to the active involvement of employees in quality improvement.
- Education and training concerns activities, which are carried out to ensure appropriate skills and experience and training in quality improvement techniques.
- Quality information, which relates to the collection of and feedback on quality results.

Regularly measuring compliance with the stated system is important to determine its effectiveness. Corrective actions derived from an internal audit will keep the system fresh and up to date (Dillon *et al.,* 2000).

The *second party audit* is an *external audit* and is carried out to audit a potential supplier against proposed requirements when a contract is to be drafted. These requirements can be prescribed in suppliers' internal documents and they may include requirements established in a quality system standard. In second party audits different approaches are used to establish supplier quality, ranging from comprehensive audits to short quality evaluations and awareness programmes. Short quality evaluations aim to check whether the supplier is following appropriate quality control procedures. Awareness programmes only try to create awareness about customers' requirements, among employees of the supplier, by providing appropriate information.

The *third party audit* is an *external audit* and implies an independent audit organisation (the third party) assessing organisations against a quality system standard. It provides a certificate, which records the results of the audit indicating that the organisation at the time of assessment had management systems, which complied with the standards. This third-party method is thought to reassure users of organisations' products and therefore reduce the need for second-party audits. In anticipation of the requirements of purchasers, suppliers try to prevent each purchaser from individually judging their quality system, by obtaining a certificate. Certification is defined as a procedure by which a third party gives a written assurance that a product and/or service, process or quality management system conforms to specified requirements (ISO 9000:2005).

Accreditation

Third party audits or certification audits are carried out by certification institutes or bodies. Before certification institutes are allowed to assess quality systems and provide certificates, they are audited as well, which is called *accreditation*. Accreditation has been defined as a procedure by which an authoritative body gives formal recognition that a body or person is competent to carry out specific tasks.

In the Netherlands, the Council for Accreditation supervises institutions in the private sector that assess products, processes, and measuring instruments on behalf of the government, companies and consumer associations. The Council for Accreditation can, for example, screen certification institutes based on established criteria and if the institute proved competence accreditation is granted. In this way the government has two lines of influence: the first by supervising the council for accreditation and the second by inspection and investigation in the company and market place.

7.6 Quality assurance from a chain perspective

The primary quality assurance guidelines/standards have been described in section 7.3. In this section we explain the principles of quality assurance standards/guidelines with an explicit chain focus and/or those that put specific assurance demands on certain actors in the agri-food production chain. Although, in all sectors along the food supply chain companies have to deal with the Codex Alimentarius hygiene codes, GMP, PRP, and HACCP principles, some sectors have defined specific standards, codes and or guidelines to support the sector in assuring quality and safety of agri and food products. For example, ISO 22000 Food Safety Management System has been developed by ISO and is aimed at assuring food safety but sets the requirements within a clear chain perspective (section 7.6.1). In section 7.6.2, we briefly describe major quality assurance initiatives/standards that are explicitly required by the retail sector like Global Food Safety Initiative (GFSI), British Retail Consortium- Global Standard for Food Safety (BRC), and International Food Standard (IFS). Sections 7.6.3-7.6.5 describe various assurance standards/guidelines that are specific for animal production, fresh produce, and flower and indoor plant production respectively.

7.6.1 Quality assurance standards with a chain focus (ISO 22000, SQF)

Both ISO 22000 and SQF (Safe Quality Food) are quality management standards that explicitly focus on ensuring food safety (and to a lesser extent quality) along the food supply chain.

ISO 22000

The growing number of national standards, codes, and guidelines for food safety management created a need to harmonise the national standards into an international standard. Therefore, ISO started in 2001 with the development of a standard with requirements for a food safety management system based on the HACCP principles. ISO 22000 specified requirements enabling an organisation to:

- plan, implement, operate, maintain and update a food safety management system aimed at providing products that, according to their intended use, are safe for the consumer;
- demonstrate compliance with applicable statutory and regulatory food safety requirements;
- evaluate and assess customer requirements and demonstrate conformity with those mutually agreed customer requirements that relate to food safety, in order to enhance customer satisfaction;
- effectively communicate food safety issues to their suppliers, customers and relevant interested parties in the food chain;
- ensure that the organisation conforms to its stated food safety policy;
- demonstrate such conformity to relevant interested parties; and
- seek certification or registration of its food safety management system by an external organisation, or make a self-assessment or self-declaration of conformity to ISO 22000.

They combined the following key elements to ensure food safety along the food supply chain: (1) prerequisite programmes, (2) HACCP principles, (3) process control, (4) system management, and (5) interactive communication.

ISO 22000 integrates the principles of the Hazard Analysis and Critical Control Point system and application steps developed by the Codex Alimentarius Commission. By means of auditable requirements, it combines the HACCP plan with prerequisite programmes. ISO 22000 requires that all hazards that may be reasonably expected to occur in the food chain, including hazards that may be associated with the type of process and facilities used, are identified and assessed. Thus it provides the means to determine and document why certain identified hazards need to be controlled by a particular organisation and why others need not. During hazard analysis, the organisation determines the strategy to be used to ensure hazard control by combining the prerequisite programmes and the HACCP plan.

Communication along the food chain is essential to ensure that all relevant food safety hazards are identified and adequately controlled at each step within the food chain. This implies communication between organisations both upstream and downstream in the food chain. Communication with customers and supplies about identified hazards and control measures will assist in clarifying customer and supplier requirements.

An effective food safety system is established, operated and updated within the framework of a structured management system and is incorporated into the overall management activities of the organisation. This provides maximum benefit for the organisation and interested parties. ISO 22000 has therefore been aligned with the structure of ISO 9001 to enhance the compatibility of the two standards.

It contains the following items:
- scope;
- normative references;
- terms and definitions;
- food safety management system;
- management responsibility;
- resource management;

- planning and realisation of safe products;
- validation, verification and improvement of the management system for safety.

Moreover, ISO is developing additional standards that are related to ISO 22000. These standards are also called the ISO 22000 family of standards. The following standards will make up the ISO 22000 family of standards:
- ISO 22000 - Food safety management systems - Requirements for any organisation in the food chain.
- ISO 22001 - Guidelines on the application of ISO 9001:2000 for the food and drink industry (replaces: ISO 15161:2001).
- ISO TS 22003 - Food safety management systems for bodies providing audit and certification of food safety management systems.
- ISO TS 22004 - Food safety management systems - Guidance on the application of ISO 22000.
- ISO 22005 - Traceability in the feed and food chain - General principles and basic requirements for system design and implementation.
- ISO 22006 - Quality management systems - Guidance on the application of ISO 9002:2000 for crop production.

ISO 22000 can be applied independently of other management system standards or can be integrated with existing management system requirements.

Safe Quality Food (SQF)

The SQF standards were developed by the Australian government in collaboration with farmer organisations since 1995. Safe Quality Food (SQF) is a *food safety standard* that specifies the quality management system requirements necessary to identify food safety and quality risks as well as validate and monitor the control measures.

The programme is made up of two codes (standards): the SQF 1000 Code for the primary production sector and the SQF 2000 Code for the food manufacturing and service sectors.

Suppliers can implement a management system which meets any of the following three levels of certification:

Level 1. Food Safety Fundamentals, which indicates that pre-requisite programmes and fundamental food safety controls have been implemented to provide a sound foundation for the further development of the supplier's management system.

Level 2. Certified HACCP Food Safety Plans, which incorporates all level 1 system requirements, and indicates that a food safety risk assessment of the product and its associated processes has been completed. This assessment must identify the hazards and the actions taken to eliminate, prevent or reduce their occurrences.

Level 3. Comprehensive Food Safety and Quality Management System incorporates all level 1 and level 2 system requirements, and indicates that a food quality risk assessment of the product and its associated process has been completed. Additionally, at level 3 an organisation must show that the actions taken to prevent the incidence of poor quality have been implemented and that the remaining quality management system procedures addressed in the standard have been implemented.

The standard combines generally recognised key elements to ensure food safety along the food chain, including:

- control of food safety hazards through prerequisite programmes and HACCP-based food safety plans;
- control of potential threats to quality through HACCP-based food quality plans;
- principles of system management;
- maintenance of the management system.

SQF Certification provides an independent and external validation that a product, process or service complies with international, regulatory and other specified standard(s) and enables a food supplier to give assurances that food has been produced, prepared and handled according to the highest possible standards. The SQF standards offer the food sector a way of managing food safety and quality simultaneously. SQF (the SQF 1000 and 2000) is recognised by the Global Food Safety Initiative as conforming to the highest international standards and utilises protocols administered by International Accreditation Forum member Accreditation Bodies.

The reader is referred to http://www.sqfi.com/ for more details about the standards.

7.6.2 Quality assurance demands in the retail sector (GFSI, BRC, IFS)

The retail sector forms the last chain before the consumers and plays an important role in the assurance of food safety by setting high assurance demands on their suppliers. Various measures have been taken by the retail sector to improve food safety and gain consumer trust. The major quality assurance initiatives and standards are described below.

Global Food Safety Initiative (GFSI)

The Global Food Safety Initiative (GFSI) was launched in May 2000. It was initiated by a group of international retailers who acknowledged the need to improve food safety, ensure consumer protection, regain consumer trust, set clear requirements for food safety schemes and reduce costs. An international tasks force (comprised of 50 retailer quality managers representing 65% of food retail revenue worldwide), worked on a guidance document against which food safety standards for manufacturing can be benchmarked. It consists of three key elements: (1) FSMS, (2) good practices for agriculture, manufacturing and distribution (GAP, GMP, GDP), and (3) HACCP. A quality management standard can be acknowledged by GFSI if it complies with all components of the key elements. Once a standard has been benchmarked successfully, the standard is acknowledged. The GFSI has now approved five standards (British Retail Consortium technical standard (BRC), Dutch HACCP code, European Food Safety Inspection Service (EFSIS), International Food Standard (IFS), Safe Quality Food (SQF 200 code)). It should be noted that GFSI does not perform any certification or accreditation activities, but encourages the use of third party audits fro benchmarked standards. For details the reader is referred to the internet (www.ciesnet.com).

British Retail Consortium global standard for food safety (BRC technical standard)

In 1998 the British Retail Consortium (BRC) published the global standard for food safety (GSFS) for producers supplying own brands goods into the UK retail market as a consequence of the 'due diligence' principle.

This principle states that *'it shall ... be a defence for the person charged to prove that he took all reasonable precautions and exercised all due diligence to avoid the commission of the offence by himself or by the person under his control'.*

So, for every private label, the technical execution, in particular production methods of all suppliers has to be verified demonstrably. This resulted in a situation where every retailer made his own standard and had an extensive QA department with a lot of auditors to audit their suppliers resulting in very high costs. To resolve this problem BRC (representing the biggest retailers in the UK like Tesco, Safeway, Sainsbury, Somerfield, etc.) compiled the requirements that the suppliers have to meet in the BRC technical standard.

The standard is regularly updated to included the latest insights into food safety. The standard has now attained usage worldwide as a framework for any business to assist the production of safe food and is nowadays available in many languages. The standard has specified safety, quality and operational criteria required to be in place within a food manufacturing organisation to fulfil obligations with regard to legal compliance and protection of consumers. The format and content of the standard allows the assessment of a company's premises, operational system, and procedures by a competent third party (certification body) against the requirements of the standard. Details about this certification process can be found on internet at http://www.brcglobalstandards.com.

The BRC standard is based on two key elements: senior management commitment and HACCP, as a step-by-step approach to managing food safety risks. The standard has issued detailed comments on the following:

- Senior management commitment and continual improvement.
- The food safety plan – HACCP.
- *Food safety and quality management system*; including, amongst others, requirements on food safety quality policy and manual, organisational structure, customer focus, audit, purchasing-supplier approval, documentation requirements, corrective actions, traceability, complaint handling and management of incidents.
- *Site standards,* including, amongst others, requirements on inclucing/external external utilities, equipment, maintenance, facilities, etc.
- *Product control*; including, amongst others, requirements on product design & improvement, foreign body detection, packaging, laboratory testing, product release, etc.
- *Process control*; including, amongst others, requirements on control of operations, quality weight, calibration and control of measuring devices.
- *Personnel*; including, amongst others, requirements on training, movement personnel, hygiene, medical screening, etc.

A major advantage of the BRC guidelines is that there is more clarity for suppliers of private labels; there is only one inspection standard and a supplier can use that standard for all retailers and so reduce auditing costs. The reader can find a more detailed table showing the specific requirement in Appendix 7.3.

European Food Safety Inspection Service (EFSIS)

The EFSIS standard is developed by the British European Food Safety Inspection Service and is based on the BRC standards with some additional requirements. For details the reader is referred to http://www.efsis.com/.

International Food Safety Standard (IFS)

In Germany too, initiatives have been undertaken by the retail sector to improve food safety and regain consumer trust. As in the UK, the German retailers (presented by the Bundesvereinigung Deutscher Handelsverbände) developed a standard to inspect their suppliers of private labels, in 1999. The objective of the initiative was to design one standard with an uniform rating system with limited interpretation of the auditor. The control must be carried out by certification bodies and auditors. The audit report should provide a reliable view on the FSMS of the company and auditing should become part of a continuous process. Critical non-conformance products should be known about. The standard and auditing protocol should be in agreement with the Global Food Safety Initiative. The working group developed an international standard based on the BRC standard but organised according to the ISO 9001:2000 format. The initiative is supported by an increasing number of European countries, such as German, France, and Italy.

The standards include five chapters: (1) management and the quality system, (2) management responsibility, (3) resource management, (4) product realisation, and (5) measurement, analyses and improvements, in accordance with ISO 9001:2000 structure. For details about the content and auditing of the standard the reader is referred to http://www.food-care.info/.

7.6.3 Quality assurance guidelines in animal production

Besides the common characteristics of primary production, animal production faces some typical quality affecting factors such as the use of veterinary medicines for growth and animal health. Living conditions for cattle and poultry, including aspects such as light, ventilation, accommodation and hygienic conditions, are also important quality factors, which need to be controlled. Another typical aspect of animal production involves animal feeding production. Animal feeding is mainly composed of (undefined) waste streams of the food industry, which complicates quality control of animal feeding production.

Due to various food crises, which originated from problems in the animal production chain (e.g. dioxin affair, BSE), much attention is paid nowadays to quality assurance in the primary part of the food production chain. In this section we briefly explain the various quality assurance guidelines/schemes that have been specifically designed for the animal production chain in Europe.

Good manufacturing practices plus (GMP+) for feed industry

The Product Board Animal Feed has developed a special quality programme for the Dutch animal feed chain that focuses on animal feed safety as a component of overall food safety. The motto of the programme, feed for food, suggests that animal feed and food for human consumption can never be viewed in isolation. In addition to safety, the nutritional quality of food is a vital concern within this common quality policy. The Dutch animal feed sector applies the same quality assurance system as the international foodstuffs industry: pro-active quality assurance based on HACCP principles. The HACCP risk management system complements the product board's existing good manufacturing, managing practice quality assurance standards (GMP). The GMP+HACCP quality assurance system (known as GMP$^+$) has been extended to the whole chain from raw materials producers up to and including the carriers and the livestock farmers. Independent auditors from certification bodies check on compliance with the regulation.

The GMP$^+$ scheme describes requirements and conditions for feed safety system including:

- ISO9001:2000: quality management systems - requirements.
- Requirements for a food safety management systems based on HACCP principles.
- ISO22000 food safety management system - requirements.

Before implementing these requirements each company should establish and implement general basic programmes (like PRP).

The quality programme of the Product Board Animal Feed also provides an early warning and response system (EWS). EWS is necessary for the timely signalling and proper solution of potential hazards to humans and/or animals which might occur in spite of the preventive measures. Moreover, tracking and tracing is another specific element which is part of quality assurance. This involves a system and a method whereby the most accurate and rapid traceability of irregularities in batches of feeds and feed materials is possible. Good registration of details on batches of raw materials and feeds in the logistical process make it possible - in the event of contamination by an undesirable substance - to organise a recall quickly and to identify the source.

For more information the reader is referred to http://www.pdv.nl/lmbiraries/inleiding-uk.pdf.

Quality assurance requirements in dairy production

Chain quality of milk (in Dutch 'keten kwaliteit melk', KKM) is another Dutch initiative in the animal production sector setting requirements on milk production. The system is included in the national regulations and is therefore compulsory. The KKM requirements are clear and simple, and farmers are supported and advised by the KKM foundation. Chain quality of milk consists of various modules (Figure 7.10), including requirements on:

- animal medicines;
- animal health and welfare;
- feed and water;
- collection of milk, storage and design of milking facilities;
- cleaning and disinfection.

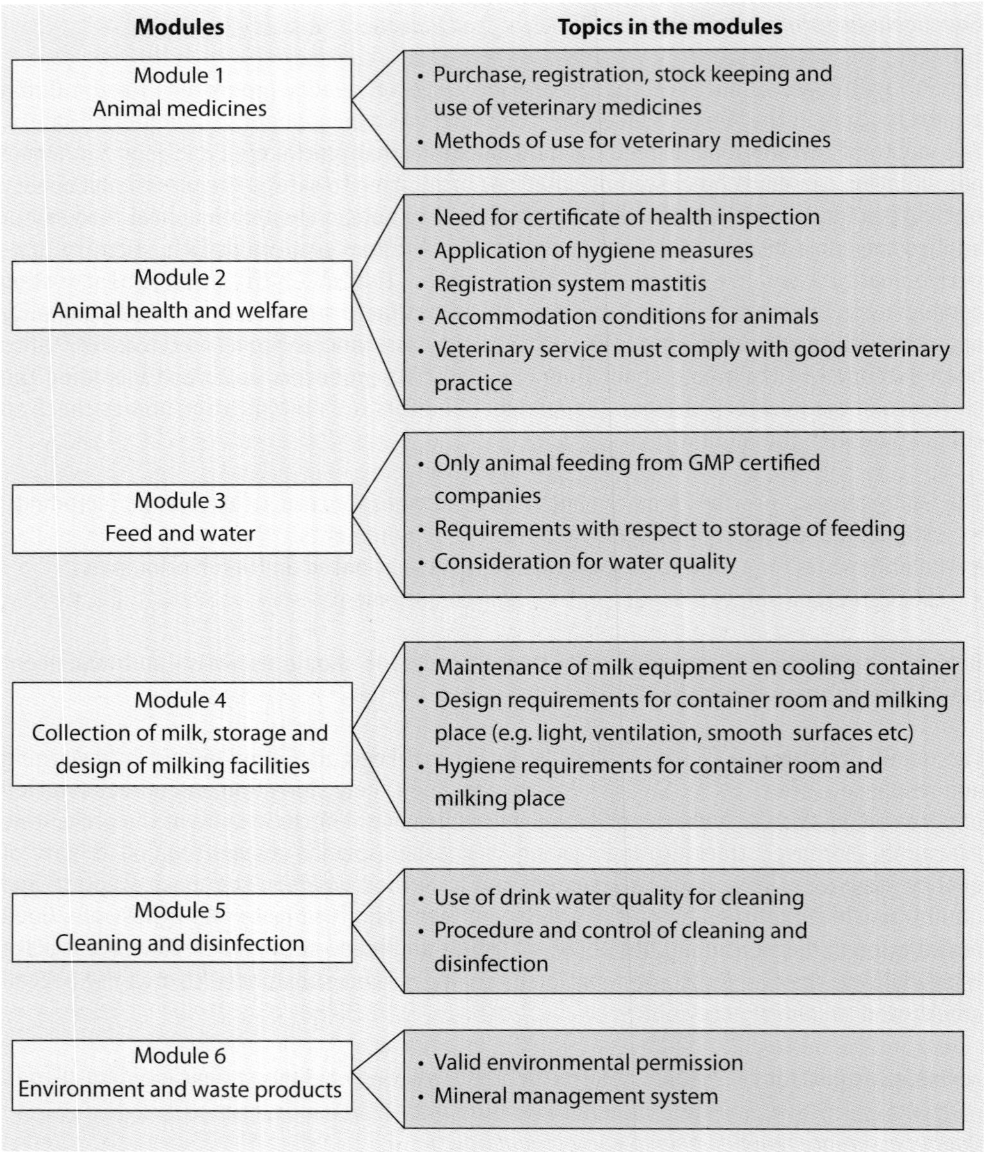

Figure 7.10. Modules of KKM (chain quality of milk).

For details the reader is referred to http://www.ketenkwaliteitmelk.nl.

Integrated chain control

In the animal production sector several integrated chain control requirements (in Dutch: 'Integrale Keten Beheersing' IKB) have been developed in the Netherlands, e.g. IKB for slaughter poultry, IKB for eggs, and IKB for cattle. The requirements are focused on the production and distribution of meat derived from pork, cattle, and poultry. Integrated chain control involves the entity of activities, which cover the whole production chain and tuning of units in the chain, in order to control quality-determining factors including those factors that affect labour conditions and environment.

IKB involves both legislative requirements and additional requirements. Typical additional requirements concern animal feed (only to be purchased from GMP certified suppliers), hygienic transportation (complying with GMP for transport), and veterinary practices according to the GVP code (good veterinary practices). IKB requirements aim at covering all steps in the production chain. Typical topics that are addressed by IKB requirements are:

- *Communication*: exchange of information between chains must be facilitated by appropriate registration of relevant data, such as results of chemical and bacterial analyses, age of animals, and vaccination schedules. These data from previous steps can be used to anticipate further in the chain.
- *Traceability:* tracing and tracking of products must be organised, by adequate administration, to simplify removal of non-conformance products.
- *Veterinary medicines* must be controlled and veterinary service must comply with good veterinary practice (GVP).
- *Animal feeding* may only be purchased from GMP-acknowledged companies.
- *Specific hygienic measures* with respect to plant design, personal hygiene, and transport must be taken to prevent microbial contamination and animal illness.
- *Independent audits/inspections* will be carried out to inspect performance of IKB.
- *Sanctions* will be executed if companies do not comply with IKB requirements.

For more information the reader is referred to http://bedrijfsnet.pve.agro.nl.

Various quality assurance initiatives

In other European countries too, specific quality assurance schemes, codes and guidelines have been developed to support the control and assurance of quality and safety of animal products for consumption, specifically addressing feed quality and safety and animal welfare. Some initiatives in the meat sector in the UK are the pigmeat quality assurance (PQA), assured British meat (ABM) assurance scheme, and quality meat Scotland assurance schemes - feed.

Companies in agribusiness and the food industry have to comply with the globally recognised standards and guidelines as well with the nationally recognised standards/schemes, which put high (and sometimes conflicting) demands on the company's own specific quality (food safety) management system.

7.6.4 Quality assurance standards for vegetables and fruits production

Also in the primary production of fresh produce (i.e. fresh vegetables, fruits, spices) various initiatives have been undertaken in Europe and worldwide to control and assure the quality and safety of fresh produce. Typical quality aspects of vegetable and fruit production relate to seed quality and disease resistance, in addition to product properties like colour, flavour, nutritional value and shelf life. Nowadays, there are also serious concerns about the safety of fresh produce (vegetables, fruits) (e.g. Matthews, 2006; Doyle and Erickson, 2008; FAO

2008). Fresh (minimally processed) produce are not usually subject to any heat treatment but can be easily contaminated by inadequate horticultural practices along the supply chain (e.g. use poor quality of irrigation water, inadequate pesticide use, contaminated soil, inadequate cleaning of equipment and environment, poor hygiene practices by food handlers) (e.g. Ingham *et al.,* 2004; Abadias *et al.,* 2008; Matthews, 2006; Beuchat and Mann, 2008). To illustrate this, several *Salmonella* outbreaks have been associated with fresh produce (e.g. Cordy *et al.,* 2005; US FDA, 2008).

GlobalGAP

In 1997 retailers started with the EurepGAP initiative (Euro-retailer produce working group Good Agricultural Practices) to create standards and procedures for good agricultural practices for the primary sector. EurepGAP has been modified to the so-called 'new GlobalGAP integrated farm assurance' or just IFA. It integrates all agricultural products into a single farm audit. Producers of different crops and livestock can now avoid multiple audits to meet various market and consumer requirements.

The standard serves as a global reference system for other existing standards and can also easily and directly be applied by all parties of the primary food sector. In other words, GlobalGAP operates like a satellite navigation system. It equips members with a reliable tool kit, which allows each partner in the supply chain to position themselves in a global market with respect to consumer requirements. The comprehensive documentation of the system is organised into five major blocks, each with a set of complementary elements. Users can select the applicable elements of each block to create a tailor-made, user-friendly manual. This selection gives direction to all relevant stages of the GlobalGAP certification process, including:
- system rules referred to as general regulations (GR);
- global GAP requirements referred to as control points and compliance criteria (CPCC);
- inspection documents referred to as checklists (CL);
- national GAP requirements referred to as approved national interpretation guidelines;
- harmonisation tools referred to as benchmarking cross reference checklist (BMCL) and other guidelines.

The ultimate objective of GlobalGAP is to grow into a world-wide acknowledged standard. Therefore it will comply as much as possible with other initiatives like the global food safety initiative (GFSI). For more information the reader is referred to http://www.globalgap.org.

Integrated quality assurance system (IKZ)

In this sector integrated quality assurance systems have been developed in the Netherlands (in Dutch: 'Integrale KwaliteitsZorg', IKZ). These systems are tailored for products e.g. IKZ sprout, IKZ apple. The systems aim at covering all steps in the whole production chain, with consumers as a point of departure. These consumer needs must be translated into requirements in the different units of the production chain, such as sowing of varieties, cultivation, harvesting, handling and packaging. Communication between units in the chain is an elementary aspect of IKZ. Furthermore, existing methods to assure quality are being integrated in IKZ systems, e.g.:

- *Seed quality control*, i.e. prior to marketing seeds varieties must be tested on purity and germinative capacity.
- *Practical value studies*, i.e. varieties are extensively tested on their illness resistance and product properties.
- *Quality attributes inspection* by auctioneers and, randomly, by other inspection authorities.
- *Residue analyses* are carried out in greenhouses to check their compliance with legal norms.
- *Established* storage and distribution *temperatures* for fruits and vegetables to improve shelf life.
- *Labelling* of packed fruits and vegetables to provide information about shelf life.

Also environmentally sound production is an important aspect of quality in vegetable and fruit production chains. Several projects have been initiated to stimulate sound production.

The *Early Warning & Response System* (EWRS) for crop protection is a system that is specifically aimed at managing the information supply about residues of crop protection agents (pesticides, fungicides, etc.). It can be used as a communication tool to and between companies. The EWRS was an initiative of the Dutch commodity board of fruits and vegetables 'Productschap Tuinbouw'. The system is aimed at:

- reducing the number of crops with residue levels that exceed the legislative norm;
- being able to obtain insight into the consequences of exceeding residue levels for human health at an early stage;
- preventing similar incidents by taking adequate measures at a sector level.

Moreover, the Dutch commodity board of fruits and vegetables has developed various hygiene codes and guidelines for developing *tracking and tracing systems* to support actors in the supply chain of fresh produce.

7.6.5 Quality assurance standards/guidelines in flower/plant sector

Flowers, plants, and indoor plants are also judged by consumers on their intrinsic and extrinsic quality attributes. Control and assurance of factors influencing final quality is thus of major importance in this sector as well, and various initiatives have been and are being taken to develop standards and guidelines for control and assurance.

Environmental project ornamental plant cultivation MPS

A Dutch quality standard has been developed for the horticultural sector, i.e. environmental project ornamental plant cultivation (MPS). MPS is basically a registration system but aims at reducing the use of pesticides, fertiliser and energy in ornamental plant production. Companies are classified depending on the extent to which they use pesticides, fertilisers and energy using the environmental scorecard (MPS-ABC). MPS works closely with MPS-ECAS (certification body) and MPS-HCS (advisory body).

Through this structure MPS profiles itself as the window for certification, knowledge, innovation and advice in the horticultural sector. The products/services of MPS are focused on entrepreneurs from the floriculture, bulb, nursery stock and vegetable sector. MPS is active in more than 50 countries.

MPS/VMS, internationally recognised environmental label

The 'Vlaams Milieuplan Sierteelt' (VMS) or Flemish environmental plan for ornamental horticulture) was set up in 1997 at the request of Flemish ornamental growers and exporters, in cooperation with the 'Nederlandse Milieu Programma Sierteelt' or Dutch organisation for sustainability in the horticultural sector (MPS, www.my-mps.com). Today, more than 150 Belgian ornamental nurseries and dealers belong to VMS-MPS. Their products can be recognised by the MPS sticker, which indicates that the required environmental standards were respected while the plant was being grown.

7.7 Quality assurance in practice

Quality assurance standards and guidelines have been widely adopted in agribusiness and the food industry (Trienekens and Zuurbier, 2008). It started in the eighties with good manufacturing practices codes. The codes provided guidelines that were aimed at assuring minimum acceptable standards and conditions for the processing and storage of food products to assure basic quality and safety. In the nineties the popularity of the ISO 9000 family increased. The 9000-series provided a framework for quality management and quality assurance. The specified quality system requirements could be used when a contract between two parties had to demonstrate that the supplier was able to prevent non-conformance at all its relevant stages (e.g. ranging from design to service). Since 1996 the application of HACCP principles has also increased rapidly because it was mandatory. Then, BRC (United Kingdom) and IFS (Germany, France) standards became popular. More recently the ISO 22000 standard has been embraced by the food industry, but it is not yet widely adopted.

Although the objective of a QMS (quality management system) or FSMS (food safety management system) is to guarantee quality and/or safety by standardised guidelines or principles, in practice the interpretation and implementation of quality assurance standards and guidelines into an internal QMS/FSMS appears to be distinctively different. For example, the requirements set in the ISO 9000 series can be elaborated differently by companies. They range from focus on establishment of procedures to focus on quality improvement. In the first situation, comprehensive procedures are made for all types of handling to control and assure quality. In the second situation, the attention is focused on improvement, whereas procedures are only prepared where control is essential for realising quality. In the latter case a more lean system can be achieved which aims at controlling and improving critical points with respect to quality.

Furthermore, in the use of PRP (PreRequisite Programmes) and HACCP principles for the development of a FSMS, different interpretations have been noticed in practice. Some companies have allocated many CCP's to be sure that safety of their products is properly controlled. Others believe that a high number of CCP reflects a poor condition of equipment and machines (insufficient PRP's). In their view technological innovations should be introduced first to improve the safety level, so reducing the number of CCP's. Moreover, international differences exist in the acknowledgement of HACCP certificates, which may hinder international trade.

Another concern is the actual performance of QMS and FSMS. Internal and external audits can be used to monitor the system, but they usually only check on compliance with the standards or requirements, giving less insight into the reasons behind poor performance. Various studies have identified and/or compiled barriers and failures in current systems (e.g. Panisello and Quantick, 2001; Luning and Marcelis, 2006), which initiated the need for systematic assessment of the internal FSM/QMS independent of the quality assurance standards/guidelines that have been implemented. Recently, a diagnostic instrument has been developed to systematically assess core control and assurance activities in a company's FSMS (Luning *et al.*, 2008, Luning *et al.*, 2009), combined with a microbial assessment scheme (MAS) to get insight into the actual microbial performance (Jacxsens *et al.*, 2009). This combined assessment gives detailed insight into which control and assurance activities and which contextual factors (e.g. organisational characteristics) need attention. The basic assumption behind the diagnostic instrument is that when a company operates in a more risky context, it requires a more sophisticated FSMS to comply with certain safety requirements. A FSMS should thus be tailored to the contextual situation.

Another bottleneck experienced by agribusiness and the food industry is the ease with which (new) requirements are set by government, retailers, and other stakeholders. The sets of requirements are not only different but they also change regularly (J. Kussaga, personal communication 2009).

In addition, it should be noted that a certificate is often considered as an ultimate object, the final quality level. This certificate is then used as tool to protect against external interference. It is used as a defence instrument instead of a consumer-oriented quality assurance tool.

From the techno-managerial perspective, QMS/FSMS should be developed on the basis of an in-depth technical and technological analysis of product, process and process equipment/facilities conditions to assess critical control points. Consensus must be reached within the company about the selection of critical control points. For this purpose, predictive modelling can be used to identify potential hazards and crucial quality parameters, whereas the Taguchi principles can be applied to determine well-considered tolerances. In addition, application of the HACCP principles and risk assessment helps in the selection of technological critical control points, which are scientifically underpinned. At the same time they provide a scientific basis for the non-critical points. It is important that identification of these critical points is executed in collaboration with the people involved in the relevant processes.

The restriction to critical points enables the development of a lean QMS or FSMS, which serves as a tool for the people involved instead of being an enforced system. Such a lean system consists of a hard core based on crucial control points and modules, which can be easily modified or extended depending on customer demands or changing environmental conditions.

From the techno-managerial approach a single focus on procedures is not advisable because it has its limits as regards directing human behaviour. The next chapter explains how the concept of total quality management (TQM) can overcome this shortcoming. Likewise, for quality assurance in the food supply chain, lean QMS/FSMS may be more suitable and more readily agreed upon by the partners involved.

Appendix 7.1. Hygiene issues that are covered in the 'Codex general principles of food hygiene' (summary of items from Codex Alimentarius, 2003)

Primary production (section III)

- Environmental hygiene
  Potential sources of contamination from environment should be considered. Primary food production should not be carried out in areas where presence of potentially harmful substances would lead to an unacceptable level of such substances in food.
- Hygienic production of food sources
  Potential effects of primary production activities on safety and suitability of food should be considered at all times. The HACCP-based approach may assist in the taking of such measures. Producers should as far as practicable implement measures to (1) control contamination from air, soil, water, feedstuffs, fertilisers (including natural fertilisers), pesticides, veterinary drugs or any other agent used in primary production; (2) control plant and animal health so that it does not pose a threat to human health through food consumption, or adversely affect the suitability of the product; and (3) protect food sources from faecal and other contamination. In particular, care should be taken to manage waste, and store harmful substances appropriately.
- Handling, storage, and transport
  Procedures should be in place to: (1) sort food and food ingredients to segregate material which is evidently unfit for human consumption, (2) dispose of any rejected material in a hygienic manner; and (3) protect food and food ingredients from contamination by pests, or by chemical, physical or microbiological contaminants or other objectionable substances during handling, storage and transport. Care should be taken to prevent, so far as reasonably practicable, deterioration and spoilage with appropriate measures which may include controlling temperature, humidity, and/or other controls.
- Cleaning, maintenance & personal hygiene
  Appropriate facilities and procedures should be in place to ensure that: (1) any necessary cleaning and maintenance is carried out effectively; and (2) an appropriate degree of personal hygiene is maintained.

Establishment: design and facilities (II)

- Location
  Establishments. Potential sources of contamination need to be considered when deciding where to locate food establishments, and the effectiveness of any reasonable measures that might be taken to protect food. Establishments should not be located anywhere where, after considering such protective measures, it is clear that there will remain a threat to food safety or suitability. Establishments should normally be located away from: (1) environmentally polluted areas and industrial activities which pose a serious threat of contaminating food, (2) areas subject to flooding unless sufficient safeguards are provided, (3) areas prone to infestations of pests, and (4) areas where waste, either solid or liquid, cannot be removed effectively.

- Equipment should be located so that it (1) permits adequate maintenance and cleaning, (2) functions in accordance with its intended use; and (3) facilitates good hygiene practices, including monitoring.
- Premises and rooms
  Design & lay-out. Where appropriate, internal design and layout of food establishments should permit good food hygiene practices, including protection against cross-contamination between and during operations by foodstuffs.
  Internal structures and fittings. Structures within food establishments should be soundly built of durable materials and be easy to maintain, clean and where appropriate, and able to be disinfected. In particular the following specific conditions should be satisfied where necessary to protect the safety and suitability of food:
  1. surfaces of walls, partitions and floors should be made of impervious materials with no toxic effect in intended use;
  2. walls and partitions should have a smooth surface up to a height appropriate to the operation;
  3. floors should be constructed to allow adequate drainage and cleaning;
  4. ceilings and overhead fixtures should be constructed and finished to minimise the build up of dirt and condensation, and the shedding of particles;
  5. windows should be easy to clean, be constructed to minimise the build up of dirt and where necessary, be fitted with removable and cleanable insect-proof screens. Where necessary, windows should be fixed;
  6. doors should have smooth, non-absorbent surfaces, and be easy to clean and, where necessary, disinfect;
  7. working surfaces that come into direct contact with food should be in sound condition, durable and easy to clean, maintain and disinfect. They should be made of smooth, non-absorbent materials, and inert to the food, to detergents and disinfectants under normal operating conditions

  Temporary/mobile premises and vending machines covered here include market stalls, mobile sales and street vending vehicles, temporary premises in which food is handled such as tents and marquees. They should be sited, designed and constructed to avoid, as far as reasonably practicable, contaminating food and harbouring pests. Any food hygiene hazards associated with such facilities should be adequately controlled to ensure the safety and suitability of food.
- Equipment
  General equipment and containers (other than once-only use containers and packaging) coming into contact with food, should be designed and constructed to ensure that, where necessary, they can be adequately cleaned, disinfected and maintained to avoid the contamination of food. Equipment and containers should be made of materials with no toxic effect in intended use. Where necessary, equipment should be durable and movable or capable of being disassembled to allow for maintenance, cleaning, disinfection, monitoring and, for example, to facilitate inspection for pests.
  Food control and monitoring. In addition to the general requirements, equipment used to cook, heat treat, cool, store or freeze food should be designed to achieve the required food temperatures as rapidly as necessary in the interests of food safety and suitability, and maintain them effectively. Such equipment should also be designed to allow temperatures to be monitored and controlled. Where necessary, such equipment should have an effective means of controlling and monitoring humidity, air-flow and

any other characteristic likely to have a detrimental effect on the safety or suitability of food. These requirements are intended to ensure that: (1) harmful or undesirable micro-organisms or their toxins are eliminated or reduced to safe levels or their survival and growth are effectively controlled, (2) where appropriate, critical limits established in HACCP-based plans can be monitored; and (3) temperatures and other conditions necessary to food safety and suitability can be rapidly achieved and maintained.
Containers for waste, by-products and inedible or dangerous substances, should be specifically identifiable, suitably constructed and, where appropriate, made of impervious material. Containers used to hold dangerous substances should be identified and, where appropriate, be lockable to prevent malicious or accidental contamination of food.

- Facilities
  Water supply. An adequate supply of potable water with appropriate facilities for its storage, distribution and temperature control, should be available whenever necessary to ensure the safety and suitability of food. Potable water should be as specified in the latest edition of WHO guidelines for drinking water quality, or water of a higher standard. Non-potable water (for use in, for example, fire control, steam production, refrigeration and other similar purposes where it would not contaminate food), shall have a separate system. Non-potable water systems shall be identified and shall not connect with, or allow reflux into, potable water systems.
  Drainage & waste disposal. Adequate drainage and waste disposal systems and facilities should be provided. They should be designed and constructed so that the risk of contaminating food or the potable water supply is avoided.
  Cleaning. Adequate facilities, suitably designated, should be provided for cleaning food, utensils and equipment. Such facilities should have an adequate supply of hot and cold potable water where appropriate.
  Personnel hygiene facilities and toilets should be available to ensure that an appropriate degree of personal hygiene can be maintained and to avoid contaminating food. Where appropriate, facilities should include: (1) adequate means of hygienically washing and drying hands, including wash basins and a supply of hot and cold (or suitably temperature controlled) water, (2) lavatories of appropriate hygienic design; and (3) adequate changing facilities for personnel. Such facilities should be suitably located and designated.
  Temperature control. Depending on the nature of the food operations undertaken, adequate facilities should be available for heating, cooling, cooking, refrigerating and freezing food, for storing refrigerated or frozen foods, monitoring food temperatures, and when necessary, controlling ambient temperatures to ensure the safety and suitability of food.
  Air quality and ventilation. Adequate means of natural or mechanical ventilation should be provided, in particular to: (1) minimise air-borne contamination of food, for example, from aerosols and condensation droplets, (2) control ambient temperatures, (3) control odours which might affect the suitability of food; and (4) control humidity, where necessary, to ensure the safety and suitability of food. Ventilation systems should be designed and constructed so that air does not flow from contaminated areas to clean areas and, where necessary, they can be adequately maintained and cleaned.
  Lighting. Adequate natural or artificial lighting should be provided to enable the undertaking to operate in a hygienic manner. Where necessary, lighting should not be such that the resulting colour is misleading. The intensity should be adequate to the

nature of the operation. Lighting fixtures should, where appropriate, be protected to ensure that food is not contaminated by breakages.
Storage. Where necessary, adequate facilities for the storage of food, ingredients and non-food chemicals (e.g. cleaning materials, lubricants, fuels) should be provided. Where appropriate, food storage facilities should be designed and constructed to: (1) permit adequate maintenance and cleaning, (2) avoid pest access and harbourage, (3) enable food to be effectively protected from contamination during storage, and (4) where necessary, provide an environment which minimises the deterioration of food (e.g. by temperature and humidity control). The type of storage facilities required will depend on the nature of the food. Where necessary, separate, secure storage facilities for cleaning materials and hazardous substances should be provided.

Control of operation (V)

- Control of hazards
  Food business operators should control food hazards through the use of systems such as HACCP. They should: (1) identify any steps in their operations which are critical to the safety of food, (2) implement effective control procedures at those steps, (3) monitor control procedures to ensure their continuing effectiveness; and (4) review control procedures periodically, and whenever the operations change. These systems should be applied throughout the food chain to control food hygiene throughout the shelf life of the product through proper product and process design. Control procedures may be simple, such as checking stock rotation calibrating equipment, or correctly loading refrigerated display units. In some cases a system based on expert advice, and involving documentation, may be appropriate (use annex HACCP).
- Key aspects of hygiene control systems
  Time and temperature control. Such controls include time and temperature of cooking, cooling, processing and storage. Systems should be in place to ensure that temperature is controlled effectively where it is critical to the safety and suitability of food. Temperature control systems should take into account: (1) the nature of the food, e.g. its water activity, pH, and likely initial level and types of micro-organisms, (2) the intended shelf life of the product, (3) the method of packaging and processing; and (4) how the product is intended to be used, e.g. further cooking/processing or ready-to-eat. Such systems should also specify tolerable limits for time and temperature variations. Temperature recording devices should be checked at regular intervals and tested for accuracy.
  Specific process steps. Other steps which contribute to food hygiene may include, for example: chilling, thermal processing, irradiation, drying, chemical preservation, vacuum or modified atmospheric packaging.
  Microbiological and other specifications. Where microbiological, chemical or physical specifications are used in any food control system, such specifications should be based on sound scientific principles and state, where appropriate, monitoring procedures, analytical methods and action limits.
  Microbiological cross-contamination. Raw, unprocessed food should be effectively separated, either physically or by time, from ready-to-eat foods, with effective intermediate cleaning and where appropriate disinfection. Access to processing areas may need to be restricted or controlled. Where risks are particularly high, access to

processing areas should only be via a changing facility. Personnel may need to be required to put on clean protective clothing including footwear and wash their hands before entering. Surfaces, utensils, equipment, fixtures and fittings should be thoroughly cleaned and where necessary disinfected after raw food, particularly meat and poultry, has been handled or processed.

Physical and chemical contamination. Systems should be in place to prevent contamination of foods by foreign bodies such as glass or metal shards from machinery, dust, harmful fumes and unwanted chemicals. In manufacturing and processing, suitable detection or screening devices should be used where necessary.

- Incoming material requirements

  No raw material or ingredient should be accepted by an establishment if it is known to contain parasites, undesirable micro-organisms, pesticides, veterinary drugs or toxic, decomposed or extraneous substances which would not be reduced to an acceptable level by normal sorting and/or processing. Where appropriate, specifications for raw materials should be identified and applied.

  Raw materials or ingredients should, where appropriate, be inspected and sorted before processing. Where necessary, laboratory tests should be made to establish fitness for use. Only sound, suitable raw materials or ingredients should be used.

  Stocks of raw materials and ingredients should be subject to effective stock rotation.
- Packaging

  Packaging design and materials should provide adequate protection for products to minimise contamination, prevent damage, and accommodate proper labelling. Packaging materials or gases where used must be non-toxic and not pose a threat to the safety and suitability of food under the specified conditions of storage and use. Where appropriate, reusable packaging should be suitably durable, easy to clean and, where necessary, disinfect.
- Water

  In contact with food. Only potable water, should be used in food handling and processing, with the following exceptions: (1) for steam production, fire control and other similar purposes not connected with food and (2) in certain food processes, e.g. chilling, and in food handling areas, provided this does not constitute a hazard to the safety and suitability of food (e.g. the use of clean sea water). Water re-circulated for reuse should be treated and maintained in such a way that no risk to the safety and suitability of food results from its use. The treatment process should be effectively monitored. Re-circulated water which has received no further treatment and water recovered from processing of food by evaporation or drying may be used, provided its use does not constitute a risk to the safety and suitability of food.

  As an ingredient. Potable water should be used wherever necessary to avoid food contamination.

  Ice and steam. Ice should be made from water that has been produced, handled and stored in a way that protects from contamination. Steam used in direct contact with food or food contact surfaces should not constitute a threat to the safety and suitability of food.
- Management and supervision

  The type of control and supervision needed will depend on the size of the business, the nature of its activities and the types of food involved. Managers and supervisors should have enough knowledge of food hygiene principles and practices to be able to judge potential risks, take appropriate preventive and corrective action, and ensure that effective monitoring and supervision takes place.

- Documentation and records
  Where necessary, appropriate records of processing, production and distribution should be kept and retained for a period that exceeds the shelf life of the product. Documentation can enhance the credibility and effectiveness of the food safety control system.
- Recall procedures
  Managers should ensure effective procedures are in place to deal with any food safety hazard and to enable the complete, rapid recall of any implicated lot of the finished food from the market. Where a product has been withdrawn because of an immediate health hazard, other products which are produced under similar conditions, and which may present a similar hazard to public health, should be evaluated for safety and may need to be withdrawn. The need for public warnings should be considered. Recalled products should be held under supervision until they are destroyed, used for purposes other than human consumption, determined to be safe for human consumption, or reprocessed in a manner to ensure their safety.

Establishment: maintenance and sanitation

- Maintenance and cleaning
  General. Establishments and equipment should be kept in an appropriate state of repair and condition to: (1) facilitate all sanitation procedures, (2) function as intended, particularly at critical steps, (3) prevent contamination of food, e.g. from metal shards, flaking plaster, debris and chemicals. Cleaning should remove food residues and dirt which may be a source of contamination. The necessary cleaning methods and materials will depend on the nature of the food business. Disinfection may be necessary after cleaning. Cleaning chemicals should be handled and used carefully and in accordance with manufacturers' instructions and stored, where necessary, separated from food, in clearly identified containers to avoid the risk of contaminating food.
  Cleaning procedures and methods. Cleaning can be carried out by separate or combined use of physical methods, such as heat, scrubbing, turbulent flow, vacuum cleaning or other methods that avoid the use of water, and chemical methods using detergents, alkalis or acids. Cleaning procedures will involve, where appropriate: (1) removing gross debris from surfaces; (2) applying a detergent solution to loosen soil and bacterial film and hold them in solution or suspension, (3) rinsing with water which complies with section 4, to remove loosened soil and residues of detergent, (4) dry cleaning or other appropriate methods for removing and collecting residues and debris; and (5) where necessary, disinfection.
- Cleaning programmes
  Cleaning and disinfection programmes should ensure that all parts of the establishment are appropriately clean, and should include the cleaning of cleaning equipment. The programmes should be continually and effectively monitored for their suitability and effectiveness and where necessary, documented. Where written cleaning programmes are used, they should specify: (1) areas, items of equipment and utensils to be cleaned, (2) responsibility for particular tasks, (3) method and frequency of cleaning, and (4)

monitoring arrangements. Where appropriate, programmes should be drawn up in consultation with relevant specialist expert advisors.

- Pest control systems
  General. Pests pose a major threat to the safety and suitability of food. Pest infestations can occur where there are breeding sites and a supply of food. Good hygiene practices should be employed to avoid creating an environment conducive to pests. Good sanitation, inspection of incoming materials and good monitoring can minimise the likelihood of infestation and thereby limit the need for pesticides (Stemerding and Souleymane, 2003).
  Preventive access. Buildings should be kept in good repair and condition to prevent pest access and to eliminate potential breeding sites. Holes, drains and other places where pests are likely to gain access should be kept sealed. Wire mesh screens, for example on open windows, doors and ventilators, will reduce the problem of pest entry. Animals should, wherever possible, be excluded from the grounds of factories and food processing plants.
  Harbourage and infestation. Potential food sources should be stored in pest-proof containers and/or stacked above the ground and away from walls. Areas both inside and outside food premises should be kept clean. Where appropriate, refuse should be stored in covered, pest-proof containers.
  Monitoring and detection. Establishments and surrounding areas should be regularly examined for evidence of infestation.
  Eradication. Pest infestations should be dealt with immediately and without adversely affecting food safety or suitability. Treatment with chemical, physical or biological agents should be carried out without posing a threat to the safety or suitability of food.
- Waste management
  Suitable provision must be made for the removal and storage of waste. Waste must not be allowed to accumulate in food handling, food storage, and other working areas and the adjoining environment except so far as is unavoidable for the proper functioning of the business. Waste stores must be kept appropriately clean.
- Monitoring effectiveness
  Sanitation systems should be monitored for effectiveness, periodically verified by means such as audit pre-operational inspections or, where appropriate, microbiological sampling of environment and food contact surfaces and regularly reviewed and adapted to reflect changed circumstances.

Establishment: personal hygiene (VII)

- Health status
  Personnel known, or suspected, to be suffering from, or to be a carrier of a disease or illness likely to be transmitted through food, should not be allowed to enter any food handling area if there is a likelihood of their contaminating food. Any person so affected should immediately report illness or symptoms of illness to the management. Medical examination of a food handler should be carried out if clinically or epidemiologically indicated.
- Illness and injuries
  Conditions which should be reported to management so that any need for medical examination and/or possible exclusion from food handling can be considered include:

jaundice, diarrhoea, vomiting, fever, sore throat with fever, visibly infected skin lesions (boils, cuts, etc.), discharges from the ear, eye or nose.

- Personal cleanliness
  Food handlers should maintain a high degree of personal cleanliness and, where appropriate, wear suitable protective clothing, head covering, and footwear. Cuts and wounds, where personnel are permitted to continue working, should be covered by suitable waterproof dressings. Personnel should always wash their hands when personal cleanliness may affect food safety, for example: (1) at the start of food handling activities, (2) immediately after using the toilet; and (3) after handling raw food or any contaminated material, where this could result in contamination of other food items; they should avoid handling ready-to-eat food, where appropriate.
- Personal behaviour
  People engaged in food handling activities should refrain from behaviour which could result in contamination of food, for example: smoking, spitting, chewing or eating; sneezing or coughing over unprotected food. Personal effects such as jewellery, watches, pins or other items should not be worn or brought into food handling areas if they pose a threat to the safety and suitability of food.
- Visitors
  Visitors to food manufacturing, processing or handling areas should, where appropriate, wear protective clothing and adhere to the other personal hygiene provisions in this section.

Transportation (VIII)

- General
  Food must be adequately protected during transport. The type of conveyances or containers required depends on the nature of the food and the conditions under which it has to be transported.
- Requirements
  Where necessary, conveyances and bulk containers should be designed and constructed so that they: (1) do not contaminate foods or packaging, (2) can be effectively cleaned and, where necessary, disinfected, (3) permit effective separation of different foods or foods from non-food items where necessary during transport, (4) provide effective protection from contamination, including dust and fumes, (5) can effectively maintain the temperature, humidity, atmosphere and other conditions necessary to protect food from harmful or undesirable microbial growth and deterioration likely to render it unsuitable for consumption, and (6) allow any necessary temperature, humidity and other conditions to be checked.
- Use and maintenance
  Conveyances and containers for transporting food should be kept in an appropriate state of cleanliness, repair and condition. Where the same conveyance or container is used for transporting different foods, or non-foods, effective cleaning and, where necessary, disinfection should take place between loads. Where appropriate, particularly in bulk transport, containers and conveyances should be designated and marked for food use only and be used only for that purpose.

Product information and consumer awareness (IX)

- Lot identification
  Lot identification is essential in product recall and also helps effective stock rotation. Each container of food should be permanently marked to identify the producer and the lot. Codex general standard for the labelling of pre-packaged foods (CODEX STAN 1-1985) applies.
- Product information
  All food products should be accompanied by or bear adequate information to enable the next person in the food chain to handle, display, store and prepare and use the product safely and correctly.
- Labelling
  Pre-packaged foods should be labelled with clear instructions to enable the next person in the food chain to handle, display, store and use the product safely. Codex general standard for the labelling of pre-packaged foods (CODEX STAN 1-1985) applies.
- Consumer education
  Health education programmes should cover general food hygiene. Such programmes should enable consumers to understand the importance of any product information and to follow any instructions accompanying products, and make informed choices. In particular consumers should be informed of the relationship between time/temperature control and foodborne illness.

Training (X)

- Awareness & responsibilities
  Food hygiene training is fundamentally important. All personnel should be aware of their role and responsibility in protecting food from contamination or deterioration. Food handlers should have the necessary knowledge and skills to enable them to handle food hygienically. Those who handle strong cleaning chemicals or other potentially hazardous chemicals should be instructed in safe handling techniques.
- Training programmes
  Factors to take into account in assessing the level of training required include: (1) the nature of the food, in particular its ability to sustain growth of pathogenic or spoilage micro-organisms, (2) the manner in which the food is handled and packed, including the probability of contamination, (3) the extent and nature of processing or further preparation before final consumption, (4) the conditions under which the food will be stored, and (5) the expected length of time before consumption.
- Instruction and supervision
  Periodic assessments of the effectiveness of training and instruction programmes should be made, as well as routine supervision and checks to ensure that procedures are being carried out effectively. Managers and supervisors of food processes should have the necessary knowledge of food hygiene principles and practices to be able to judge potential risks and take the necessary action to remedy deficiencies.
- Refresher training
  Training programmes should be routinely reviewed and updated where necessary. Systems should be in place to ensure that food handlers remain aware of all procedures necessary to maintain the safety and suitability of food.

Appendix 7.2. Overview of prerequisite programmes (compiled from Jacxsens *et al.*, 2009, Escriche *et al.*, 2006)

PRP

- Cleaning & disinfection
  Cleaning and disinfection are crucial for prevention of cross contamination via machines, equipment, tools, cutting boards, conveyor belts, utensils, etc. Cleaning and disinfection is based on the following basic principles:
  - use of authorised products for food industry;
  - suitable for type of dirt;
  - applied at recommended dose;
  - following recommended procedure.
- Pest control
  Presence of insects, rodents and other animals constitutes a serious threat to food safety.
  - The company building design is one of the main barriers to pest entrance.
  - Another aspect is adequate control of food storage and food residues to prevent animals getting food.
  - Outside of building free of dense vegetation that could be a pest source.
- Water & air quality
  Water from unsafe sources has been the cause of enteric infections. Water for food use should therefore comply with microbiological and chemical criteria. Bacteriological quality should be tested on a regular basis, for chemical analysis a certificate from the water supplier is sufficient. Non-potable water (used for e.g. steam) should not come into contact with food.
  Air flow can provide a route for micro-organisms and can contribute to the dissemination through the company.
  - Rooms for aseptic filling require devices for sterile filtration of air.
  - Ventilation must be controlled (and regularly cleaned) to prevent large differences between air and product temperature.
- Temperature control & registration
  Temperature and moisture are critical factors influencing growth of micro-organisms (including psychotropic ones that can grow close to 0 °C).
  - Cold rooms must be carefully monitored (preferably automatically with alarm).
  - Capacity of cooling and heating equipment should not be exceeded.
  - Chilled transport must be equipment with thermometers for registration.
  - Product temperature should be measured on regular basis.
  - Temperature measuring equipment should be regularly calibrated.
- Personnel
  Food handlers form a potential risk for transferring pathogens to food.
  - They can contaminate products by poor food handling (contact with raw food and subsequently processed food).
  - They can contaminate food by being a source of contamination themselves due to illness (bacterial and or viral infections.
  - They can become ill due to working with infected food (typically zoonosis).

Important measures are:
- health inspection;
- hygienic handling of food (special care for raw and cooked products);
- personal hygiene (hand washing, clothing, hair covers, etc.);
- training of personnel (prevention of bad habits, poor routine handling).

- Structure & infrastructure
  The circumstances for food production are very important for safety.
  - With respect to *location and surrounding* area one should consider possibility of potential sources of contamination, sufficiency/quality water supply, waste removal, availability of transport, climate, actual location of company, etc.
  - With respect to *buildings* one should consider aspects like: special rooms, areas and processes (e.g. low risk zone, high care zone, etc.), structural components (like smooth no-slip floor surfaces, smooth walls, ceilings where dirt and dust cannot accumulate, windows (as few as possible), tight-fitting doors, smooth pipelines, etc.).
  - With respect to *production environment* one should consider aspects like lighting, effluent disposal, storage depots and recipients, storage of toxic materials.
  - With respect to *sanitary facilities* one should consider aspects like clothes-changing facilities, toilets, and hand-washing facilities).
  - With respect to *equipment* one should consider aspects like cleanability, food contact materials (non-toxic, durable, resistant to corrosion, etc.), devices for monitoring connected to equipment, installation and lay-out of equipment.
- Technical maintenance and calibration
  In order for food equipment to be used safely it needs to be inspected and maintained on a regular basis.
  - A maintenance plan should be developed to repair defect machines and equipment, and to prevent defects.
  - Food grade lubricants must be used.
  - Calibration programme/schedule for measuring equipment must be developed.
- Waste management
  Waste, and in particular animal waste, often contains a large number of spoilage micro-organisms and possibly pathogens. The following aspects must be considered:
  - different types of waste (chemical, paper, food waste);
  - how different types are collected (which recipients, colour of containers);
  - who is responsible for removal of waste during and after production activities;
  - storage of different types of waste;
  - removal of different types of waste (contracts with specialised companies).
- Control of raw material
  Poor raw materials can lead to a serious risk of contamination. Raw materials refer to those used in food production, auxiliary materials (like colouring agents), and packaging. Important aspects here are purchasing specifications, supplier evaluation, receipt control (how to deal with defects) and storage
- Traceability, recall, returned goods, rejections/no conform products
  Taking poor quality (unsafe) products off the market is important for food safety (and quality) as well.
  Every company must have a procedure for:
  - recalls (in case of severe calamities);
  - returned goods (returned out the market due to complaints, other reasons);

  - non-conformance products (not complying with specifications within own process).
- Allergens
  Since an increasing group of people appears to be allergic to various compounds, also a general allergy policy should be outlined. It should be determined:
  - which allergens can be associated with the produced products;
  - what are the consequences for raw material selection, if allergens are kept out of the company;
  - which precautionary measures are taken to prevent cross contamination, if allergens are present in the production process.
- Physical and chemical contamination
  In this PRP it is described how contamination with *physical* hazards is controlled in the company. A procedure for breakage of glass or hard plastic should also be included. Likewise for *chemical* hazards; clear policy (use of food-grade chemicals only).
- Management of product information
  The methods of preparation are company specific but are important for food safety and quality. Aspects that need to be considered here are:
  - establishing of handling methods of preparation;
  - label information;
  - final product specifications.
- Work methodology (procedures)
  Working in a hygienic way is crucial for food safety. Working instructions should be straightforward and simple and must:
  - indicate what is expected from the personnel in a certain production place;
  - ensure that different employees work and carry out checks in the same way.

Appendix 7.3. Overview of items for which requirements are described in global standard for food safety (GSFS) from the British Retail Consortium (BRC)

*Senior management commitment and continual improvement**

This section describes requirements for the *demonstration* of *commitment* to implement GSFS requirements by senior management, by providing *adequate resources* (human and financial), *effective communication, systems of review* and actions taken to effect *continuous improvement*, which must be implemented and fully documented.

*The food safety plan HACCP**

This section describes requirements for a systematic, comprehensive, thorough and fully implemented and maintained food safety plan based on HACCP principles. It explicitly addresses requirements (e.g. *what* to specify and that it must be *documented*) for:

- forming an HACCP food safety management team;
- describing product description;
- constructing and verifying flow diagram;
- listing potential hazards, analysing hazards, and considering measures to control identified hazards;
- determining CCP´s and establishing critical limits;
- establishing monitoring system and corrective action plan;
- establishing verification procedures;
- documenting and record keeping;
- reviewing the HACCP plan.

Food safety and quality management

This section describes requirements for the development and documentation of a food safety and quality policy, authorised, reviewed and signed and dated by an appropriate senior manager. It explicitly addresses requirements for:

- food safety and quality manual;
- organisational structure, responsibilities and management authority (responsible persons affecting safety-quality, their relationships, etc.);
- contract review and customer focus (identification of customer requirements, expectations, etc.);
- *internal audit** (what and how to audit compliance with GSFS requirements);
- purchasing supplier approval and performance monitoring (control of purchasing processes critical to safety/quality);
- general documentation requirement (about control of documentation, setting specifications, record completion and maintenance);

---

* Items indicated with a * are called fundamental requirements, which means that these requirements relate to systems that are crucial to the establishment and operation of an effective food quality and safety operation.

- *corrective and preventive action plan** (about procedures to record, investigate, analyse and correct non-conformity with standards);
- *traceability** (about a system to identify and trace product lots through all raw materials, processing stages, and distribution);
- complaints handling (about how to handle complaints effectively);
- management of incidents, withdrawals, recalls (about how to effectively manage these aspects).

Site standards

This section describes requirements on external and internal environmental aspects of the site (factory, company). It explicitly addresses requirements for:
- external standards (about outside environment of the site);
- security (about prevention of unauthorised persons entering production/storage areas);
- internal standards (about procedures to minimise risk of contamination on premises and in plant);
- lay out, product flow and segregation*;
- building fabric, walls, floors, ceiling, lay-overs, windows, lightning, air conditioning-ventilation;
- utilities and equipment (should be designed, constructed, and maintained to control risk of product contamination);
- maintenance (about maintenance of all equipment and plant critical to safety/quality);
- staff facilities (about accommodation so that personnel and staff can operate with minimum of risk of product contamination), e.g. hand-washing facilities, toilets, clothing, etc.;
- chemical and physical contamination control (about procedures to control the risks of chemical or physical contamination);
- *housekeeping and hygiene** (about housekeeping and cleaning systems to reduce the risk of contamination), e.g. requirements on cleaning procedures, personnel to carry out cleaning, chemicals and equipment for cleaning, verification of effectiveness of cleaning, revalidation of effectiveness;
- waste disposal (about collecting and disposing of waste);
- pest control (about preventing risk of pest infestation);
- storage and transport (about facilities used for transportation of products, movement around the site, and dispatch of finished products).

Product control

This section describes requirements for typical issues related to the product. It explicitly addresses requirements on:
- product design and development (about design procedures to ensure that manufacturing processes are capable of producing safe and compliant products);
- *handling specific materials, materials containing allergens and identity of preserved materials** (about special procedures for handling specific materials to ensure product quality;
- foreign body detection (about appropriate and effective equipment for detecting foreign bodies);

- product packaging (about appropriateness of packaging for intended use);
- product inspection and laboratory testing (about appropriate inspection and analyses critical to confirm product safety, legality and quality requirements);
- control of compliant and non-compliant product (about identification, labelling and quarantining of compliant and non-compliant products).

Process control

This section describes requirements for demonstrating the effective control of all operations. It explicitly addresses requirements for:
- *control of operations** (about verifying that processes and equipment are capable of producing consistently safe and compliant product with desirable quality characteristics);
- quantity, weight, volume and number control (about quantity control systems to comply with legal requirements and sector specific codes or consumer requirements);
- calibration and control of measuring and monitoring devices (about identification and calibration (based on recognised international standards) of monitoring equipment.

Personnel

This section describes requirements for competences, skills, safety practices, etc., of personnel performing work that could affect product safety, legality and quality. It explicitly addresses requirements on:
- *training** (about training of people to enable them to perform their tasks properly, which may affect product safety, legality and quality);
- personal hygiene (to reduce risk of contamination from poor hygiene behaviour);
- medical screening;
- protective clothing (to reduce risk of contamination from poor hygiene behaviour).

Chapter 8

Quality policy and strategy

Quality policy and strategy is the food quality management function that structurally embeds food quality management in the company's strategy and environment. The policy and strategy development process will be described in section 8.2. Then the focus of this chapter will be on quality policy. There will be an elaboration of its influencing factors and content with special attention to the policy of total quality management (sections 8.3 and 8.4). In order to verify the performance of quality policy, sets of criteria have been developed. These sets are enclosed in different awards and prizes as explained in section 8.5. Then strategic collaboration will be analysed (section 8.6), and finally practical aspects of quality policy will be discussed (section 8.7).

8.1 Introduction

Customers today are becoming increasingly critical about quality. Moreover competition is much more sophisticated. The following trends are on-going and result in changing market strategies (Johnson and Scholes, 2006; Hellriegel *et al.*, 2005; Noori and Radford, 1995):

- Rapid advances in technology and in biological and physical sciences are providing industries with new materials, new products and new market needs.
- Escalation of new product introduction is shortening the life span of most products.
- Continuous reductions in transportation and communication costs are increasing global competition.
- Global competition is intensifying as countries are moving to eliminate protectionism and create large free-trade zones.
- Markets in general are fragmenting, not expanding. Customers are demanding high-quality products, at a reasonable price, that meet their individual needs. Unsatisfied customers are switching from one product to another.
- Growing international concern about the environment is being translated into stricter legislation. Conventional energy supplies are not limitless, nor is the world's ability to absorb the disposal of products and process by-products.

Being proactive on developments in the environment, the task of top management is to formulate a vision, long-term goals, and strategies for how to achieve these goals. In management today, quality issues are of great significance in this process. To make this clear many companies have formulated a specific quality policy. In other companies the quality values are completely integrated into other strategic documents. However, what is important is that quality values are firmly established in the company's strategy process (Bergman and Klefsjö, 1994).

Strategy refers to the determination of a mission and long-term objectives of a company and the adoption of courses of action and allocation of resources necessary to achieve these aims. Therefore, objectives are a part of strategy formulation.

Policies are general statements or understandings that guide managers' thinking in decision-making. They ensure that decisions fall within certain boundaries. They do not usually require action but are intended to guide managers in their commitment to the decision they ultimately make.

In Figure 8.1 the quality policy and strategy function is described in its main elements. It includes a decision-making process wherein strategic quality objectives are translated into quality policy statements that serve as boundaries and guidelines for decision-making in quality by determining long-term food quality goals and objectives, and how to achieve them by the quality system. It includes determining quality levels of products, and technological and human resources, and the type of quality system.

Strategic decision-making is complicated by different factors. While decision-making heavily depends on the prediction of future developments, it is typically in the strategic area in which prediction is very difficult. Both food products and food production processes are making rapid technological advancements, and despite all efforts in scientific research future technological developments are not really clear, partly due to the technological complexity (e.g. novel foods with health claims). In addition, the food supply chain as a whole is quite unpredictable. Take, for example, developments in the international trade of raw materials and final products, national and international legislation, and competition in the market. And finally, one has to deal with restrictive policies in areas such as environmental protection, animal welfare, and international political developments.

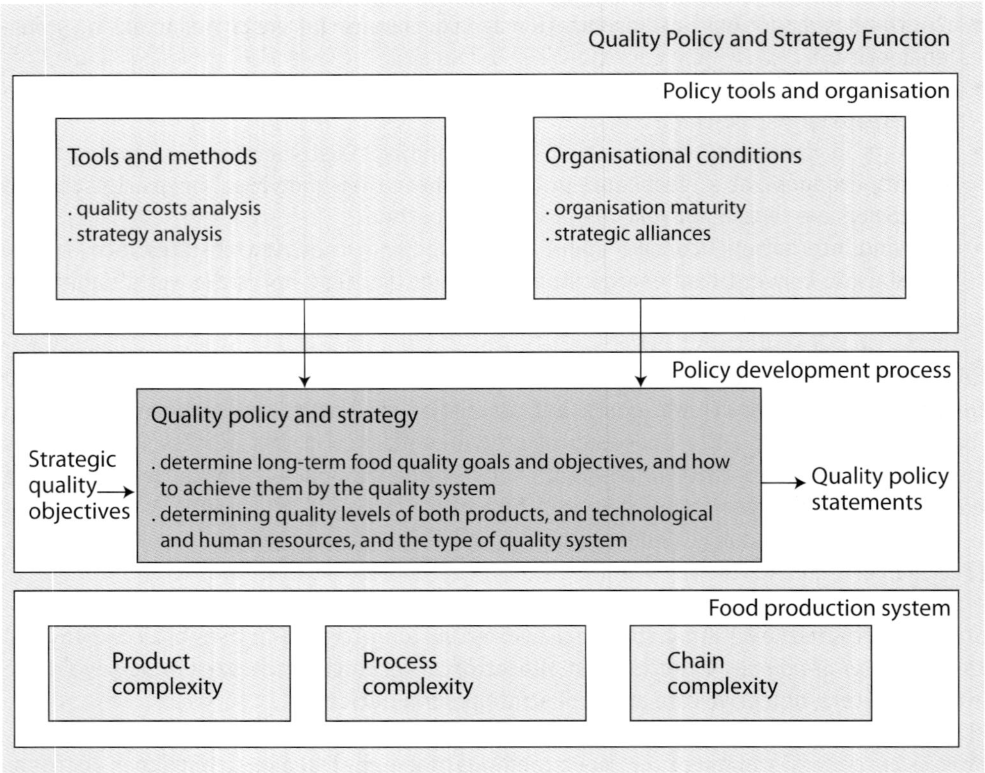

Figure 8.1. Quality policy and strategy function.

Typical tools, methods, and organisational conditions support effective decision-making in the quality strategy and policy process, like the SWOT analysis and costs analysis. They will be described within the topics in the following sections.

8.2 The quality policy and strategy process

Quality policy and strategy is an important food quality management function because it relates the quality objectives to the goals and strategy of the company. The importance of a quality policy and clearly established quality values in the company's strategy process is increasingly being acknowledged (Bergman and Klefsjö, 1994; Hellsten and Klefsjö, 2000).

> *The core mechanism of quality policy and strategy is to relate quality objectives to the goals and strategy of the company by setting basic requirements on product and resource quality levels and on the quality management system, and by translating them into policy statements that can serve as boundaries and guidelines for decision-making in quality.*

In this section the steps in the quality policy and strategy process, as well as the decisions involved, are described.

8.2.1 Quality policy and strategy activities

Both policy and strategy are developed within the strategic management process. The essence of policy is vision, whereas strategy concerns the direction in which human and material resources will be applied in order to achieve selected objectives (Weihrich and Koontz, 1993).

Strategic management is the process of formulating and implementing strategies to accomplish an organisation's mission and objectives and to ensure competitive advantage. The essence of strategic management is to look ahead, understand the environment, and effectively position an organisation for competitive success in changing times (Schermerhorn, 2007). The strategic management process entails both formulation activities, which are the major focus of planning, and implementation activities. These activities will actually realise the devised strategy (Johnson and Scholes, 2006, Schermerhorn, 2007). Both activities are shown in Figure 8.2 and will be discussed in more detail.

The first step in strategic management is the careful assessment and classification of organisational mission and goals.

The *mission* or purpose of any organisation may be described as its reason for existence. In today's quality-conscious and highly competitive environment, the sense of purpose must be directly centred on serving the needs of customers or clients. After all, their satisfaction and continued support are the ultimate keys to organisational survival. In general the firm's mission reflects information such as, what types of products or services it produces, who its customers tend to be, and what important values it holds. Values are broad beliefs about what is or is not appropriate. The corporate culture is the predominant value system for the

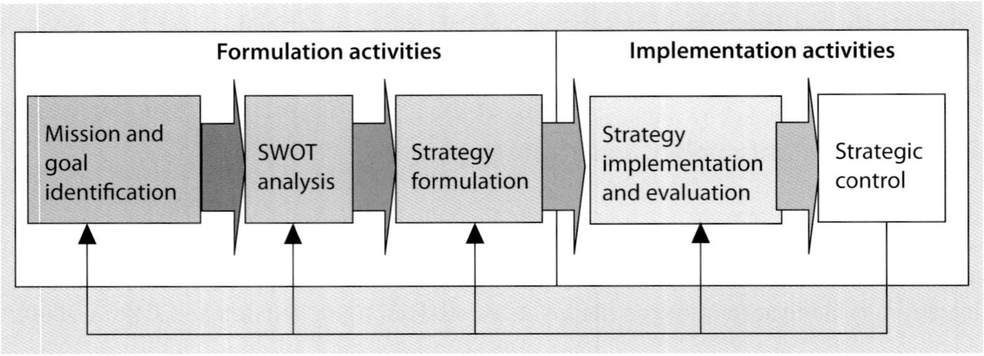

Figure 8.2. Strategic management process.

organisation as a whole. In strategic management the presence of strong core values helps to build corporate identity and supports the mission statement.

Goals direct activities toward key and specific results. Typical goals of a firm are profitability, market share, product quality level, and social responsibility.

The second step is the evaluation of the firm's internal strengths and weaknesses and opportunities and threats (SWOT) associated with the firm's environment. The SWOT analysis emphasises that the fit between firm and its environment is of paramount importance. A firm's strategy should be built around this match. A *strength* is an ability or attribute, which gives the firm a distinctive competence, which may be reflected in for example the firm's product quality offered or cost advantages due to the firm's overall quality performance. Conversely, a *weakness* is an attribute or skill that a firm lacks or has not developed and at which it performs poorly. An *opportunity* is an environmental circumstance that is potentially beneficial for a firm, such as convenience food due to the fact that households want to spend less time on cooking. A *threat* on the other hand, is an environmental factor that could be potentially harmful to the firm, such as a law banning the use of battery housing systems for chickens.

The third step is to formulate the *strategy*. As a rule, a firm will formulate its strategy on the basis of its greatest strengths and to keep its weaknesses from being exploited by competitors and other factors in the environment. Opportunities and threats are often the same for all firms in an industry. However, a successful strategy will better exploit the opportunities and limit the business environmental threats. One can use two criteria to select potential strategies, i.e. its content and the procedure followed. The *content criteria* consist of elements such as:

- Suitability: to what extent does the strategy meet the strategic problem and objectives.
- Feasibility: to what extent is the strategy economically and technically feasible.
- Acceptability: to what extent are relevant partners, departments, management and workers willing to commit to the strategy.

Procedural criteria relate to the way in which the selection of strategies is carried out. In other words the conditions under which the decisions are made, such as knowledge of those involved, availability of information, and authority of decision-makers (Johnson and Scholes, 2006).

Then, the strategy has to be *implemented*. This requires the involvement of the entire organisation. It means initiating and directing activities to ensure that all activities are focused on implementing the firm's strategy. Finally, strategic control is the feedback mechanism in the strategic management process. It compares the firm's actual performance to its intended performance. Strategic control enables evaluation of the way in which the strategy was planned, formulated and implemented. As a result, the firm may find that a potentially successful strategy was formulated well but implemented poorly, and *vice versa*.

8.2.2 Quality policy and strategy decisions

The policy and strategy development process is a decision-making process in which an analysis of the environment as well as the competencies of the company provides alternative strategic choice possibilities.

The food quality decisions grid (Table 3.1) shows which decisions can be seen as the result of quality policy and strategy activities, i.e. decisions on:

- Quality category of food products (influencing food dynamics), e.g. choosing branded products (which require a high quality level), or choosing differentiated products for a certain market segment like food products for medical purposes.
- Competence level of processes, equipment and buildings (influencing technological conditions), e.g. the decision to invest only in hygienic designed equipment.
- Quality category of people's actions (influencing human dynamics), e.g. the decision to accept qualified people only (with certain diplomas) for production work.
- Competence level of people, organisation, information system (influencing administrative conditions), e.g. the decision to invest in training, or the decision to decentralise quality responsibilities in the organisation.

These decisions are prominent in quality decision-making in the entire organisation, when they are communicated well and supported by a clear demonstration of quality commitment from top management.

8.2.3 SWOT analysis and strategic alternatives

The importance of step 2, the SWOT analysis, in the strategic management process, becomes clear, when the literature is analysed. It offers a range of strategic analysis models and techniques that can be applied to analyse the firm's position in its environment. Besides general techniques for analysing business environment (like PEST analysis: political, economical, social and technical trends) and general techniques for market and consumer research, several forms of analysis are frequently used in the SWOT analysis:

1. the portfolio analysis for determining the firm's product position in the market;
2. the five forces analysis for determining the firm's competitive position;
3. the value chain analysis; and
4. the core competence analysis, both for determining the relative strength of internal resources.

The *portfolio analysis* (Boddy and Paton, 1998), developed by the Boston Consulting Group (BCG), is a useful model for examining the market share of individual products or services in relation to growth rate in their particular markets (Figure 8.3).

The model's assumptions are underpinned by the concept of the *experience curve,* which holds that companies already in a particular market for a significant period of time have learned from experience how to produce and deliver their product or service more efficiently and effectively. They are likely to have built up a strong market share, and on the basis of both the associated economies of scale and their 'experience', to have achieved relatively low unit costs. Thus the matrix trends to associate a high market share with an ability to generate large, positive cash flows. The matrix descriptors are explained as follows:

- *Cash cows* have high market share in a mature market. They occupy a leading position with little need for investment in production capacity or marketing. They are likely to be cash generators which can be 'milked', contributing to the development of future stars.
- *Stars* have high market share within a growing market, but expenditure may be high so that the product or service both generates and uses cash. They are likely to be the cash cows of the future.
- *Question marks,* are also found in growing markets, but have a low market share. They are typically products or services that are at an early stage in life requiring substantial amounts of cash to be spent in an effort to increase market share. However, the payback from high spending is uncertain.
- *Dogs* experience low market share in low-growth markets, and are likely to drain the organisation of cash by using a disproportionate amount of management time.

		Competitive position (market share)	
		High	Low
Market growth	High	Stars (develop)	Question marks (investigate)
	Low	Cash cows (milk)	Dogs (divest)

Business type	Profitability	Growth opportunities	Typical net cash flow
Star	High	Yes	Balanced to negative
Cash cow	High	Limited	Positive
Dog	Low	Limited	Balanced to negative
Question mark	Low	Yes	Very negative

Figure 8.3. The Boston portfolio matrix.

A portfolio of products and/or services, which is balanced in terms of this matrix, is crucial to the survival of the business because investment for growth must be internally generated, which is the role of the cash cow. The star and perhaps the question mark are essential to future success. Although the dog is often regarded as a candidate for speedy elimination, Johnson and Scholes (2006) warned against immediate action without further consideration. A product should be considered, in the light of whether it is a necessary element in a complete product range, or whether dropping a service will be politically acceptable. In other words, there might be a good reason to retain a loss-making product or service.

The *five forces model* of Porter (Porter, 1980; Dessler, 2001) is a well-known tool for analysing the firm's competitive position. Essentially, Porter's model outlines the primary forces that determine competitiveness within an industry and illustrates how those forces are related. The model suggests that in order to develop effective organisational strategies, managers must understand and react to these forces within that industry. Porter's model is presented in Figure 8.4. According to the model, the following factors determine competitiveness within an industry:

- products that might act as a substitute for goods or services that companies within the industry produce;
- new companies entering the market threatening the market share of the company;
- ability of suppliers to control issues like costs of materials that industry companies use to manufacture their products;

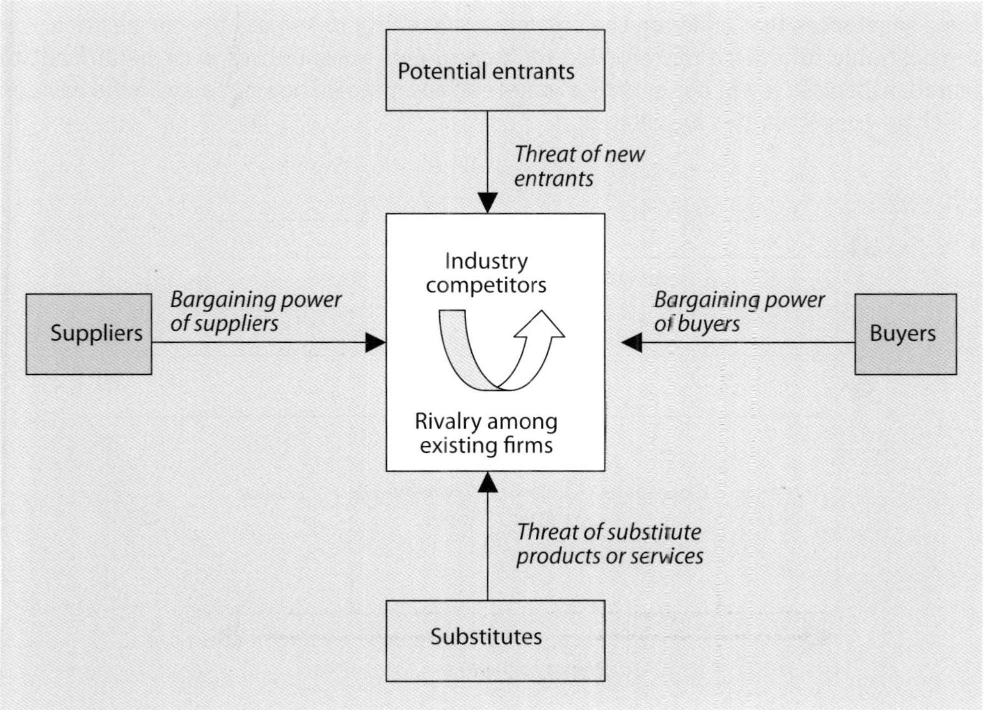

Figure 8.4. Forces driving industry competition.

- bargaining power that buyers possess within the industry;
- general level of rivalry or competition among firms within the industry.

According to the model, buyers, product substitutes, suppliers, and potential new companies within an industry all contribute to the level of rivalry among industry firms.

The *concept of the value chain*, introduced by Porter (1985), is derived from an established accounting practice of calculating the value added to a product by individual stages in a manufacturing processes (Figure 8.5). Porter applied this idea to the activities of an organisation as a whole, arguing that it is necessary to examine activities separately in order to identify sources of competitive advantage. The usefulness of value chain analysis is that it recognises that individual activities in the overall production process play a part in determining e.g. cost, quality, and image of the end-product or service. That means that each can contribute to a firm's relative position and create a basis for differentiation' (Porter, 1985). Porter argues that key factors, which drive cost efficiency and value, the so-called cost drivers and value drivers, should be identified.

Hamel and Prahalad (1994) developed the *core competence philosophy*. They argued that the difference in performance of organisations in the same 'industry' can rarely be fully explained by differences in their resource base *per se*. Superior performance is also determined by the way in which resources are *deployed* to create competences in the organisation's separate activities. Moreover, it is determined by the processes of linking these activities together to sustain excellent performance. Although the organisation needs to achieve a threshold level of competence in all of its activities, only some will be core competences. These are the competences that underpin the organisation's ability to surpass the competition – or demonstrably to provide better value for money. Core competences must be difficult to imitate otherwise it will not provide a long-term advantage. They may also be the basis on which new opportunities are created.

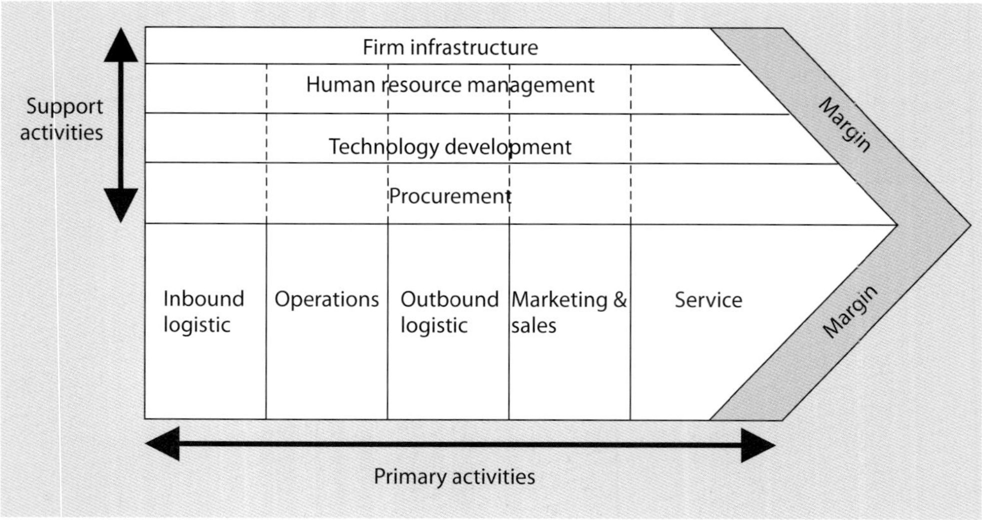

Figure 8.5. The value chain.

Johnson and Scholes (2006) derived from both models the bases on which an organisation's core competence can be built:

- cost efficiency;
- added value, i.e. effectiveness in matching customer requirements;
- managing linkages within the organisation's value chain and linkages into the supply and distribution chains;
- robustness, i.e. competences must be difficult to imitate.

Strategic alternatives

In considering possible strategies to achieve its objectives, management faces three choices (Johnson and Scholes, 2006):

- On what basis do they wish to compete with, or differentiate themselves from, their competitors? This is referred to as choosing a *positioning* or *generic* strategy.
- Should they develop new products or services, new markets or a combination of those two, and should they drop certain activities? This concerns choosing a *strategic direction*.
- Having made the first two sets of decisions, how should they develop the new services, products and/or markets? This refers to choosing the *means of developing* a strategy.

The choices available can be categorised as shown in Figure 8.6.

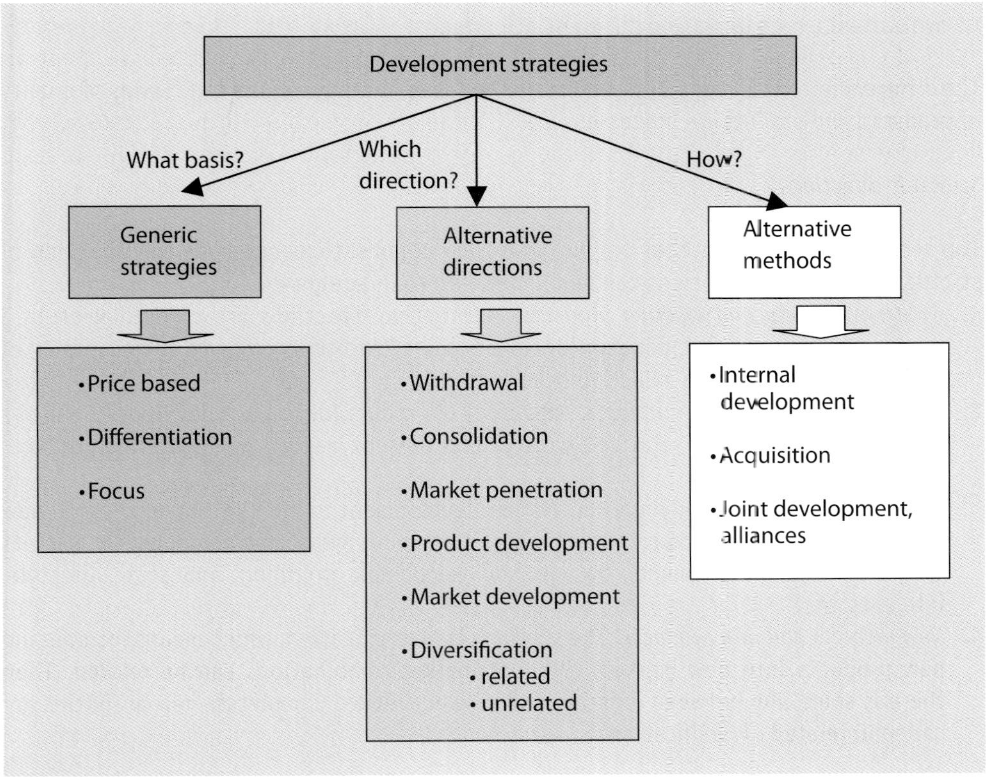

Figure 8.6. Strategic alternatives.

Generic strategies

Porter (1980, 1985) identified two basic types of competitive advantage that a firm can possess, i.e. low costs or differentiation. From this concept Porter (1980, 1985) classified three generic strategies or means by which companies can identify their position to develop and maintain competitive advantage.

The first, *cost leadership*, is a strategy whereby a firm aims to produce its product or service at a lower cost than its competitors. This could mean developing a large market share to attain economies of scale in production costs. Cost leadership does not necessarily mean low price.

The second generic strategy is *differentiation*, whereby a company aims to offer a product or service that is distinctive – and valued as such by customers – from those of its competitors. Porter argued that differentiation is 'something unique beyond simply offering a low price'. It allows firms to demand a premium price or to retain buyer loyalty during cyclical or seasonal downturns in the market. For example, firms that develop a reputation for reliability may be able to enhance their market share or to achieve higher margins by charging slightly higher prices, which customers are willing to pay.

The third, *focus strategy* is a strategy whereby management applies either costs leadership or differentiation to a narrow marker segment, rather than to a whole market. Porter refers to two variants, cost focus and differentiation focus.

The differentiation strategy, and even more the focus strategy require high levels of quality of products and production processes.

Strategic directions

The second set of choices that are important for organisations concerns the direction of strategy development. Options can be divided into four categories:

1. *Existing markets and existing products.* Alternative directions are consolidation in a growing market, market penetration in a mature market in order to increase market share, and withdrawal in a declining market.
2. *Existing markets and new products.* A strategy of product (or service) development allows a company to retain the relative security of its present markets while altering products or developing new ones.
3. *New markets and existing products.* Market development is the strategy of seeking new markets for existing products or services. This could mean new demographic markets or new groups of customers, by age, socio-economic grouping, education, lifestyle, interests, etc.
4. *New markets and new products.* The strategy is diversification, which means introducing new products into new markets. Product/market combinations can be related. Then there is some link between existing and new activities. Completely new activities are called unrelated diversification.

The more a strategy includes new products and new markets, the higher the demand on the organisation for quality management, due to the risk of making mistakes through a lack of knowledge and understanding.

Alternative methods

A final set of decisions concerns the means by which an organisation pursues the chosen strategy. Internal development means that all aspects of the development of new products are totally under the organisation's control and under the influence of internal management and expertise. Acquisitions allow rapid entry into new product/market areas. They may also be appropriate when the acquiring company lacks the necessary skills and resources.

Joint development and alliances mean turning to partners to co-operate in developing products or services. Arrangements can vary from highly formal contractual relationships to looser forms. Examples are franchising, licensing and subcontracting.

8.3 Factors influencing quality policy

In this section quality policy will be analysed. Firstly, a characterisation of the quality gurus' philosophies will be given. Secondly, Crosby's maturity-grid, which relates quality policy to stage of development of the company, will be described. Then attention is paid to the costs of quality and their influence on quality policy. Finally, the influence of technology will be discussed.

8.3.1 Quality philosophies

As stated earlier, policies or general statements are understandings that guide manager's thinking in decision-making. A quality policy must be firmly established in the decision-making processes and be the guiding principle for all activities whenever quality is concerned, with respect to external as well as internal customers and also as regards suppliers to the company (Bergman and Klefsjö, 1994).

Quality policy in most companies is more or less influenced by the ideas of the four major quality gurus Deming, Juran, Feigenbaum and Crosby (see Chapter 3 for a description).

Their ideas about quality management can be illustrated by the more or less philosophical approach of Deming. With this approach he believed that a commitment to quality requires a transformation of the entire organisation. The Deming philosophy is based on a system known as the 'Fourteen Points':

1. Create and disseminate to all employees a statement of the aims and purposes of the company or other organisation. The management must constantly demonstrate their commitment to this statement.
2. Learn the new philosophy, top management and everybody else in the organisation.
3. Understand the purpose of inspection, for the improvement of processes and reduction of cost.
4. End the practice of awarding business on the basis of price tag alone.
5. Improve constantly and forever the system of production and service.

6. Institute training.
7. Teach and institute leadership.
8. Drive out fear. Create trust. Create a climate for innovation.
9. Optimise the efforts of teams, groups, staff areas toward the aims and purposes of the company.
10. Eliminate exhortations for the workforce.
11. a. Eliminate numerical quotas for production. Instead, learn and institute methods for improvement.
    b. Eliminate MBO (management by objective). Instead, learn the capabilities of processes and how to improve them.
12. Remove barriers that rob people of pride and workmanship.
13. Encourage education and self-improvement for everyone.
14. Take action to accomplish the transformation.

All gurus agree that the ultimate focus should be on quality, customer orientation and a company-wide approach to the prevention of quality problems. However, they differ in their opinion as to how to achieve this. These differences can be illustrated with the aid of the administrative concepts (see also Chapter 3).

Management approaches are described in a matrix with two axes: centralisation versus decentralisation and high versus low degree of formalisation. In Figure 8.7, the gurus are positioned in the same matrix, indicating that Feigenbaum and Crosby, more than Deming and Juran, start their programmes at the top of the company. Feigenbaum and Deming believe, more than Crosby and Juran, in a formalised way of doing things, whereas the latter focus heavily on everyone's attitude towards quality, an attitude that refuses to accept today's quality levels as good enough. Note that these differences between the quality philosophies are not so explicit in practice. Deming, for instance, has also been stressed the importance of attitude. Where they all agree is that quality policy implies an obligation for management (Bergman and Klefsjö, 1994). Unless management acts in line with the quality policy, no other employee in the company will take it seriously.

		Formalisation high, focus on: Methods	Formalisation low, focus on: Attitude
Centralisation low, focus on:	Decentralised decisionmaking	Deming	Juran
Centralisation high, focus on:	Centralised decisionmaking	Feigenbaum	Crosby

Figure 8.7. Approaches from different quality philosophies.

8.3.2 Organisational maturity

Quality policy evolves over the course of time. A systematic approach is policy deployment, which is a top-down and bottom-up process in order to align objectives through all levels (Bergman and Klefsjö, 1994). Objectives are identified at each level by top-down and bottom-up consultation and communication, thus translating the overall goals of the company into specific targets for departments and groups. The outcome of policy deployment processes depends greatly on the *stage of development of quality understanding and management.*

Like Deming and Juran, Crosby viewed the path to quality as a never-ending journey (Crosby 1979, Melnyk and Denzler, 1996). Crosby typified the stages of development in quality management in his maturity grid (Table 8.1).

In this grid, corporate quality awareness and quality maturation from a level of uncertainty to one of certainty is traced. The stages of development are:

- *Uncertainty:* the firm remains unaware of the importance of quality as a strategic necessity.
- *Awakening:* managers understand the importance of quality but they have not yet acted to improve it.
- *Enlightenment*: managers openly acknowledge the importance of quality and the corporate need for improvement, and they have started to take concrete action to improve overall quality by establishing a formal programme.
- *Wisdom:* managers have established a quality management programme that is working well, identifying problems early and routinely pursuing corrective actions. The programme emphasises prevention rather than inspection.
- *Certainty:* quality has become an unavoidable component of the current management structure. The entire system is designed to ensure that the firm can attain its goal of zero defects. Any problems are both infrequent and truly random events.

Referring to quality costs Crosby states that quality costs can fall from 20% of sales to 2.5% of sales.

When developing a quality policy, the stage of organisational maturity has to be taken into account, because it shows which measures will have good organisational conditions for success, e.g. starting quality circles in a company in stage 1 will not be successful due to the lack of awareness of quality and its revenues.

8.3.3 Technological complexity

The characteristics of technology used in food production do have consequences for the choice of a quality policy. Simple and well-known technology is easier to control and assure than advanced and new technology. The technology that can be used depends on the production organisation. Unit or small-batch producers do use basic multi-purpose technology. Continuous-process producers like vegetable oil factories do use complex one-purpose technology. Large batch manufacturers, which are very common in the food industry, do have an intermediate position.

Table 8.1. Crosby's quality management maturity grid.

Measurement categories	Stage I Uncertainty	Stage II Awakening
Management understanding and attitude	No comprehension of quality as management tool. Tend to blame quality department for 'quality problems'	Recognised that quality management may be of value but not willing to provide money or time to make it all happen
Quality organisation status	Quality is hidden in manufacturing or engineering departments. Inspection is probably not part of organisation. Emphasis on appraisal and sorting	A stronger quality leader is appointed but main emphasis is still on appraisal and moving the product. Still part of manufacturing or other
Problem handling	Problems are tackled as they occur; no resolution; inadequate definition; lots of yelling and accusations	Teams are set up to tackle major problems. Long-term solutions are not solicited
Cost of quality as % of sales	Reported unknown. Actual 20%	Reported 3%. Actual 18%
Quality improvement actions	No organised activities. No understanding of such activities	Trying obvious 'motivational' short-term efforts
Summation of company quality posture	'We don't know why we have problems with quality'	'Is it absolutely necessary to always have problems with quality?'

Perrow (1967) introduced a concept for comparing organisations by the characteristics of their technologies (Figure 8.8), categorising them according to two dimensions. The first is task variability: the number of exceptional cases encountered in the work, which is the degree to which stimuli are perceived as familiar or unfamiliar. This varies on a scale from a few to many exceptions. Low task variability requires a low level of skill, limited originality, high repetition, high uniformity, and little training.

The second is the type of problem that arises when exceptions occur with technology: they could be well-defined and easy to analyse or ill-defined and difficult to analyse. The search process that is undertaken by people is typified differently in these situations. The first type refers to the well-defined technological problems, and involves a search that can be conducted on a logical, systematic, and analytical basis. There are often programmes showing how to deal with such problems. If a programme exists, only a very trivial search is involved in switching from one programme to another when the stimuli change. The second type of search process occurs when the problem is so vague and poorly conceptualised

Stage III Enlightenment	Stage IV Wisdom	Stage V Certainty
While going through quality improvement programme learn more about quality management; becoming supportive and helpful	Participating. Understand absolutes of quality management. Recognise their role in continuing emphasis	Consider quality management as an essential part of company system
Quality department reports to top management; all appraisal is incorporated and quality manager has role in management of company	Quality manager is an officer of company; effective status reporting and preventive action. Involved with customer affairs and special assignments	Quality manager on board of directors. Prevention is main concern. Quality is a true leader
Corrective action communications established. Problems are faced openly an resolved in an orderly way	Problems are identified early in development. All functions are open to suggestions and improvements	Except in the most unusual cases, problems are prevented
Reported 8%. Actual 12%	Reported 6.5%. Actual 8%	Reported 2.5%. Actual 2.5%
Implementation of the 14-step programme with thorough understanding and establishment of each step	Continuing the 14-step programme and starting. Make certain	Quality improvement is a normal and continuous activity
Through management commitment and Q-improvement problems are identified and solved	Defect prevention is a routine part of our operation	We know why we do not have problems with quality

that it is virtually impossible to analyse. In this case, no 'formal' search is undertaken, but instead one draws upon experience or intuition, or relies upon chance and guesswork.

Figure 8.8 shows four types of technology from these two dimensions. Perrow labelled them as follows:

- *Routine technology* is well-known and reliable technology to be dealt with in a standardised way, e.g. standard technology for mass production.
- *Craft technology* is characterised by reliable technology but difficult when problems occur, e.g. existing technological modules used in new applications.
- *Engineering technology* is technology with many exceptions but they can be handled in a systemised way, e.g. existing technology changed for new products.
- *Non-routine technology* is new technology that shows unexpected behaviour, e.g. new technology for new products and processes.

The type of technology is an important factor when developing the quality policy, as will be discussed in the next section.

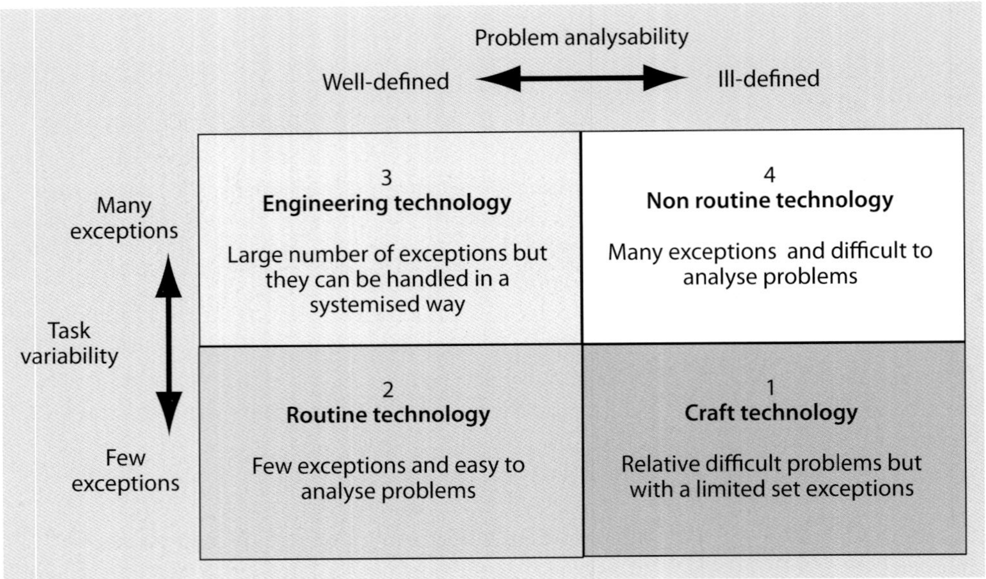

Figure 8.8. Matrix of technologies by Perrow.

8.3.4 Quality costs

In most firms cost accounting is an important function. All organisations measure and report costs as a basis for control and improvement. Quality cost information serves a variety of purposes. Those costs which are associated with avoiding poor quality or have been incurred as a result of poor quality, help management to evaluate the relative importance of quality problems and thus identify opportunities for cost reduction. It can aid in budgeting and cost control activities. Finally, it can serve as a scoreboard to evaluate the organisation's success in achieving quality objectives with regard to the firm's strategy. To establish a cost of quality programme, one must identify the activities that generate cost, measure them, report them in a way that is meaningful to management and analyse them to identify areas for improvement.

Quality costs (Figure 8.9) can be organised into four major categories: prevention costs, inspection costs, internal failure costs and external failure costs (Evans and Lindsay, 2005).

Prevention costs are incurred in an effort to keep non-conformance products from occurring and reaching the customer, and include the following specific costs:

- Quality planning costs, such as salaries of individuals associated with quality planning and problem-solving teams, the development of new procedures, new equipment design and reliability studies in order to improve quality or to prevent poor quality.
- Process control costs, which include costs spent on analysing production processes and implementing process control plans.
- Information systems costs incurred to develop data requirements and measurements.
- Training and general management costs, including internal and external training programmes, clerical staff expenses and miscellaneous supplies.

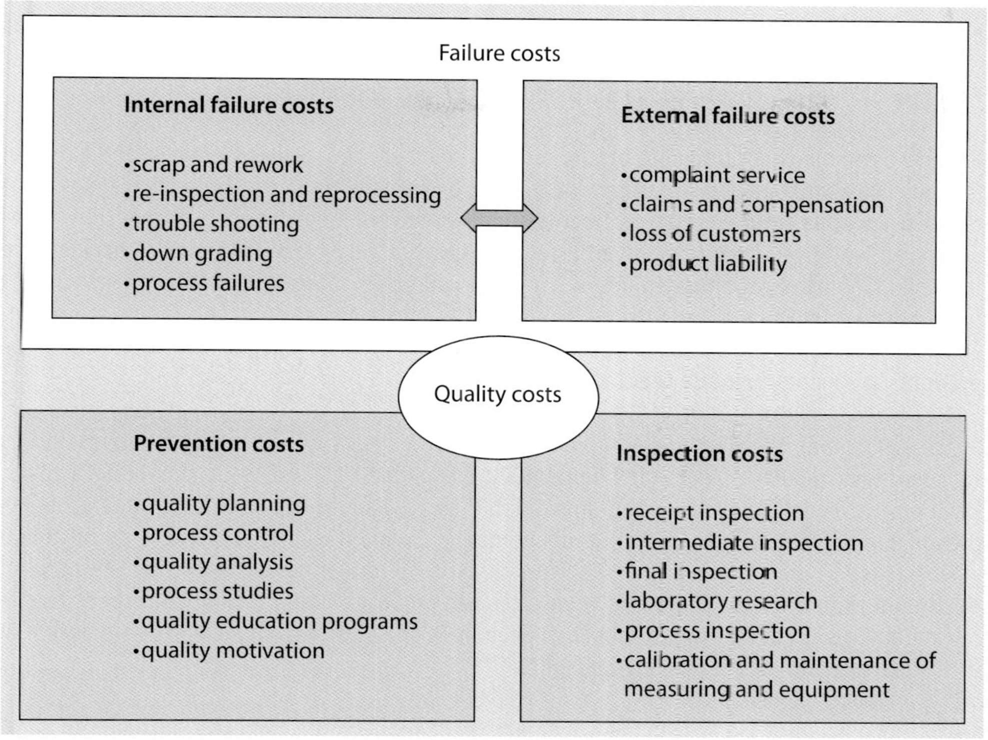

Figure 8.9. Quality costs.

Inspection costs are incurred to ascertain quality levels through measurement and analysis of data in order to detect and correct problems. Categories of inspection costs include:

- Test and inspection costs associated with incoming materials, work in process and finished goods, including equipment costs and salaries.
- Instrumental maintenance costs, due to calibration and repair of measuring instruments.
- Process measurement and control costs, which involve the time spent by workers gathering and analysing quality measurements

Internal failure costs are a result of unsatisfactory quality detected before delivery of a product to the customer; some examples include:

- Scrap and rework costs, including material, labour and overheads.
- Costs of corrective action, arising from time spent determining the causes of failure and correcting production problems.
- Downgrading costs, such as revenue lost when selling a product at a lower price because it does not meet specifications.
- Process failures, such as unplanned machine downtime or unplanned equipment repair.

External failure costs occur after poor quality products reach the customer, specifically:

- Costs due to customer complaints and returns, including rework on returned items, cancelled orders, freight premiums, and loss of customers.

- Product recall costs and warranty claims, including the cost of repair or replacement as well as associated administrative costs.
- Product liability costs, resulting from legal actions and settlements.

Experts estimate that 60 to 90% of total quality costs are the result of internal and external failure. These costs are not easily controllable by management. In the past management reacted to high failure costs by increasing inspection. Consequently inspection costs increased while very often the overall result improved very little. In practice an increase in prevention usually generates larger savings in all other categories. Better prevention of poor quality clearly reduces internal failure costs, since fewer defective items are made. External failure costs will also decrease. In addition, less inspection is required, since products are made correctly the first time.

Juran (1951) was the first one to give managers the means of answering the question as to how much quality is enough (Figure 8.9). He divided the costs of achieving a given level of quality into avoidable and unavoidable costs. Failure costs are avoidable, whereas prevention and inspection costs are unavoidable. Juran stated that there is an optimum where a rise in prevention and inspection costs is no longer compensated by the parallel decline in failure costs. There has been criticism on Juran's approach with respect to the optimum. Crosby (1979) argued that the optimum is at a 100% quality level (= zero defects), as is illustrated in Figure 8.10 at the right.

Taking into account costs from the perspective of food production, the optimum of total quality costs will not necessarily be at 100% quality. It is almost impossible to produce products of constant and perfect quality because of the nature of agri- and food products, e.g. biological variation, climate fluctuations and limited shelf life qualities. This is in accordance with Juran's approach to the proven reality that poor quality is uneconomical and the unproved theory that perfect quality is uneconomical. However, supporters of Crosby's 'zero-defects' programme argue that with a careful stepwise approach the level of zero-defects can be accomplished. However, nowadays higher and higher demands are put on food production in terms of safety, freshness, health and sustainability. Furthermore, firms are confronted with increasing competition and quickly changing consumer markets. Therefore,

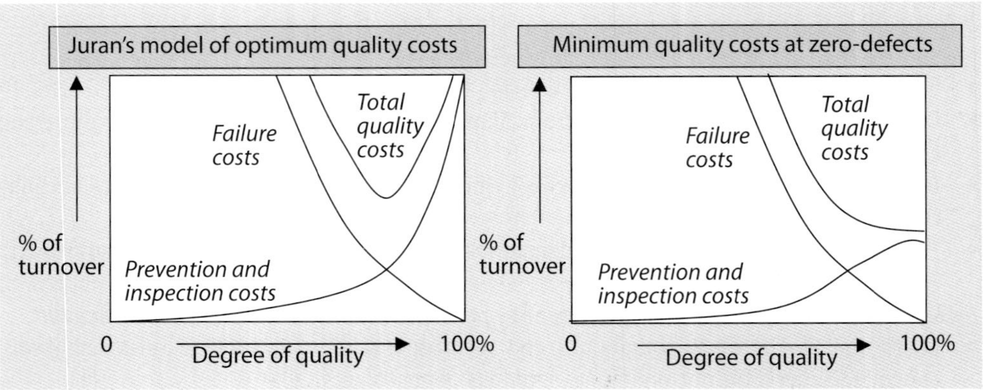

Figure 8.10. Two quality costs models.

firms have to improve their products and processes continuously. As a consequence the optimum of total quality costs will decrease, while the general quality level will increase (Zuurbier *et al.*, 1996).

Quality costs themselves provide little information, since they may vary due to factors such as production volume and seasonal effects. Some common measurement indicators are labour, sales, manufacturing costs and unit of products. By using these indices one should be aware of the time horizon included, the possible need to weigh to a standard, price fluctuations and interaction with other factors. This illustrates that all of these indices, although often used in practice, have a fundamental problem. A change in the indicator can appear to be a change in the level of quality or productivity alone. For instance, if the selling price is decreased due to increased competition, the direct sales-based index will increase even if the quality does not change. Nevertheless, use of such indexes is widespread and useful for comparing quality costs over time. In general, sales bases are the most popular, followed by cost, labour and unit bases. It has to be mentioned that this approach to quality costs is limited to efficiency and effectiveness and does not fit in a TQM philosophy. For instance, if direct labour is decreased due to managerial improvements, the direct labour-based index will increase even if the quality does not change. However, the overall firm's competitiveness (total quality) has increased because it succeeded through improvements in reducing direct labour costs. Thus it succeeded in cutting its total costs. Thus a different approach is required (Evans and Lindsay, 2005).

8.4 Total quality management

This section starts with the concept of different policy orientations and the conditions necessary to apply them. Total quality management is a company policy, that fits with the later stages of Crosby's maturity grid. In this section total quality management will be explained. Briefly, the implementation of this policy will be discussed.

8.4.1 Policy orientations

For quality policy various orientations are used depending on the organisation's culture and the individual beliefs of top management. In companies where technology plays an important role, managers tend to introduce technological solutions first for any emerging problems, such as the automation of processes and redesign of products and processes. In other companies managers prefer changing the culture and attitude of employees, because they strongly believe that this is the only way to accomplish improvements in future. In general, three policy orientations can be distinguished (Figure 8.11):

1. *Product/process orientation.* Technological solutions are preferred in order to make product quality, processes, and materials predictable and controllable.
2. *Procedure/organisation orientation.* Solutions are looked for within the quality system. The underlying belief is that a clear assignment of responsibilities accompanied by a broad description of procedures and work instructions form the basis for controlling quality performance.
3. *People orientation.* Problems are analysed in the field of human behaviour. Solutions are sought in changing people by means of training and education and by improving commitment.

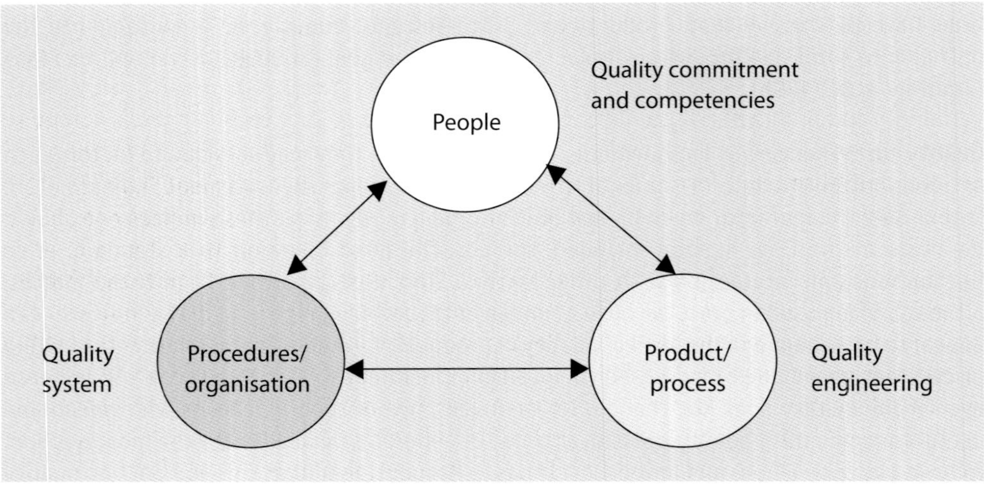

Figure 8.11. Different quality policy orientations.

Experience among food companies reveals that integrating the three policy orientations is very difficult. Only a few companies succeed in doing so. Some companies combine people and procedure orientation but more companies integrate product/process and procedure orientation. Most of the companies only use a single orientation in resolving quality problems and controlling the quality system.

Frequently, using one policy orientation won't be sufficient for solving problems (Edge, 1990). For instance, the introduction of new technology, such as modified atmosphere packaging, will give rise to problems in traditionally organised companies unless they question long-held belief and practices. New manufacturing methods can also shock commercial departments because automated processes can produce products with more narrow specifications, which in turn requires another market approach. Another example is the tightness of the procedures that are required for automated processes. These automated operations might magnify harmful effects that defective upstream processes might have on downstream processes. Without machine operators, who are physically handling the products, there is no one to correct the process. Process engineering might try to provide expert systems to replicate an operator's talent for recognising errors or to build in sensors and programmed control. However, there remains the input of pure craftsmanship.

So, new manufacturing systems will work best in an organisation where technical and managerial aspects are integrated with the contributions of various people. Such organisations have flat structures (Edge, 1990). In the administrative concepts model (Figure 3.9) this situation would be typified as the flexible concept, in which decentralisation is combined with a low degree of formalisation. The organisation is based on the teamwork structure (Figure 3.20) and only a limited set of procedures exists.

However, applying this concept is often not possible in practice. First of all, the functional structure is often used in food companies and changing such an organisational structure to a teamwork or network structure requires adequate change strategies and a lot of time and effort (Robbins and Coulter, 2007). Secondly, the type of technology and or the characteristics of people may restrict the introduction of flat structures. From Perrow's matrix of technologies (Figure 8.8) it can be seen that technologies are different in task variability and problem analysability. As a consequence, different technologies will require different processes of analysis and decision-making. Routine technology, for instance, will require centralised and formalised management processes, while conversely non-routine technology will be better managed with decentralised and non-formalised processes.

Comparable influences originate from people characteristics. In Chapter 3 section 3.5 it has already been noted that people differ in commitment and competencies and that they show dynamic quality behaviour. Flat structures can only be built on people who are sufficiently competent and committed. In addition, complex technology puts demands on people, e.g. hidden safety risks (e.g. contamination from the growth of pathogens due to unhygienic conditions or improper processing) require a rather profound knowledge of microbial processes in food. But, in agri-food production processes people with rather low education levels are often employed, which complicates the implementation of flat structures.

In summary, developing a quality policy should include a careful alignment of the three policy orientations in order to integrate technology, organisation and people characteristics.

Total quality management (TQM) is a policy that simultaneously embraces the three policy orientations, thus striving for the broadest range of analysing methods and possibilities for solutions. TQM is elaborated on further in the following section.

8.4.2 Total quality management

Total quality management is a management view that strives for the creation of a customer-focused culture, which defines quality for the organisation and establishes the foundation for activities aimed at attaining quality-related goals. TQM is not merely a technique, but a philosophy anchored in the belief that long-run success depends on a uniform and firm-wide commitment to quality, which includes all activities that a firm carries out. TQM can become the core of a firm's strategy to remain competitive (Schermerhorn, 2007; Evans and Lindsay, 2005; Ross, 1999).

According to Dean and Evans (1994) the term *total quality management* (TQM) covers a total, company-wide effort including all employees, suppliers, and customers, that seeks to continuously improve the quality of products and processes to meet the needs and expectations of customers. TQM has become the basic *business strategy* for firms that aspire to meet and exceed the needs of their customers.

There are probably as many different approaches to TQM as there are businesses. Although no TQM programme is ideal, successful programmes share many characteristics. According to Dean and Evans (1994) the basic attributes of TQM are: (1) customer focus, (2) strategic planning and leadership, (3) continuous improvement, and (4) empowerment and teamwork (Figure 8.12).

Total quality management	**Customer focus** . sensitivity to customer/consumer requirements + measurement customer/consumer satisfaction . awareness of market developments + rapid and flexible response
	Strategic planning and leadership . top commitment and preventive measures planning + regular review and assessment . demonstrating high quality values and expectations + recognition and reinforcement
	Continuous improvement . improvement is a part off all activities + regular cycles of planning, execution and evaluation
	Empowerment and teamwork . all employees Q-focused and able to make decisions + self control and learning . sharing knowledge and collaboration + team control and rewarding

Figure 8.12. Principles of total quality management (TQM).

Customer focus

The customer is the final judge of quality. Quality systems must address all product and service attributes that provide value to the customer, and which lead to customer satisfaction and loyalty. Many factors influence value and satisfaction throughout the customer's overall purchasing, ownership, and service experiences. This includes the relationship between the company and customers, i.e. the trust and confidence in products and services, that leads to loyalty. This concept of quality includes not only product and service attributes that meet basic requirements, but also those features that enhance the product and differentiate it from competing offers. This is often referred to as the 'service bundle' or 'service package'. A business can achieve success only by understanding and fulfilling the needs of customers.

TQM is an important element of business strategy that is directed toward market-share gain and customer retention. It demands constant sensitivity to emerging customer and market requirements, and measurement of the factors that drive customer satisfaction. TQM also requires an awareness of developments in technology and a rapid and flexible response to customer and market needs. Such requirements extend beyond merely reducing defects and complaints or meeting specifications. Nevertheless, defect and error reduction and the elimination of causes of dissatisfaction contribute significantly to the customers' views of quality and so are important parts of TQM. In addition, the company's approach to recovering from defects and errors is crucial to its improving both quality and relationships with customers.

Also crucial to assuring quality for external customers are the *internal customers,* such as the next department in a manufacturing process. Failures in meeting the needs of internal customers will likely affect external customers as well. Employees must view themselves as customers of other employees and suppliers to others.

Strategic planning and leadership

Achieving quality and market leadership requires a long-term strategy. Improvements do not occur overnight. Planning and organising of improvement activities require time and a major commitment from all members of the organisation. Strategies, plans and budget allocations need to reflect long-term commitments to customers, employees, stockholders and suppliers. They must also address training, employee development, supplier development, technology evolution, and other factors that support quality. A key part of the long-term commitment is regular review and assessment of progress in long-term goals.

Leadership for quality is the responsibility of top management. Senior leadership must create clear quality values and high expectations. They must implement these quality values and expectations into the way companies operate. This requires substantial personal commitment and the involvement of senior management. The leaders must take part in the creation of strategies, systems, and methods for achieving excellence. The systems and methods must guide all activities and decisions of the company and encourage participation and creativity by all employees. The senior leaders must serve as role models through their regular personal involvement in visible activities, such as planning, reviewing company quality performance, supporting improvement teams and recognising employees for quality achievement. They must reinforce the values and encourage leadership in all levels of management.

Continuous improvement

Continuous improvement is part of the management of all systems and processes. Achieving the highest levels of quality and competitiveness requires a well-defined and well-executed approach to continuous improvement. Such improvement needs to be part of all operations and all work unit activities in a company. Improvements may be of several types:

- enhancing value to the customer through new and improved products and services;
- reducing errors, defects, and waste;
- improving responsiveness and cycle time performance; and
- improving productivity and effectiveness in the use of all resources.

Thus, improvement is driven not only by the objective to provide better quality, but also by the need to be responsive and efficient. Both activities result in additional marketplace advantages. To meet all these objectives, the process of continuous improvement must contain regular cycles of planning, execution, and evaluation. This requires a, preferably quantitative, basis for assessing progress and for deriving information for future cycles of improvement. Quality must be *measured*. A company should select data and indicators that best represent the factors that determine customer satisfaction and operational performance. A system of indicators tied to customer and company performance requirements will provide a clear and objective basis for directing all activities of the company toward common goals.

Empowerment and teamwork

All functions at all levels of an organisation must focus on quality to achieve corporate goals. Teamwork can be viewed in three ways:

1. *Vertical view,* i.e. teamwork between top management and lower-level employees. Employees are empowered to make decisions that satisfy customers without a lot of bureaucratic hassle, and barriers between levels are removed.
2. *Horizontal view,* i.e. teamwork within work groups and across functional lines. An example is a cross-functional product development team, that might consist of designers, manufacturing personnel, suppliers, salespeople, and customers.
3. *Interorganisational view,* i.e. organisational partnerships with suppliers and customers. Rather than dictating specifications for purchased parts, a company might develop specifications jointly with suppliers to take advantage of the suppliers' manufacturing capabilities.

The person who best understands his or her job is the one performing it. In addition, everyone must participate in quality improvement efforts. Thus employees must be empowered to make decisions that affect quality and to develop and implement new and better systems. This often requires a radical shift in the philosophy of senior management. In the traditional philosophy the workforce should be 'managed' to comply with existing business systems, which is contrary to what is required in TQM.

Companies can encourage participation by recognising team and individual accomplishments. So they share success stories throughout the organisation and encourage risk taking by removing the fear of failure. Moreover, they encourage the formation of employee involvement teams, implementing suggested systems that act rapidly, provide feedback, and reward implemented suggestions. Moreover, they provide financial and technical support to employees to develop their ideas.

Employees need training in quality skills related to performing their work, and to understanding and solving quality-related problems. Training brings all employees to a common understanding of goals and objectives and the means to attain them. Training usually begins with awareness of quality management principles and is followed by specific skills in quality improvement. Training should be reinforced through on-the-job applications of learning, involvement, and empowerment.

Firms applying TQM successfully achieve a decrease in long-term costs by building a more satisfied customer base. However, because short-term costs for new equipment, materials and additional labour are sometimes required to implement TQM processes, many firms hesitate to adopt TQM. Time is another bottleneck to the adoption of TQM. Many firms do not want to wait the two to ten years typically required to obtain the initial benefits from TQM.

TQM processes versus QA control processes

Jabnoun (2002) labels regular quality management processes as QA control processes. Usually, these processes are focused on prevention by measuring inputs in a way that enables employees to conform to standards. It also includes the analysis of performance data in order to detect special causes of variations. TQM processes focus on customer satisfaction and continuous improvement. They include the role of management providing and altering the necessary input of structure, skills, systems and shared values in order to meet customers' varying demands. Table 8.2 shows the differences between the regular and TQM control processes.

Table 8.2 shows that there is a major difference between the QA control and TQM control focus. Prevention is a shared goal by emphasising the role of management in providing the input that will build quality into the system. TQM control does this by delegating authority and reinforcing TQM values while it is restricted to systems in QA control. As a result of this, in QA only the managers have the authority to correct performance while in TQM empowered employees check performance and take corrective actions.

8.4.3 Total quality control and total quality learning

With similar underlying TQM concepts, organisations can apply two different orientations, TQC (total quality control) and TQL (total quality learning) with TQC being associated with quality in terms of conformance, and TQL being related to innovation (Prajogo, 2005; Prajogo and Sohal, 2001; Sitkin, 1994).

TQC and TQL share the same TQM principles but translate them into different sets of operating principles (Table 8.3), e.g.:

- *Customer satisfaction*: TQC will monitor and assess customer needs to better understand them, and will improve response to customer needs. TQL will search for an understanding of new customer needs and stimulate new customer need definitions.
- *Continuous improvement*: TQC will exploit existing skills and resources, and will improve control and reliability. TQL will explore new skills and resources, and will increase learning and understanding.

Table 8.2. Differences between the TQM and QA control processes (Jabnoun, 2002).

	QA control	TQM
Focus	Conformance	Customer satisfaction, continuous improvement
Nature	Static	Dynamic
Management role	Establish documented systems, provide training and resources, supervise employees	Empower employees by infusing TQM values and providing resources, know-how, systems and delegating authority
Employees' role	Conform with documented standards	Participate in setting standards, searching for new improvement, checking performance and correcting actions
Main enabler	Systems and process documentation	TQM cultural values

Table 8.3. Linking the distinctive principles associated with TQC and TQL to common underlying TQM percepts (Sitkin *et al.*, 1994).

Shared TQM concepts	Control-oriented principles (TOC)	Learning-oriented principles (TQL)
Customer satisfaction	Monitor and assess known customer needs Benchmark to better understand existing customer needs Respond to customer needs	Scan for customer needs or issues Test customer need definitions Stimulate new customer need definitions and levels
Continuous improvement	Exploit existing skills and resources Increase control and reliability	Exploit new skills and resources Increase learning and resilience
Organisation as an open system	First-order learning (cybernetic feedback) Participation enhancement focus	Second-order learning Diversity enhancement focus

Sitkin (1994) suggested that, when uncertainty is low, the assumptions and practices associated with a TQC approach make sense. In contrast, when uncertainty is high, the only reasonable goal may be to try to do a good job of exploration and learning through the use of a TQL approach. He also concluded that most situations include mixtures of subtasks that vary in their level of certainty and, thus, mandate that the two distinct approaches can be simultaneously in some appropriately balanced way. Before implementation the primary purpose for which TQM is being adopted should be very clear as the approach used should also consider the fact that the organisation changes and matures.

Another approach to distinguish TQM is 'hard' and 'soft' TQM, which is in line with the TQC and TQL perspective. The so-called 'hard' indicators of TQM focus on total quality control, and will focus more on food quality by conformance. On the other hand, the set of 'soft' indicators to TQM concentrate on total quality learning. It stresses the development of new skills, exploration of new areas, and other innovative-like activities (Sitkin, 1994). Those indicators are focused on executive commitment, loyalty, teamwork & empowerment, and training & education.

Both TQC and TQL are TQM alternatives, but due to their typical differences TQC may be perceived as the first phase of TQM in a further development towards TQL. Companies with a TQM focus strive for continuous improvement and such companies have much in common with learning organisations.

8.4.4 Implementing TQM

In most companies it will be evident that implementing a TQM programme implies a long-term fundamental change process. These organisational change processes and strategies have been previously discussed in Chapter 6. Shared power strategies and, in particular, organisation development (OD) were proposed as adequate for complex, long-range change efforts, involving improvement of internal problem-solving capacities.

When implementing these change efforts, senior management, middle management, and the workforce each have a critical role to play. Senior managers must ensure that their plans and strategies are successfully executed within the organisation. Middle managers provide leadership by which the vision of senior management is translated into the operation of the organisation. In the end, it is the workforce that delivers quality and must have not only empowerment, but also a true commitment to quality in order for TQM to succeed (Dean and Evans, 1994).

Senior managers must ensure that the organisation focuses on the needs of the customer. They must promote the mission, vision, and values of the company throughout the organisation. Senior managers must identify critical processes that need attention and improvement. They also have to recognise the resources and trade-offs that must be made to fund the TQM activity. They must improve progress and remove barriers to implementation. Finally, they must improve the processes in which they are involved (strategic planning, for example), both to improve performance of the process and to demonstrate their ability to use quality tools for problem-solving (Tenner and De Toro, 1992).

Middle management has been viewed by many as a direct obstacle to creating a supportive environment for TQM. Middle managers are often seen as feeding territorial competition, stifling information flow, not developing and/or preparing employees for change, and feeling threatened by continuous improvement efforts. However, middle management's role in creating and sustaining a TQM culture is critical. Middle managers improve the operational processes that are the foundation of customer satisfaction; they can make or break co-operation and teamwork. They are the principal means by which the workforce prepares for change.

The workforce must develop ownership of the quality process. Ownership and empowerment gives employees the right to have a voice in deciding what needs to be done and how. It is based on a belief that what is good for the organisation is also good for the individual and vice versa. Training, recognition and better communication are the key success factors for transferring ownership to the workforce.

8.4.5 Quality policy evaluation

A company's quality policy should be evaluated from a strategic point of view. Mostly this is executed within the normal strategy evaluation processes, which are related to overall business performance. It should be noted, however, that business performance evaluation is frequently restricted to financial criteria. There are a limited number of frameworks for evaluating (total) quality management. The most prominent frameworks are management awards like the United States' Malcolm Baldrige Award, the Japanese Deming's price and the European Foundation for Quality Management Award (Evans and Lindsay, 2005).

The Malcolm Baldrige Award

The criteria of the Malcolm Baldrige Award are built upon a set of core values and concepts that include:
- customer driven quality;

- leadership;
- continuous improvement;
- employee participation and development;
- quick response;
- design quality and prevention;
- long-range view of the future;
- management by fact;
- partnership development;
- corporate responsibility and citizenship;
- results orientation.

These core values are embodied in seven categories, which form the basis for the assessment (Figure 8.13):

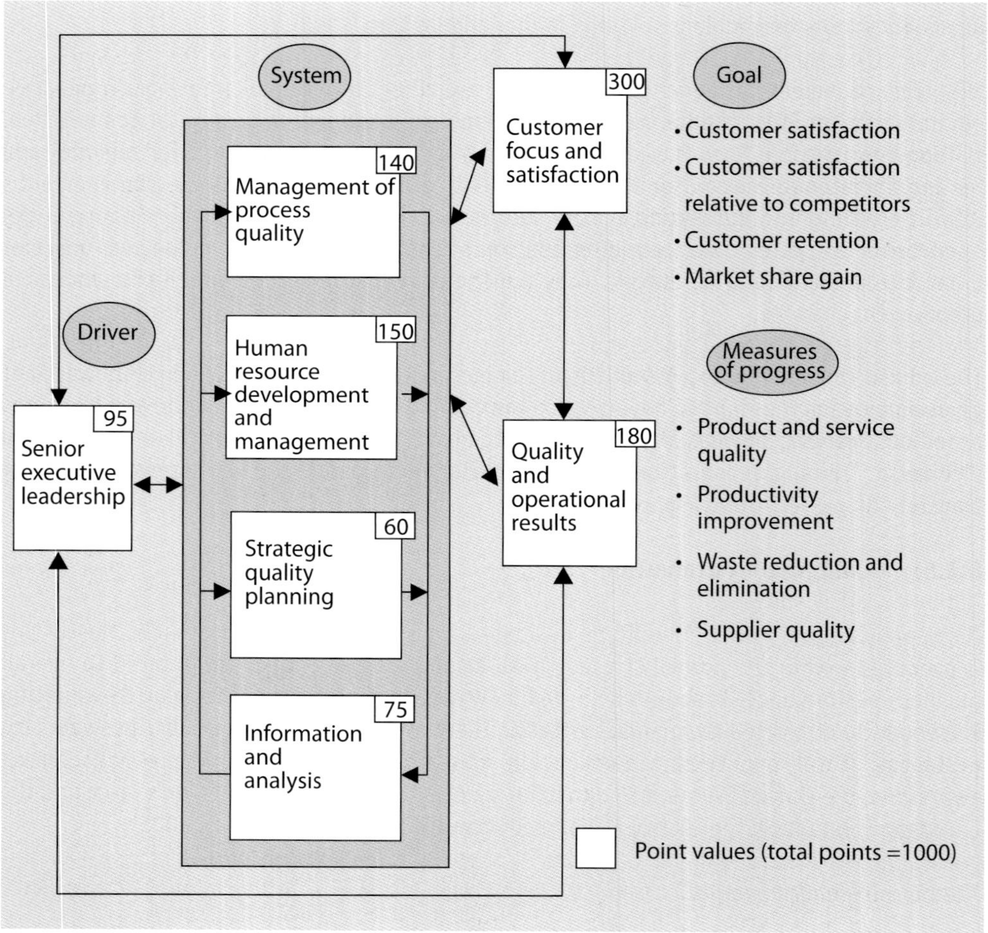

Figure 8.13. Malcolm Baldrige National Quality Award criteria framework.

The seven categories include leadership, information and analysis, strategic planning, human resource development and management, business results and customer focus and satisfaction.

1. Leadership: this category examines senior executives' leadership and personal involvement in setting strategic directions and building and maintaining a leadership system conducive to high performance, individual development and organisational learning.
2. Information and analysis: this category addresses the information and analysis requirements for performance improvements based upon the improvement of key processes.
3. Strategic quality planning: this category focuses on the firm's strategic and business planning and development of plans, along with the firm's attention to the customer and operational performance requirements.
4. Human Resource Development and management: the areas addressed in this category are concerned with how well human resource practices tie into and are aligned with the firm's strategic directions.
5. Management of process quality: in this category the key issues of process management-effective design, a prevention orientation, evaluation and continuous improvement, linkage to suppliers and overall high performance are assessed.
6. Business results: the company's performance in key business areas. It provides an appropriate orientation for all processes and process improvement activities.
7. Customer focus and satisfaction: the final category addresses how the company determines current and emerging customer requirements and expectations, provides effective customer relationship management and determines customer satisfaction.

The framework has four basic elements. Senior executive leadership sets (1) directions, (2) values, (3) goals and (4) systems. It guides the pursuit of customer value and firm performance improvement. The system comprises a set of well-defined and well-designed processes for meeting the firm's customer and performance requirements. Measures of progress establish a results-oriented basis for channelling actions to deliver ever-improving customer value and company performance.

The basic aims of the system are delivery of ever-improving value to customers and success in the market place.

The European Quality Award

The European Quality Award assessment is based on customer satisfaction, business results, processes, leadership, people satisfaction, resources, people management, policy and strategy and impact on society (Figure 8.14).

In comparing these elements with the Baldrige Award framework many similarities and several differences are apparent. Results, including customer satisfaction, people (employee) satisfaction and impact on society, constitute a higher percentage of the total score. Leadership and the quality system drive these results, as in the Baldrige Award. The three criteria of people satisfaction, the impact on society, and business results are somewhat different. People satisfaction refers to how employees feel about the organisation, including

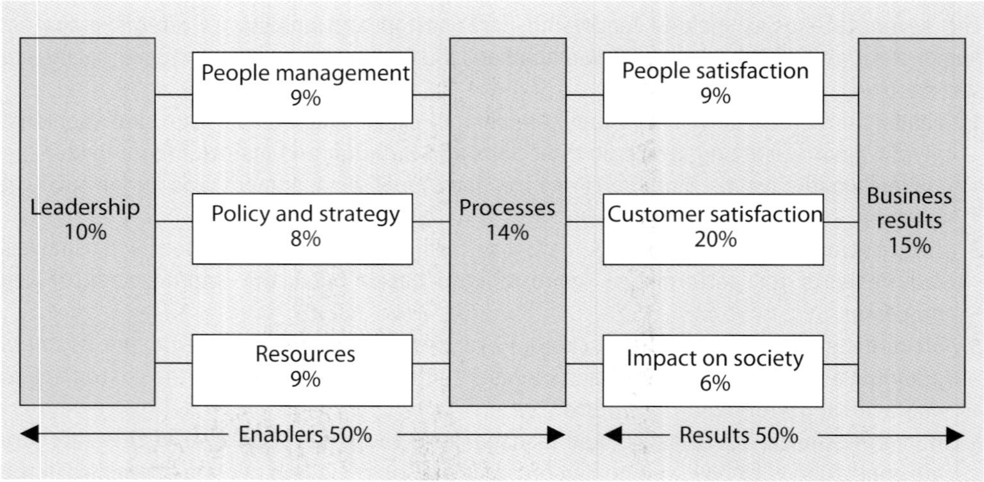

Figure 8.14. The European Quality Award criteria.

the work environment, perception of management style, career planning and development and job security. Unlike category 4 of the Baldrige Award (human resource development and management), the category people satisfaction is an independent result. The category impact on society focuses on the perception of the company by the community at large and the company's approach to quality of life, environment and the preservation of global resources. The European Quality Award criteria put more emphasis on this category than on the public responsibility item in the Baldrige Award criteria. The business results criterion explicitly addresses the financial performance of the firm, its market competitiveness and its ability to satisfy shareholders' expectations. Other non-financial areas of performance, such as order processing time, new product design lead-time, and time to break-even are also considered.

8.5 Quality policy from a chain perspective

Quality management in the agri- and food industry is strongly dependent on co-operation, due to the mutual dependencies between elements of the food chain. In this section the concept of the value chain is followed by an analysis of collaboration in the food chain.

8.5.1 The value system

In section 8.2.3 the concept of the value chain is mentioned. One of the key features of most industries is that a single organisation almost never undertakes all of the value activities from product design to delivery of the final product or service to the final consumer. There is usually specialisation and any organisation is part of the wider *value system,* which creates a product or service (Figure 8.15).

In understanding the basis of an organisation's strategic capability, it is not sufficient to consider the organisation's internal position alone. Much of the value creation will occur in the supply and distribution chains, and this *whole process* needs to be analysed and

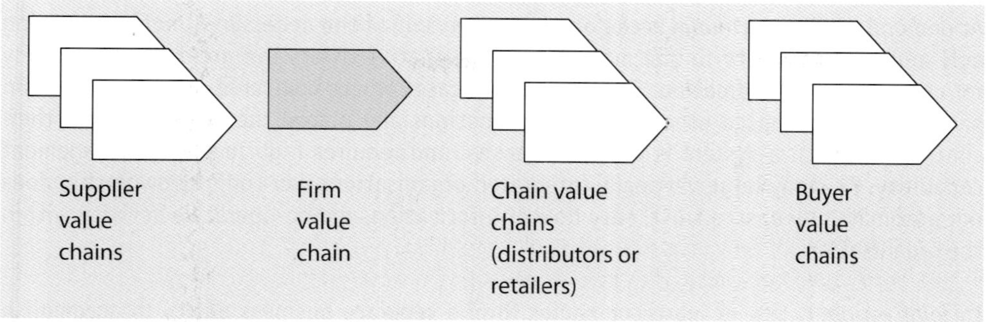

Figure 8.15. The value system.

understood. The quality of a food product when it reaches the final purchaser is not only influenced by the activities within one manufacturing company itself. It is also determined by e.g. quality of raw materials, half-fabricates and ingredients, and performance of the distributors and retailers. The ability of an organisation to influence the performance of other organisations in the value chain is crucially important for competence and a source of competitive advantage (Porter, 1985; Johnson and Scholes, 2006).

In order to be able to assure the highest quality levels in product quality and food safety, food companies and other food chain members feel a strong incentive to co-ordinate. However, in some cases individual companies value independence and autonomy so highly that they are unwilling to allow the needs of the system to dictate their decisions. Some agribusiness firms feel they are better off by retaining the ability to change sources at their discretion or find new outlets for their products without worrying about the implications for a channel relationship. Even in a single vertically integrated firm there are often significant challenges in getting each level of the chain to consistently perform at the required levels. Clearly, the longer the controlled chain, the greater the management challenges to make it work effectively. There are also external factors that affect the degree of co-ordination that may exist in any industry. The public desire to protect small independent farms or agribusinesses sometimes results in the development of legal barriers to contractual or vertically integrated co-ordination.

In some cases, the savings or market advantages of co-ordination may not offset the costs associated with co-ordinating the activities. Here, a more traditional system of independent suppliers bringing commodities to an open market, sorting product into various categories, and allowing prices to be negotiated based on available supply and demand serves the needs of the market quite adequately (Downey, 1996).

8.5.2 Collaboration in the food supply chain

There is a continuum of ways of organising the agricultural production/processing/distribution system. All alternatives along this continuum are forms of collaboration where different processes are bound together with some form of co-ordination. It is likely that no single co-ordination mechanism will evolve across the agricultural products industries because of the wide differences in markets and products (Downey, 1996).

At one end of the continuum is the outright *ownership* of the necessary functions to bring agri and food products to the market. This alternative offers the greatest opportunity for control and co-ordination. There are many cases where companies in the food chain have built or purchased other food chain functions and placed them totally under their control. But such a system is capital intensive and requires high levels of management capability. As many large vertically integrated organisations can confirm, ownership does not automatically ensure trust, easy flow of information, and co-operative activity within the organisation.

In *joint ventures*, two or more companies form a separate business entity to accomplish mutually beneficial functions. These joint ventures have become relatively common, but are not always successful. The long-term commitment of joint venture partners is often met by significant differences in philosophy and strategy that might result in a less successful relationship.

Contractual relationships between agribusiness entities can also accomplish the objective of co-ordination if they are carefully negotiated with the proper motivation. An infinite set of contractual arrangements is possible. Contractual arrangements are usually short-term with less commitment than joint ventures, and so more attractive to a larger number of agri- and food businesses.

Informal agreements or simply developing a 'preferred supplier' relationship with a buyer are the least formal type of buyer/seller alliances and possibly the most common. 'Preferred supplier' arrangements are based totally on informal relationships and proven dependability. With no legally binding commitment, this alliance is the least restrictive, but many will argue that it is still a very effective business arrangement.

Strategic relationships

As the complexity of markets has evolved, a new form of collaboration has emerged: the strategic relationship. It is based on trust and disclosure. A strategic relationship can be viewed as a collaborative agreement between two or more firms, where each partner remains independent, while cooperating on specific projects.

The strategic relationships can be either vertical or horizontal. Vertical relationships between suppliers and customer/buyer are known as *partnerships* and horizontal relationships, such as those between two suppliers are known as *alliances*. Figure 8.16 identifies four different types of strategic relationship, and identifies the objectives of each (Gattorna, 1996). Those are combined downstream activity, combined upstream activity, partnership sourcing and third party. A clear graph show how they are formed and the relationship they have with the customers.

Strategic relationships go beyond contractual arrangements and generate close working relationships that can enhance the operating efficiency and marketing effectiveness of the combined business entities.

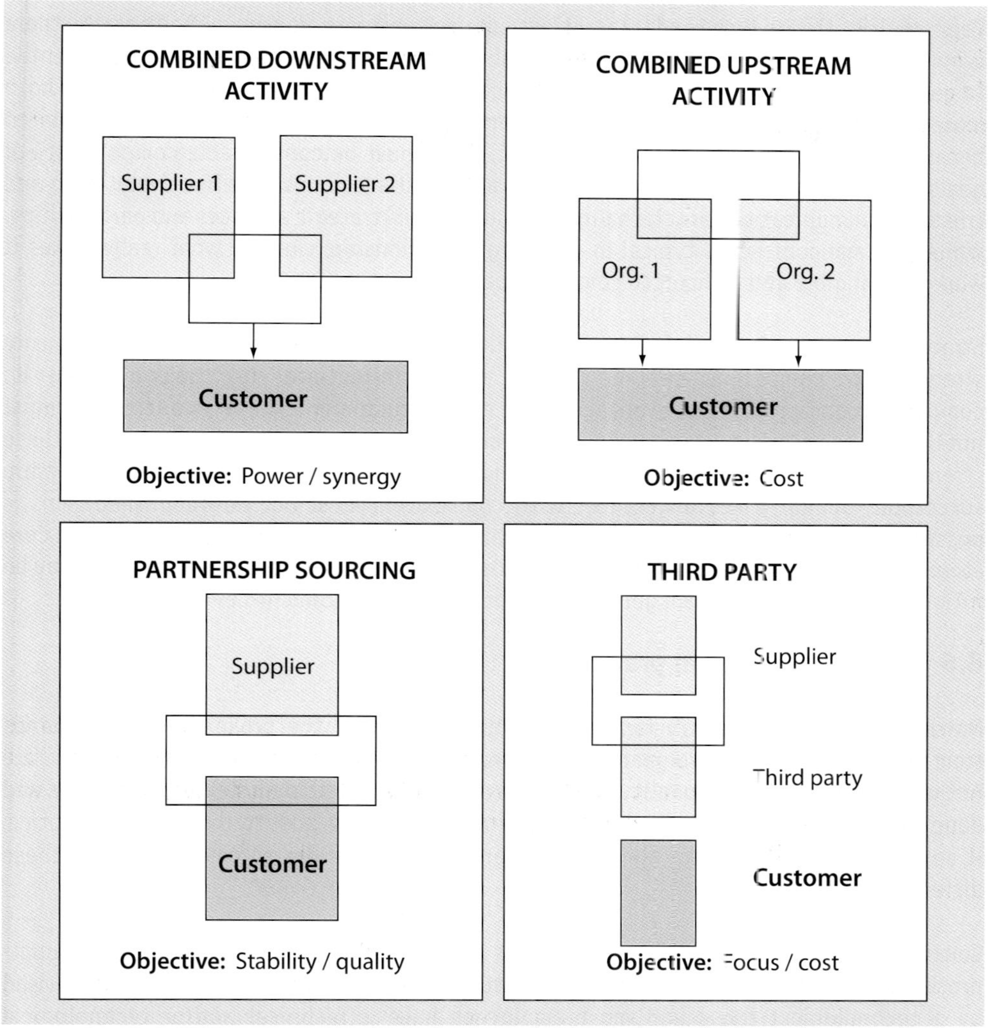

Figure 8.16. Four different types of strategic relationships and their corresponding objectives (Gattorna, 1996).

A partnership or alliance is a long-term relationship with common goals, mutual understanding, and a depth of commitment that goes beyond what can easily be negotiated into a contract. There must be a non-adversarial attitude from all parties involved. Moreover, all parties must perceive the relationship as beneficial. True strategic relationships generally exceed simple single person relationships, so they will survive personal changes. The details of the relationship have to be negotiated with several levels of management involved in order to seek areas of mutual benefit. A major advantage is that joint solutions to system problems can be found more easily and the benefits can be shared among all parties. Often these benefits are improvements in quality and result in competitive advantage in the market place that is reflected in customer satisfaction.

Paperwork is often reduced and transactions between partners are streamlined because there is mutual understanding and agreement. Usually companies can be managed more efficiently. In general, the level of customer service is enhanced because of a system that operates both more efficiently and effectively. The number of strategic relationships is usually limited because of the intensity of the relationship. There must be considerable compatibility of goals, philosophy, and operations for true strategic alliances to work effectively. Open and trusting communications are the hallmark of successful strategic alliances and partnerships. While a formal contract is typical in a strategic relationship, much of what really makes it work is intangible and difficult to commit to paper (Downey, 1996).

Strategic partnerships and alliances have their downsides. Each company has its core strengths or competence-specific talents that differentiate the company from its competitors and gives it an advantage in the eyes of customers. These core strengths must not be weakened by the relationship, which can happen if resources are diverted from these strengths or if technological or strategic strengths are compromised to make the partnership successful. Similarly, key differences with competitors must not be diminished. This is possible if key technology is shared or if entry barriers are reduced for the competition. Core strengths do not necessarily correspond to large investments or resources, and they may be intangible items such as management skills or brand image (Simchi-Levi *et al.*, 2000).

8.6 Quality policy in practice

Nowadays the food industry faces the challenge to meet the broad range of demands from the market and environment (e.g. other companies, legislation, etc.). Demands are not only made on product quality, cost and availability, but also on flexibility, service and dependability of the business as reflected in the extended quality triangle. In practice, it appears to be difficult for agri-business and food companies to comply with all these different requirements.

Considering the different policy orientations reveals that in food production the product-process orientation is predominant. As a matter of fact, the food industry is characterised by a technological focus and most employees have a technical and/or technological education. Next to the product-process orientation is the procedure-orientation. The latter orientation is also widely accepted in the food industry because procedures are considered as a powerful tool to provide assurance in technological processes. However, in practice the procedure-orientation often gives rise to complications because the effects of human behaviour and decision-making processes on the accurate execution of procedures are underestimated. The people-orientation is not widely applied in food production because it is often considered as a soft and intangible approach.

Considering the policy orientations in agribusiness and food companies, it is not surprising that QA-systems are commonly perceived as technological support tools with a predictable outcome. This might explain why these systems (e.g. HACCP, ISO) are rather appreciated by the food industry. Nevertheless, it is clear that these QA systems have their limitations. In practice, employees and operators experience that these QA systems are forced and that they are too dominant in the internal quality system, while the benefits are not clear. Those systems that are developed by operators and employees themselves and which are considered as useful, are the most successful and well accepted. This is one of the reasons why systems should be as simple and practical as possible. Overly complex quality management systems will not be effective in practice. Pressing operators to comply with the enforced guidelines will be experienced as unpleasant and can result in a discrepancy between operators and the quality department. External certification will only serve to increase the enforcement idea.

Conflicting requirements

Another typical problem for the food industry originates from the demands from the market situation. Considering the administrative concepts model (Figure 3.9) the market situation puts conflicting demands on food companies. On the one hand quality demands are set which make a strong appeal to the *bureaucratic administrative* concept such as:

- demands on high and advanced technological product quality, like for functional foods;
- legal obligations and market demands with respect to safety;
- increased requirements with respect to costs and price level.

This set of demands requires a consistent quality approach at a high level with a *high degree of assurance*. In practice, this is reflected in a high level of centralisation and bureaucratic systems with many rules based on legislative obligations.

However, on the other hand the continuously changing market and environment require a more *flexible concept*:

- high demands are put on service accompanied by a customer-oriented attitude;
- high demands on flexibility due to e.g. unexpected happenings, seasonal influences, etc.;
- the increasing reduction of life cycle of products, which leads to much focus on product development;
- increased mass customisation in the food industry leads to an enormous broadening of and major changes in product assortments.

Al these aspects require a *responsive quality approach* that places responsibilities low down in the organisation and that is not hindered by detailed rules and guidelines.

From the techno-managerial view point, the above-mentioned challenge should be approached from the following starting points:

- The technological discussion must be focused on the technological core competence of the business. If a company has a competitive position in the market for specific products (and the corresponding technology) then it cannot allow any failures.
- Technological core competence should be supported by a suitable quality management system, based primarily on GMP and HACCP and to a lesser extent ISO.
- Integration of the three policy orientations in a concept of total quality management (TQM), which is build upon strategic consumer values and core competence of the company.
- The quality management system must be a tool for all employees and is subservient to the TQM principles.
- The core competence of each unit in the production chain should take advantage of intensive collaboration and chain development. This process can be started by extensive exchange of information between units of the production chain.

In conclusion, the total food production chain needs to focus on the technological core competence coupled to the principles of total quality management in order to comply with customer value and consumer satisfaction.

Chapter 9

Food quality management in perspective

The current situation in the food market and the agri-food production chain is still very turbulent due to changing consumer requirements, environmental issues, governmental interests, and internationalisation. In this chapter, firstly the current dynamics of the agri-food production chain will be described. Then, the approach in this book is reflected into a concise picture of core developments in future food quality management. Finally, the principles behind food quality management research are very briefly described.

9.1 Dynamics in the agri-food production chain

Agribusiness and the food industry have changed from a productivity/costs focus towards a customer focus resulting in a dynamic situation in the agri-food chain. From a chain-oriented perspective three major cycles of change can be distinguished:

- The first cycle encompasses *developments in the market*. The decreasing life cycle of products and rapidly changing preferences of consumers lead to more impulsive behaviour by consumers who have become a moving target for product developers.
- The second cycle refers to the *technologies associated with processing and production systems*. In general, innovation in technologies is more slow than changes in the market situation.
- The third cycle deals with *plant breeding and primary production* and is actually the slowest cycle even with the use of modern biotechnology. Short-term changes in the market are impossible to follow.

From the chain perspective it is of the utmost importance for those actors active in the primary production cycle or the processing cycle, to have a strategic view of market developments and to identify market niches, where they can be strong and ahead of their competitors. On the other side, it is of paramount importance to have a clear view of technological possibilities and limitations from the processing and the primary production perspective to be able to follow the most effective quality design strategy.

One other development, relevant in this context, is the globalisation of food production systems. Raw materials are taken from all over the world, as are finished food products. In Western countries consumers are used to the fact that all kind of food products are available all year round. Trade barriers are disappearing, reinforcing this type of development. From this perspective it is obvious that strategic collaboration within the supply chain is of paramount importance.

Considering the cycles in more detail reveals that each actor in the chain faces different requirements and environmental influences. On the *consumer* side, the many changes in the market for new food products call for a repositioning of existing food production systems and raise the question as to whether the concepts currently used can survive the challenges of the future. Apart from market saturation, a number of other developments have a large influence on market conditions. In general, consumers are becoming better educated and more demanding. They have become less predictable in their purchasing behaviour, they eat out more and are more conscious about health-related aspects. As a result there is a continuous need for new products and a more differentiated food product assortment. One typical example is the ongoing developments in the functional food area. Related to these developments product life cycles have become shorter, and the efficiency and flexibility of food production systems have become even more important.

In addition to the continuously changing consumer demands, *retailers* have become more powerful and put high demands on food companies (with respect to safety, quality, profits, and circulation rate). Retailers are in fact the major link between consumers and food companies. Their knowledge about consumer purchases is even increased due to the use of information technology, which puts them in a very powerful position. Food *manufacturing companies* attempt to anticipate these dynamics by concentrating on product development, intensifying their knowledge potential, geographic expansion and differentiation by brands.

The *primary sector* also faces dynamic demands. Due to several serious food safety problems, which originated in the primary sector, much attention has to be paid to improvement of quality systems. Furthermore, they have increased costs due to environmental regulations, labour and soil prices and there is pressure on prices due to liberalisation of the market.

The various technological and organisational changes in the food supply chain put great demands on food quality management. More specifically, typical technological changes, such as the trend towards less intensively processed food, higher level of preparation (like, ready-to-eat components and meals), incorporation of active health improving compounds, and replacement of products and ingredients places huge demands on quality control and assurance. In addition, more flexible supply chains, higher frequency of supply, new requirements on tracking and tracing, and an increasing number of food outlets have their consequences. The trend towards 'fresh on demand' with higher supply frequencies has increased the demands on quality management systems along the whole food supply chain. The increase in food outlets selling prepared, and therefore more vulnerable food products, puts increasing demands on hygienic behaviour, hygiene knowledge, and compliance with safety control procedures. Last but not least, the ongoing concern of public and government with regard to food safety and wholesome food production has also placed increasing demands on food quality management (Luning and Marcelis, 2009).

9.2 Core developments in food quality management

There are no immediate ready-to-implement solutions to the challenges that the agribusiness and food industries are confronted with. A profound analysis, based on adequate models, theories and different viewpoints, is necessary. Thinking along the lines of the techno-managerial approach, several models and methods have been presented in this book, which may help in an analysis of the problem situations and the development of solutions for them. Furthermore, in the last sections of Chapters 4 to 8, a characterisation of common problems in the food industry has been given. Summarising these sections, the following highlights arise:

The *functional organisation,* which is common in the food industry, provides too much of a basis for departments to meet their own interests and responsibilities, for centralisation of knowledge and for a top-down culture.

Although there is clear progress, food companies and agribusiness still face difficulties with fast *responsiveness* and full *market orientation*. The conflicting requirements, responsiveness on the one hand and a high degree of assurance on the other, tend to be solved in the direction of assurance. This seems to favour the formal quality assurance systems.

In addition, the food industry does not seem to handle *knowledge* with care. Much of the knowledge remains with the suppliers of basic materials and equipment and is not available for the users. Internal knowledge is also poorly maintained and stored and feedback to production and other places in the organisation, where quality can be improved, is not always organised. This is a problem when one realises that knowledge plays an important role in developing core competencies of the organisation.

Chain collaboration is still difficult due to lack of mutual trust and belief in the advantages of co-operation. In addition, the fact that actors consider quality from different viewpoints often results in a variable understanding of the concept of product quality.

In the previous chapters it has been made clear that these problems may result in inadequate fulfilment of the different quality management functions. Since these problems commonly occur simultaneously negative effects are enhanced. Considering the dynamics in the agri-food production chain as mentioned before, it can be concluded that the situation is becoming worse. Modern developments, like customer orientation, market competition, chain collaboration and advanced technology, typically require flexible organisations and more use of knowledge.

In this book possible solutions have been proposed from the techno-managerial viewpoint. The different directions of solutions have been summarised in Figure 9.1, which is a simplified representation of the core model of this book (Figure 1.5), for a series of companies in the food supply chain. In Figure 9.1, the core developments in food quality management have been depicted based on considerations described in the book and experiences of the authors.

The three core developments as depicted in Figure 9.1 are:

1. Focus on *critical quality points* (CQP). In order to comply with the conflicting requirements: assurance on the one hand and responsiveness on the other hand, the points in the process that should be assessed are those that have a decisive influence on final product quality. These points are often related to the technological core competence of the company. These critical quality points require an in-depth analysis with respect to necessary investment, research and control activities based on detailed consideration of costs, environmental impact, and meeting consumer demands. The methods and principles of risk analysis are expected to play a major role. Moreover, the use of predictive modelling can accelerate and optimise design processes, which will support the technological core competence of the company. Developments in information and communication technology will also support above developments.

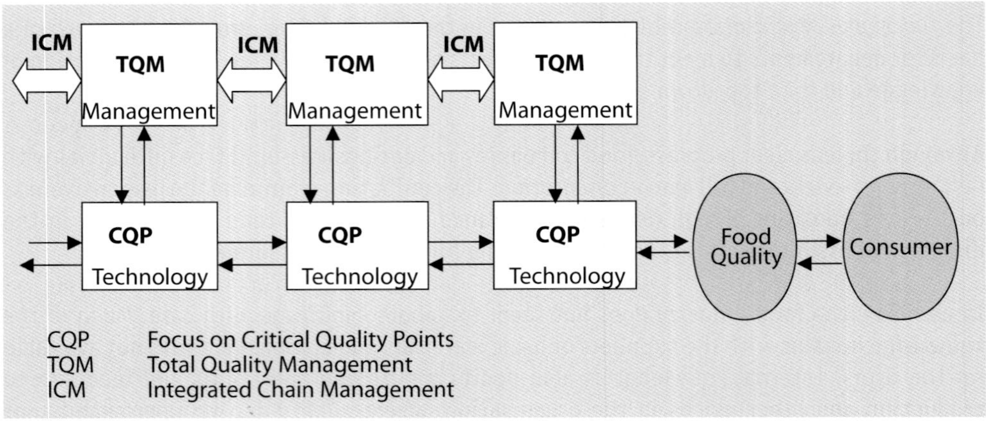

Figure 9.1. Core developments in food quality management.

2. *Total quality management* (TQM). The principles of total quality management should be embraced in agribusiness and the food industry. To avoid the disadvantages of the functional organisation, the principles of empowerment in combination with customer focus will result in a transformation from a vertical into a horizontal organisation. A potential risk of this transformation is that knowledge development and management do not receive sufficient attention and will be dispersed throughout the organisation. Consequently, management has to create an organisational structure, in which knowledge-intensive departments, like for instance the quality department, the engineering department and the maintenance department, will have a strong position. Of course, good co-operation between these departments and the production departments is vital. Advanced knowledge management systems are useful tools in achieving this.
   Teambuilding will be more important for the responsiveness of the company as well as knowledge integration in product design and business development and will therefore become an important tool in management decision-making. In this context the use of current quality systems such as ISO should no longer be a top-down assignment of procedures but a tool that assists and guides everyone involved in quality management. As a result systems will become lean and bureaucratic overkill will be avoided.
   Commitment of top management will be crucial, and they will have to take the initiative to integrate the technological and managerial aspects. A climate of quality consciousness should be created in which continuous improvement is a daily challenge. The use of discipline-integrating methods and concepts like QFD and HACCP will be very important as a catalyst for this type of collaboration and knowledge development.
3. *Integrated chain management* (ICM). Food quality is increasingly dependent on the production and supply chain as a whole to create superior quality. Using this concept, technological developments will be triggered in such a way that each actor in the chain will be supported in improving areas of core competence and at the same time superior quality of the final product will be achieved. In addition, communication with consumers about all technological aspects throughout the food chain will be vital to obtain and improve trust. Transparency from the chain perspective remains one of the big issues for the near future. The complexity of the concept of food quality requires a consumer-oriented approach and as a consequence a commonly developed quality

notion all along the chain. Collaboration is crucial and will be extended to innovation and design processes.

Quality systems will have to be redesigned into chain quality systems from the perspective of efficiency as well as assurance as a future consumer demand.

9.3 Food quality management research

The combined use of the core elements, mentioned above, offers opportunities for policy-making in food quality. However, policy-making is very complex and has to be supported by in-depth research.

Food quality management research is rather specific because it deals with paradigms from natural (food technology) and social sciences (management), and it deals with two research types, respectively managerial problem-solving and fundamental scientific research.

Commonly two distinctive research types are distinguished based on the different ways they deal with theory and practice, i.e. *managerial problem-solving* and *fundamental scientific research* (Gill and Johnson, 1997; Teale *et al.*, 2003). In managerial problem solving the researcher starts in practice with a problem and stays there, using theory mainly unconsciously. Information is gathered to analyse the situation and to develop alternative solutions in a process of continuously matching goals and ways to reach them. It is a process of *selection* of alternative choice options, resulting in a decision at a certain moment, and is thus finite. In fundamental scientific research the researcher starts in theory and basically stays there. It starts with a question, driven by a feeling of curiosity or amazement. Then more and more information is gathered (e.g. by analyses, observations), followed by continuous matching of objective facts and creative ideas. It is a process of *cognition* resulting in knowledge, and is more or less infinite (Luning and Marcelis, 2009).

Food quality management research should be positioned between both types of research. It must result in solutions for the problem situation, while at the same time high requirements are put on the dependability of the provided solutions, due to customer, public, and or governmental concerns. Food quality management research differs from managerial problem-solving in consciously using theory, and it differs from fundamental scientific research in being aimed at solving problems in practice.

As mentioned above, food quality management issues need to be analysed from a technological and managerial perspective using theories from both disciplines. As a consequence, *paradigms* from food technology and management sciences must be acknowledged. It should be noted that both disciplines represent different worlds of science. Paradigms of technological disciplines basically fit with a positivistic viewpoint, in which the world is external and objective, the observer is independent, and science is value-free. On the other hand, paradigms of managerial disciplines essentially fit with a phenomenological viewpoint, in which the world is socially constructed and subjective, and the observer cannot detach him/herself from what is observed, and human interests drive science (Teale *et al.*, 2003).

For FQM research it means that *researchers* in relation to the technological and managerial theories deal with questions like: to what extent can I work in diverse paradigms, what is my level of critical awareness of possible theories, what are my skills/experiences in using them, and what is my personality, cognitive style? Researchers dealing with the situation in practice should ask questions like: who do I see as problem owners, what is my commitment to them, what are the expectations about my role (e.g. facilitator or expert), and what resources are available (access, time, participants)? Finally, the researcher should critically analyse the relationship between practice and theory, dealing with questions such as, what is the culture of the organisation with regard to using certain theories, and what is the history of past theory use, and to what extent are certain values embedded in the theories appropriate to the situation (Mingers and Gill, 1997; Luning and Marcelis, 2009).

Despite the difficulties mentioned, food quality management researchers are required to execute profound analysis and we hope that the methods, concepts and models as proposed in this book might contribute to this process.

References

Abadias, M., Usall, J., Anguera, M., Solsona, C. and Vinas, I. (2008). Microbiological quality of fresh, minimally – processed fruit and vegetables, and sprouts from retail establishments. *International Journal of Food Microbiology*, 123, 121-129.

Abu-Ruwaida, A.S., Sawaya, W.N., Dashti, B.H., Murad, M. and Al-Othman, H.A. (1994). Microbiological quality of broilers during processing in a modern commercial slaughterhouse in Kuwait. *Journal of Food Protection*, 57 (10), 887-892.

Aladjadjiyan, A. (2006). Physical hazards in the agri-food chain. In: Luning, P.A., Devlieghere, F. and Verhé, R. (eds.) Safety in the agri-food chain. Wageningen Academic Publishers, Wageningen, the Netherlands, pp. 209-222.

Alwitt, L. and Berger, I. (1993). Understanding the link between environmental attitudes and consumer product usage: measuring the moderate role of attitude strength. In: McAlister, L. and Rothschild, M.L. (eds.) Advances in consumer research, Volume 20. Association for Consumer Research, Provo, UT, USA, pp. 189-194.

Analoui, F. (1999). Effective human resource development: a challenge for developing countries. Ashgate Publishing Ltd., Aldershot, UK. 320 pp.

Andersen, H.J., Oksbjerg, N., Young, J.F. and Therkildsen, M. (2005). Feeding and meat quality - a future approach. *Meat Science*, 70 (3), 543-554.

Antony, J., Kumar, M. and Madu, C.N. (2005). Six sigma in small and medium-sized UK manufacturing enterprises: some empirical observations. *The International Journal of Quality and Reliability Management*, 22 (8), 860-874.

Anttonen, M.K., Hoppula, K.I., Nestby, R., Verheul, M.J. and Karjalainen, R.O. (2006). Influence of fertilization, mulch color, early forcing, fruit order, planting date, shading, growing environment, and genotype on the contents of selected phenolics in strawberry (*Fragaria × ananassa* Duch.) fruits. *Journal of Agricultural and Food Chemistry*, 54 (7), 2614-2620.

Argyris, C. (1982). The executive mind and double-loop learning. *Organisational dynamics*, 11 (Autumn), 5-22.

Armstrong, M. (1996). A handbook of personnel management practice. Kogan Page Ltd, London, UK. 928 pp.

Arsenos, G., Banos, G., Fortomaris, P., Katsaounis, N., Stamataris, C., Tsaras, L. and Zygoyiannis, D. (2002). Eating quality of lamb meat: effects of breed, sex, degree of maturity and nutritional management. *Meat Science*, 60 (4), 379-387.

Auldist, M.J. and Hubble, I.B. (1998). Effects of mastitis on raw milk and dairy products. *Australian Journal of Dairy*, 53 (1), 28-36.

Azanza, M.P.V., Ortega, M.P. and Valdezco, R.G. (2001). Microbial quality of rellenado milkfish (*Chanos chanos*, Forskal). *Food Control*, 12 (6), 365-371.

Azanza, M.P.V. and Zamora-Luna, M.B.V. (2005). Barriers of HACCP team members to guideline adherence. *Food Control*, 16, 15-22.

Banens, P.J.A., Does, R.J.M.M., Van Dongen, G.B.W., Engel, J., Hasselaar, M.M.A., Lieshout, R.A.J.M., Praagman, J., Schriever, B.F., Trip, A. and Van der Veen, H. (1994). Industriële statistiek en kwaliteit. Kluwer Bedrijfswetenschappen BV., Deventer, the Netherlands. 432 pp. (in Dutch).

Baner, A.L., Franz, R. and Piringer, O. (1999). Alternative fatty food stimulants for polymer migration testing. In: Mathlouthi, M. (ed.) Food packaging and preservation. Blackie Academic Professional, London, UK, pp. 23-47.

Barbut, S., Zhang, L. and Marcone, M. (2005). Effects of pale, normal, and dark chicken breast meat on microstructure, extractable proteins, and cooking of marinated fillets. *Poultry Science*, 84 (5), 797-802.

Barlow, S.M., Greig, J.B., Bridges, J.W., Carere, A., Carpy, A.J.M.,Galli, C.L., Kleiner, J., Knudsen, I., Koëter, H.B.W.M., Levy, L.S., Madsen, C., Mayer, S., Narbonne, J.-F., Pfannkuch, F., Prodanchuk, M.G., Smith, M.R. and Steinberg, P. (2002). Hazard identification by methods of animal-based toxicology. *Food and Chemical Toxicology*, 40 (2-3), 145-191.

Bartz, J.A. and Brecht, J.K. (2002). Postharvest physiology and pathology of vegetables. 2nd edition. CRC press, Boca Raton, FL, USA. 744 pp.

Beckley, J.H., Foley, M.M. and Prinyawiwatku, W. (2007). Accelerating new food product design and development. Blackwell Publishing, London, UK. 400 pp.

Benner, M., Jongen, W.M.F., Linnemann, A.R. and Folstar, P. (2003). Quality function deployment (QFD). Can it be used to develop food products? *Food Quality and Preferences*, 14, 327-339.

Bergman, B. and Klefsjö, B (1994). Quality, from customer needs to customer satisfaction. McGraw-Hill, New York, NY, USA. 606 pp.

Bessant, J. and Francis, D. (1999). Developing strategic continuous improvement capability. *International Journal of Operations & Production Management*, 19, 1106-1119.

Beuchat, L.R. and Mann, D.A. (2008). Survival and growth of acid-adapted and unadapted *Salmonella* in and on raw tomatoes as affected by variety, stage of ripeness and storage temperature. *Journal of Food Protection*, 71 (8), 1572-1579.

Bhaskaran, S., Polonsky, M., Cary, J. and Fernandez, S. (2006). Environmentally sustainable food production and marketing - Opportunity or hype? *British Food Journal*, 108 (8), 677-690.

Blake, R.R., Mouton, J.S., Barnes, B.L. and Greiner, L.E. (1964). Breakthrough in organization development. *Harvard Business Review*, 42 (6), 133-155.

Boddy, D. and Paton, R. (1998). Management, an introduction. Prentice Hall, London, UK. 728 pp.

Boom, R.M., Dekker, M. and Esveld, D.C. (2005). Food production: trends in system innovation. In: Jongen, W.M.F. and Meulenberg, M.T.G. (eds.) Innovation in agri-food systems. Product quality and consumer acceptance. Wageningen Academic Publishers, Wageningen, the Netherlands, pp. 173-206.

Bossert, J.L. (2004). The supplier management handbook. 6th edition. ASQ Customer-Supplier Division, Milwaukee, WI, USA. 296 pp.

Botta, R. (1995). Evaluation of seafood freshness quality. VCH Publishers, New York, NY, USA. 180 pp.

Botterbrodt, S. (2003). Quality standards in food and feed field. *Getreide, Mehl und Brot*, 75 (5), 311-316.

Boulding, K. (1956). General systems theory, the skeleton of science. *Management Science*, 2 (3), 197-208.

Bounds, G., Yorks, L., Adams, M. and Ranney, G. (1994). Beyond total quality management: towards the emerging paradigm. McGraw-Hill, New York, NY, USA. 832 pp.

Bourne, M.C. (2002). Food texture and viscosity: concept and measurement. Academic Press, Oxford, UK. 427 pp.

Bover-Cid, S., Izquierdo-Pulido, M. and Vidal-Carou, M.C. (2001). Effectiveness of a *Lactobacillus sakei* starter culture in the reduction of biogenic amine accumulation as a function of the raw material quality. *Journal of Food Protection*, 64 (3), 367-373.

Branen, A.L., Davidson, P.M., Salminen, S. and Thorngate III, J.H. (2002). Food additives. 2nd edition. CRC Press, Boca Raton, FL, USA. 952 pp.

Bremmer, H.A. (2000). Toward practical definitions of quality for food science. *Critical Reviews in Food Science and Nutrition*, 40 (1), 83-90.

Brinkman, J.H.M. and Hogendoorn, F. (2006) Proeven van succes: sensorisch onderzoek: technieken, procedures en toepassingen. 2nd editie. Keesing Noordervliet, Houten, the Netherlands. 230 pp.

Brown, M. and Stringer, M. (2002). Microbiological risk assessment in food processing. Woodhead Publishing Ltd., Cambridge, UK. 301 pp.

Bruhn, C.M., Diaz-Knauf, K., Feldman, N., Harwood, J., Ho, G., Ivans, E., Kubin, L., Lamp, C., Marshall, M., Osaki, S., Stanford, G., Steinbring, Y., Valdez, I., Williamson, E. and Wunderlich, E. (2007). Consumer food safety concerns and interest in pesticide-related information. *Journal of Food Safety*, 12 (3), 253-262.

Brul, S., Van Gerwen, S. and Zwietering, M. (2007). Modelling microorganisms in food. Woodhead Publishing Limited, Cambridge, UK. 294 pp.

Bryan, F.L. (2002). Where we are in retail food safety. How we got to where we are and how do we get there. *Journal of Environmental Health*, 65(2), 29-36.

Cardello, A.V. (1996). The role of human senses in food acceptance. In: Meiselman, H.L. and MacFie, H.J.H. (eds.) Food choice, acceptance and consumption. Blackie Academic & Professional, London, UK, pp. 1-81.

Cartea, M.E. and Velasco, P. (2007). Glucosinolates in *Brassica* foods: bioavailability in food and significance for human health. *Phytochemistry Reviews*, 7, 213-229.

Chawla, S.P. and Chander, R. (2004). Microbiological safety of shelf-stable meat products prepared by employing hurdle technology. *Food Control*, 15 (7), 559-563.

Clayton, D.A., Griffith, C.J. Price, P. and Peters, A.C. (2002). Food handlers´ beliefs and self-reported practices. *International Journal of Environmental Health Research*, 12, 25-39.

Clements, M.D., Lazo, R.M. and Martin, S.K. (2008). Relationship connectors in NZ fresh produce supply chains. *British Food Journal*, 110 (4-5), 346-360.

Codex Alimentarius Commission (1998). A joint FAO/WHO food standards programme. Report of 31st session of the codex committee on food hygiene, Alinorm 99/13A, 1998.

Codex Alimentarius Commission (2003a). Hazard analysis and critical control point (HACCP) system and guidelines for its application. ANNEX to Recommended International Code of Practice/General Principles of Food Hygiene. CAC/RCP 1-1969, Rev 4. FAO/WHO Codex Alimentarius Commission.

Codex Alimentarius Commission (2003b). Good hygienic practices according to the Codex General Principles of. Food Hygiene. CAC/RCP 1-1969, Rev.4-2003.

Cohen, L. (1995). Quality function deployment. How to make QFD work for you. Engineering Process Improvement Series. Addison-Wesley Publishing Company, Reading, MA, USA. 348 pp.

Cooper, R.G. (2001). Winning at new products: accelerating the process from idea to launch. 3rd edition. Perseus Books, Reading, MA, USA. 416 pp.

Cordy, R., Lanni, V., Kistler, V., Dato, V., Weltman, A., Yozviak, C., Waller, K., Nalluswami, K., Moll, M., Lockett, J., Lynch, M., Braden, C., Gupta, S.K. and DuBois, A. (2005). Outbreak of *Salmonella* infections associated with eating Roma tomatoes. United States and Canada, 2004. *Morbidity and Mortality Weekly Report*, 54, 325-328.

Costa, A.I.A., Dekker, M. and Jongen, W.M.F. (2000). Quality function deployment in the food industry: a review. *Trends in Food Science and Technology*, 11, 306-314.

Crisosto, C.H., Mitchell, F.G. and Johnson, S. (1995). Factors in fresh market stone fruit quality Review. *Postharvest News and Information*, 6 (2), 17N-21N.

Crosby, P.B. (1979). Quality is free. The art of making quality certain. McGraw Hill, New York, NY, USA. 270 pp.

Crowther, J., Herd, T. and Michels, M. (1993). Food safety education and awareness: a model training programme for managers in the food industry. *Food Control*, 4 (2), 97-100.

Cunningham, E.P. (2005). Gene-based technologies for livestock industries in the 3rd millenium. In: H.P.S. Makkar and G.J. Viljoen (eds.) Applications of gene-based technologies for improving animal production and health in developing countries. Springer, Dordrecht, the Netherlands, pp 1-21.

Davies, M.B., Austin, J. and Partridge, D.A. (1991). Vitamin C, its chemistry and biochemistry. Royal Society of Chemistry, Cambridge, UK. 164 pp.

De Meulenaer, B. (2006). Chemical hazards. In: P.A. Luning, F. Devlieghere and R. Verhe. Safety in the agri-food chain. Wageningen Academic Publishers, Wageningen, the Netherlands, pp. 145-208.

De Sitter, H. (1998). Regulatory aspects of food production systems. In: Jongen W.M.F. and Meulenberg, M.T.G. (eds.). Innovation of food production systems. Wageningen Pers, Wageningen, the Netherlands, pp. 159-175.

Dean, J.W. and Evans, J.R. (1994). Total quality: management, organization and society. West Publishing Company, New York, NY, USA, 376 pp.

Dekker, M., Verkerk, R. and Jongen, W.M.F. (2000). Predictive modelling of health aspects in the food production chain: a case study on glucosinolates in cabbage. *Trends in Food Science & Technology*, 11 (4-5), 174-181.

Deming, W.E. (1986). Out of the crisis. MIT Centre for advanced Engineering Studies, Cambridge, MA, USA. 507 pp.

Deming, W.E. (1993). The new economics for industry, government, education. MIT Centre for Advanced Engineering study, Cambridge, MA, USA. 266 pp.

Denzau A.T. and North D.C. (1994). Shared mental models: ideologies and institutions. *Kyklos: International Review for Social Sciences*, 47, 3-31.

Dessler, G. (2001). Management: leading people and organisation in 21st century. Prentice Hall, London, UK. 688 pp.

Devlieghere, F., Francois, K., De Meulenaer, B. and Baert, K. (2006). Modelling food safety. In: P.A. Luning, F. Devlieghere and R. Verhe (eds.) Safety in the agri-food chain. Wageningen Academic Publishers, Wageningen, the Netherlands, pp. 397-438.

Dijksterhuis, G.B. and Piggott, J.R. (2000). Dynamic methods of sensory analysis. *Trends in Food Science & Technology*, 11(8): 284-290.

Dillon, M. and Griffith, C. (2000) Auditing in the food industry, from safety and quality to environmental and other audits. Woodhead Publishing Ltd., Cambridge, UK. 240 pp.

Doyle J.K. and Ford D.N. (1998). Mental models concepts for system dynamics research. *System Dynamics Review*, 14 (1), 3-29.

Doyle, M.P. and Erickson, M.C. (2008). Summer meeting 2007 - the problems with fresh produce: an overview. *Journal of Applied Microbiology*, 105, 317-330.

Downey. W.D. (1996). The challenge of food and agri product supply chains. In: J.H. Trienekes and P.J. Zuurbier (eds.) Proceedings of the 2nd International Conference on chain management in agri-and food business. Wageningen Agricultural University, Wageningen, the Netherlands, pp. 3-13.

Dufey, P.A. (1995). Fleisch- und Fettqualitat bei Schweinemast mit Weidegang. *Agrarforschung*, 2, 453-456.

Dymnicka, M., Klupczynski, J., Lozicki, A., Micinski, J. and Strzetelski, J. (2004). Polyunsaturated fatty acids in M-longissimus thoracis of fattening bulls fed silage of grass or maize. *Journal of Animal and Feed Sciences*, 13, 101-104.

Edge J. (1990). Quality improvement. In: Lock, D. (ed.) Handbook of quality management. Gower Publishing Company, Aldershot, UK, chapters 18, 19 and 20.

Edwards, S.A. and Casabianca, F. (1997). Perception and reality of product quality from outdoor pig systems in Northern and Southern Europe. In: Sorensen, J.T. (ed.) Livestock farming systems. More than food production. Proceedings of the 4th International Symposium on Livestock farming systems. EAAP Publication No 89. Wageningen Pers, Wageningen, the Netherlands, pp. 145-156.

Edwards, D.B., Bates, R.O. and Osburn, W.N. (2003). Evaluation of Duroc- vs. Pietrain-sired pigs for carcass and meat quality measures. *Journal of Animal Science*, 81, 1895-1899.

Edwards, S.A. (2005). Product quality attributes associated with outdoor pig production. *Livestock Production Science*, 94 (1-2), 5-14.

Efstratiadis, M.M., Karirti, A.C. and Arvanitoyannis, I.S. (2000). Implementation of ISO 9000 to the food industry: an overview. *International Journal of Food Sciences and Nutrition*, 51 (6), 459-473.

Ellis, M.J. and Man, C.D.M. (2000). The methodology of shelf life determination. In: Man, C.M.D. and Jones, A.A.(eds.) Shelf life evaluation of foods. 2nd edition. Aspen Publishers, Gaiterburgh, MA, USA, pp. 23-33.

Emblem, A. (2000). Predicting packaging characteristics to improve shelf life. In: Kilcast, D. and Subramaniam, P. (eds.) The stability of shelf life of food. Woodhead Publishing Ltd., Cambridge, UK, pp. 145-169.

Erkmen, O. and Dogan, C. (2004). Effects of ultra high hydrostatic pressure on *Listeria monocytogenes* and natural flora in broth, milk and fruit juices. *International Journal of Food Science & Technology*, 39 (1), 91-97.

Escriche, I., Domenech, E. and Baert, K. (2006). Design and implementation of an HACCP system In: P.A. Luning, F. Devlieghere and R. Verhe (eds.) Safety in the agri-food chain. Wageningen Academic Publishers, Wageningen, the Netherlands, pp. 303-354.

Evans, J.R. and Lindsay, W.M. (2005). The management and control of quality. 6th edition. Thomson Corporation, South Western, Ohio, USA. 848 pp.

Evans, C.R. and Packer, L. (2003). Flavonoids in health and disease. CRC press, Boca Raton, FL, USA. 504 pp.

FAO/WHO Food Standards Programme (2001). Codex Alimentarius - Food Hygiene - Basic Texts - Second Edition. FAO, Rome, Italy. 72 pp. Available at: ftp://ftp.fao.org/docrep/fao/005/Y1579e/Y1579e.pdf.

FAO (2008): Microbiological hazards in fresh fruits and vegetables. Available at: http://www.fao.org/ag/agn/agns/files/FFV_2007_Final.pdf.

Feigenbaum, A. (1961). Total quality control. McGraw-Hill, New York, NY, USA. 896 pp.

Feil, B. and Stamp, P. (1993). Sustainable agriculture and product quality: a case study for selected crops. *Food Reviews International*, 9, 361-388.

Fennema, O.R. (1996). Food chemistry. 3rd edition. Marcel Dekker. New York, NY, USA. 1088 pp.

Ficarella, A., Milanese, M. and Laforgia, D. (2006). Numerical study of the extrusion process in cereals production: part II. Analysis of variance. *Journal of Food Engineering*, 72 (2), 179-188.

Fielding, L.M., Ellis, L. Beveridge, C. and Peters, A.C. (2005). An evaluation of HACCP implementation status in UK small and medium enterprises in food manufacturing. *International Journal of Environmental Health Research*, 15 (2), 117-126.

Figura, L.O. and Teixeira, A.A. (2007). Food physics: physical properties - measurement and applications. Springer-Verlag, Berlin, Germany. 550 pp.

Flament, I. (2002). Coffee flavor chemistry. John Wiley & Sons, Chichester, UK. 424 pp.

Food and Nutrition Board (2005). Dietary reference intake for energy, carbohydrate, fibre, fat, fatty acids, cholesterol, protein, and amino acids (macronutrients). Report of the panel of macronutrients. The National Academic Press, Washington DC, WA, USA. 1357 pp.

Forsythe, S.J. and Hayes, P.R. (1998). Food spoilage. In: Forsythe, S.J. and Hayes, P.R. (eds). Food hygiene, microbiology and HACCP. Aspen Publishers Inc., Gaithersburg, MA, USA, pp. 86-149.

Forza, C. and Salvador, F. (2001). Information flows for high-performance manufacturing. International *Journal of Production Economics*, 70, 21-36.

Foster, C., Green, K., Bleda, M., Dewick, E., Evans, B., Flynn, A. and Mylan, J. (2006). Environmental impact of food production and consumption. A report to the Department of Environment, Food and Rural Affairs. Manchester Business School. Defra, London, UK. 198 pp. Available at: http://www.heartland.org/custom/semod_policybot/pdf/20893.pdf.

French, J. and Raven, B. (1960). The bases of social power. In: D. Cartwright and A. Zanders (eds). Group dynamics: research and theory. Harper and Row, New York, NY, USA. 826 pp.

Frewer, L.J., Howard, C., Hedderley, D. and Shepherd, R. (1997). Consumer attitude towards different food-processing technologies used in cheese production. The influence of consumer benefit. *Food Quality and Preferences*, 8, 271-280.

Fuller, G.W. (2004). New food product development: from concept to marketplace. CRC Press, Boca Raton, FL, USA. 388 pp.

Gade, P.B. and Christensen, L. (1998). Effect of different stocking densities during transport on welfare and meat quality in Danish slaughter pigs. *Meat Science*, 48 (3-4), 237-247.

Galema, S.A. (1997). Microwave chemistry. *Chemical Society Reviews*, 26, 233-238.

Garvin, D.A. (1984). What does product quality really mean? *Sloan Management Review*, 26, 25-43.

Garvin, D. (1993). Building a learning organization. *Harvard Business Review*, 71 (4), 78-91.

Gattorna, J.L. and Walters, D.W. (1996). Managing the supply chain - a strategic perspective. Palgrave Macmillan, Basingstoke, UK. 384 pp.

George, S. and Weimerskirch, A. (1994). Total quality management: strategies and techniques proven at today's most successful companies. John Wiley and Sons, New York, NY, USA. 254 pp.

Gerats, G.E.C. (1990). Working towards quality (in Dutch with a summary in English). Dissertation, University of Utrecht, Utrecht, the Netherlands.

Gill, J. and Johnson, P. (1997). Research methods for managers. 2nd edition. Paul Chapman Publishing, London, UK. 234 pp.

Gonzalez, A.R., Mauromoudtakos, A., Prokakis, G. and Aselage, J. (1992). Influence of year, cultivar and fruit maturity on quality of peach puree. *Journal of Food Quality*, 15 (2), 97-109.

Gordon, S.R. (2008). Supplier evaluation and performance excellence: a guide to meaningful metrics and successful results. ASQ Customer-Supplier Division, Milwaukee, WI, USA. 232 pp.

Gould, W.A. and Gould, R.W. (2001) Total quality assurance for the food industries. 3rd edition. CTI Publications Inc., MA, USA. 500 pp.

Griffin, A and Hauser, J.R. (1996). Integrating R&D and marketing: a review and analysis of the literature. *Journal of Product Innovation Management*, 13 (3), 191-215.

Grunert, K.G. (2005). Food quality and safety: consumer perception and demand. *European Review of Agricultural Economics*, 32 (3), 369-391.

Haard, N.F. (1985). Characteristics of edible plants tissues. In: Fennema, O. (ed.) Food chemistry. Marcel Dekker Inc., New York, NY, USA, pp. 857-911.

Hagg, H., Ahvenainen, R., Evers, A.M. and Tiilikkala, K. (1999) Agri-food quality II: quality management of fruit and vegetables. Woodhead Publishing Limited, Camebridge, UK. 378 pp.

Hamel, G. and Prahalad C.K. (1994). Competing for the future. Harvard Business School Press, Boston, USA. 357 pp.

Hammer, M. and Champy, J. (1993). Re-engineering the corporation: a manifesto for business revolution. Harper Collins, New York, NY, USA. 272 pp.

Harding, F. (1996). Milk quality. Springer Verlag, Berlin, Germany. 166 pp.

Harris, S.R. (1988). Production is only half the battle - a training manual in fresh produce marketing for the Eastern Caribbean. FAO, Rome, Italy. Available at: http://www.fao.org/wairdocs/X5014E/X5014e00.htm.

Heeschen, W.H. and Harding, F. (1995). Contaminants. In: Harding, F. (ed.) Milk quality. Springer Verlag, Berlin, Germany, pp. 133-150.

Heizer, J. and Render, B. (2006). Operations management. 8th edition. Prentice Hall, Englewood Cliffs, NJ, USA. 672 pp.

Hefle, S. (1999). Impact of processing on food allergens. In: Jackson, L.S., Knize, M.G. and Morgan, J.N. (eds.) Impact of food processing on food safety. Advances in experimental medicine and biology, volume 459. Kluwer Academic, New York, NY, USA, pp. 107-120.

Hellriegel, D., Jackson, S. and Slocum, J.W. (2005). Management. a competency-based approach. Addison-Wesley, New York, NY, USA. 608 pp.

Hellsten, U. and Klefsjö, B. (2000). TQM as a management system consisting of values, techniques and tools. *The TQM Magazine*, 12 (4), 238-244.

Henning, H. (1988). Sozialpolitik, III: Geschichte. Handwörterbuch der Wirtschaftswissenschaften. Stuttgart u.a., Vol. 7, pp. 85-110.

Hersey, P. and Blanchard, K. (1988). Management and organizational behavior. Prentice Hall, Englewoods Cliffs, NJ, USA. 368 pp.

Hogarth, R.H., Kahneman, D. and Tversky, A. (1994). In: M. Bazerman (ed.) Judgement in managerial decision making. John Wiley and Sons, New York, NY, USA.

Hollowood, T.A. (2002). Taste-aroma-matrix interactions determine flavour perception. PhD thesis, University of Nottingham, Nottingham, UK.

Holt, S.H.A., Brand Miller, J.C, Petocz, P. and Farmakalidis, E. (1995). A satiety index of common foods. *European Journal of Clinical Nutrition*, 9, 675-690.

Hopkins, D.L., Hegarty, R.S. and Farrell, T.C. (2005). Relationship between sire estimated breeding values and the meat and eating quality of meat from their progeny grown on two planes of nutrition. *Australian journal of Experimental Agriculture*, 45, 525-533.

Howard, L.W. and Foster, S.T. (1999). The influence of human resource practices on empowerment and employee perceptions of management commitment to quality. *Journal of Quality Management*, 4 (1), 5-22.

Howes, M.S., McEwen, S., Griffiths, M. and Harris, L. (1996). Food handler certification by home study: measuring changes in knowledge and behaviour. *Dairy Food and Environmental Sanitation*, 16, 737-744.

Hubbard, M.R. (1996). Statistical quality control for the food industry. Chapman and Hall, New York, NY, USA. 346 pp.

Imai, M. (1986). Kaizen, the key to Japan's competitive success. Random House, New York, NY, USA. 260 pp.

Ingham, S.C., Losinski, J.A., Andrews, M.P., Breuer, J.E., Breuer, J.R, Wood, T.M. and Wright, T.H. (2004). *Escherichia coli* contamination of vegetables grown in soils fertilized with non composted bovine manure: garden-scale studies. *Applied Environmental Microbiology*, 70, 6420-6427.

Ishikawa, K. (1985). What is total quality control? The Japanese way. Prentice Hall, Englewood Cliffs, NJ, USA. 215 pp.

International Standardisation Organisation (ISO) (1999). ISO/CD2 9001:2000. Quality management systems - Requirements. Document ISO/TC/SC 2/N 434. Available at: http://www.iso.org/iso/home.htm.

International Standardisation Organisation (ISO) (2005). ISO 9000. Quality management and quality assurance – Vocabulary. Available at: http://www.iso.org/iso/home.htm.

International Standardisation Organisation (ISO) (2005). ISO 9000. Quality management systems - Fundamentals and vocabulary. Available at: http://www.iso.org/iso.

International Standardisation Organisation (ISO) (2004). ISO 10002:2004. Customer satisfaction - Customer complaint procedure. Available at: http://www.iso.org/iso.

International Standardisation Organisation (ISO) (2006). ISO 2859-10:2006 Sampling procedures for inspection by attributes - Part 10: introduction to the ISO 2859 series of standards for sampling for inspection by attributes. Available at: http://www.iso.org/iso/iso_catalogue.

Ivancevich, J.M., Lorenzi, P. and Skinner, S. (1994). Management, quality and competitiveness. Irwin, Boston, USA. 656 pp.

Jabnoun, N. (2002). Control processes for TQM. *Work study,* 51 (4), 182-190.

Jacxsens, L., Kussaga, J., Luning, P.A., Van der Spiegel, M., Uyttendale, M. and Devlieghere, F. (2009). A microbial assessment scheme to support microbial performance measurements of food safety management systems. *International Journal of food Microbiology.* DOI: 10.1016/j.ijfoodmicro.2009.02.018.

Jay, J.M. (1996). Modern food microbiology. 5th edition. In: Heldman, D.R. (ed.). Chapman and Hall, New York, NY, USA. 661 pp.

Jay, J.M., Loessner, M.J. and David A. Golden, D.A. (2005). Modern food microbiology. 7th edition. Springer, Berlin, Germany. 790 pp.

Johnson, G. and Scholes, K. (2006). Exploring corporate strategy. Prentice Hall, London, UK. 672 pp.

Jones, A.A. (2000). Ambient-stable sauces and pickles. In: Man, C.M.D. and Jones, A.A. (eds). Shelf life evaluation of foods, 2nd edition. Aspen Publishers, Inc., Gaiterburgh, MA, USA, pp. 211-226.

Jones, P., Clarke-Hill, C., Shears, P. and Hillier, D. (2001). Retailing organic foods. *British Food Journal,* 103 (5), 358-365.

Jones, T.F. and Angulo, F.J. (2006). Eating in restaurants: a risk factor for foodborne disease? *Clinical Infectious Diseases,* 43 (10), 1324-1328.

Jongen, W. (ed.) (2005). Improving the safety of fresh fruit and vegetables. Woodhead Publishing Ltd., Camebridge, UK. 656 pp.

Jongen, W.M.F. and Meulenberg, M.T.G. (2005). Innovation in agri-food systems. Product quality and consumer acceptance. Wageningen Academic Publishers, Wageningen, the Netherlands. 399 pp.

Jongen, W.M.F. (2002). Fruit and vegetable processing: improving quality. Woodhead Publishing Ltd., Cambridge, UK. 388 pp.

Juran, J.M. (1951). Quality control handbook. McGraw Hill, New York, NY, USA. 1774 pp.

Juran, J.M. (1964). Managerial breakthrough. McGraw Hill, New York, NY, USA. 396 pp.

Juran, J.M. (1988). Juran on planning for quality. The Free Press, New York, NY, USA. 384 pp.

Juran, J.M. (1990). Juran on leadership for quality. The Free Press, New York, NY, USA. 384 pp.

Juran, J.M. (1992). Juran on quality by design. The new steps for planning quality into goods and services. The Free Press, New York, NY, USA. 538 pp.

Käferstein, F.K. (2004). Food safety - a worldwide public health issue. Actions to reverse the upward curve of foodborne illness. In: Smulders, F.J.M. and Collins, J.D. (eds.) Food safety assurance and veterinary public health: safety assurance during food processing. Wageningen Academic Publishers, Wageningen, the Netherlands, pp. 19-32.

Kamann, D.J.F. (1989). Actors in networks. In: F.W.M. Boekema and D.J.F. Kamann (eds.) Social-economical networks. Wolters-Noordhoff, Groningen, the Netherlands. (in Dutch).

Kampfraath, A.A. and Marcelis, W.J. (1981). Management and organisation. Kluwer, Deventer, the Netherlands. (in Dutch).

Karel, M. and Lund, D.B. (2003). Physical principles of food preservation. 2nd edition. Marcel Dekker, Inc., New York, NY, USA. 400 pp.

Kaufman, M. (2000). Antibiotics in animal feed. A growing public health hazard worries rise over effect of antibiotics in animal feed. Available at: http://www.organicconsumers.org/toxic/animalfeed.cfm.

Kays, S.J. (1991). Postharvest physiology and handling of perishable plant products. Kluwer Academic Publishers, Deventer, the Netherlands. 532 pp.

Kehoe, D.F. (1995). The fundamentals of quality management. Chapman and Hall, New York, NY, USA. 304 pp.

Kilcast, D. and Subramaniam, P. (2000). The stability and shelf life of food. Woodhead Publishing Ltd., Camebridge, UK. 350 pp.

Kjærnes, U. and Lavik, R. (2007). Farm animal welfare and food consumption practices. Results from surveys in seven countries. In: Kjærnes, U. Miele, M. and Roex, J. (eds.) Welfare quality reports No.2. Attitudes of consumers, retailers and producers to farm animal welfare. Cardiff University Press, Cardiff, UK, pp. 1-30.

Knee, M. (2002). Fruit quality and its biological basis. Sheffield Academic Press, Sheffield, UK. 279 pp.

Kolb, D.A. (1984). Experiential learning: experience as the source of learning and development. Prentice Hall, Englewood Cliffs, NJ, USA. 267 pp.

Korteweg, L.I. (1991). Complaints: compliment and warning. *Voedingsmiddelentechnologie*, 5 (in Dutch).

Knura, S., Gymnich, S., Rembialkowska, E. and Petersen, B. (2006). Agri-food production. In: P.A. Luning, F. Devlieghere and R. Verhe (eds.) Safety in the agri-food chain. Wageningen Academic Publishers, Wageningen, the Netherlands, pp. 19-66.

Krajewski, L., Ritzman, L. and Malhotra, M. (2007). Operations management, process and value chains. Prentice Hall, Englewood Cliffs, NJ, USA. 752 pp.

Kroes, R., Muller, D., Lambe, J., Lowik, M.R.H., Van Klaveren, J., Kleiner, J., Massey, R., Mayer, S., Urieta, I., Verger, P. and Visconti, A. (2002). Assessment of intake from the diet. *Food Chemistry and Toxicology*, 40 (2-3), 327-385.

Lancaster, G. and Massingham, L. (1999). Essentials of marketing. Mc-Graw-Hill, London, UK. 512 pp.

Lang, T. and Rayner, G. (2005). Obesity: a growing issue for European policy? *Journal of European Social Policy*, 15 (4), 301-327.

Lawrence, P.R. and Lorsch, J.W. (1967). Organization and environment. Harvard University, Boston, USA. 296 pp.

Lechevalier, V., Perine, E., Jeantet, R., Lesaffre, C., Croguennec, T., Guerin-Dubiard, C. and Nau, F. (2005). Statistical analysis of effects of industrial processing steps on functional properties of pasteurised liquid egg white. *Journal of the Science of Food and Agriculture*, 85 (5), 757-769.

Lee, D.S., Haggar, P.E., Lee, J. and Yam, K.L. (2006). Model for fresh produce respiration in modified atmospheres based on principles of enzyme kinetics. *Journal of Food Science*, 56 (6), 1580-1585.

Legrenzi P., Girotto V. and Johnson-Laird P.N. (1993). Focusing in reasoning and decision-making. *Cognition*, 49, 37-66.

Leistner, L. and Gould, G.W. (2002). hurdle technologies: combination treatment for food stability, safety and quality. Food Engineering Series. Springer, Berlin, Germany. 208 pp.

Lelieveld, H.L.M., Notermans, S. and De Haan, S.W.H. (2007). Food preservation by pulsed electric fields: from research to application. Woodhead Publishing Ltd., Cambridge, UK. 363 pp.

Lelieveld, H., White, E.B. and Holah, J. (2003). Hygiene in food processing: principles and practice. Woodhead Publishing Ltd., Cambridge, UK. 392 pp.

Lelieveld, H.L.M., Mostert, M.A., Holah, J. and White, B. (2003). Hygiene in food processing. Routledge, New York, NY, USA. 379 pp.

Lelieveld, H., White, E.B. and Holah, J. (2005). Handbook of hygiene control in the food industry. Woodhead Publishing Ltd., Cambridge, UK. 744 pp.

Lévy, C.M., MacRae, A. and Köster, E.P. (2006). Perceived stimulus complexity and food preference development. *Acta Psychologica*, 1223 (3), 394-413.

Lewin, K. (1952). Group decision and social change. In: Swanson G.E., Newcomb, T.M. and Hartley, E.L. (eds.) Readings in social psychology. Holt Rinehart, New York, NY, USA, pp. 459-473.

Lin, C.T.J., Jensen K.L. and Yen, S.T. (2005). Awareness of food borne pathogens among US consumers. *Food Quality and Preferences*, 16 (5), 401-412.

Linssen, J.P.H. and Roozen, J.P. (1999). Food flavour and packaging interactions. In: Mathlouthi, M. (ed.) Food packaging and preservation. Blackie Academic Professional, London, UK, pp 48-61.

Little, C.L., Omotoye, R. and Mitchell, R.T. (2003). The microbiological quality of ready-to-eat foods with added spices. *International Journal of Environmental Health Research*, 13 (1), 31-42.

Loftus, R. (2005). Traceability of biotech-derived animals: application of DNA-technology. *OIE Revue Scientifique et Technique*, 24 (1), 231-242.

Lopez Gomez, A. and Barbosa Canovas, G.V. (2004). Food plant design. CRC Press, Boca Raton, FL, USA. 388 pp.

Luning, P.A., Marcelis, W.J. and Jongen, W.M.F. (2002). Food quality management; a techno-managerial approach. Wageningen Academic Publishers, Wageningen, the Netherlands. 320 pp.

Luning, P.A. and Marcelis, W.J. (2005). Food quality management and innovation. In: Jongen, W.M.F. and Meulenberg, M.T.G. (eds.) Innovation in agri-food systems. Product quality and consumer acceptance. Wageningen Academic Publishers, Wageningen, the Netherlands, pp. 293-341.

Luning, P.A. and Marcelis, W.J. (2006). A techno-managerial approach in food quality management research. *Trends in Food Science and Technology*, 17, 378-385.

Luning, P.A., Marcelis, W.J. and Van der Spiegel, M. (2006). Quality assurance systems and food safety. In: P.A. Luning, F. Devlieghere and R. Verhe (eds.) Safety in the agri-food chain. Wageningen Academic Publishers, Wageningen, the Netherlands, pp. 249-303.

Luning, P.A. and Marcelis, W.J. (2007). A food quality management functions model from a techno-managerial perspective. *Trends in Food Science & Technology*, 18 (3), 159-166.

Luning, P.A., Van der Spiegel, M. and Marcelis, W.J. (2007). Colorants and food quality management. In: C. Socaciu (ed.) Food colorants, chemical and functional properties. CRC Press, Boca Raton, FL, USA.

Luning, P.A., Bango, L., Kussaga, J., Rovira, J. and Marcelis, W.J. (2008). Comprehensive analysis and differentiated assessment of food safety control systems: a diagnostic instrument. *Trends in Food Science & Technology*, 19, 522-534.

Luning, P.A. and Marcelis, W.J. (2009). A food quality management research methodology integrating technological and managerial theories. *Trends in Food Science and Technology* 20 (1), 35-44.

Luning, P.A., Marcelis, W.J., Rovira, J., Van der Spiegel, M., Uyttendaele, M. and Jacxsens, L. (2009). Systematic assessment of core assurance activities in company specific food safety management systems. *Trends in Food Science & Technology*. doi:10.1016/j.tifs.2009.03.003.

Luten, J.B., Jacobsen, C., Bekaert, K., Sæbø, A. and Oehlenschläger, J. (2005). Seafood research from fish to dish - Quality, safety & processing of wild and farmed fish. Wageningen Academic Publishers,Wageningen, the Netherlands. 400 pp.

Maarse, H. (1991). Volatile compounds in foods and beverages. Marcel Dekker, New York, NY, USA. 784 pp.

Mack, A., Schmitz, T., Schulze Althoff, G., Devlieghere, F. and Petersen, B. (2006). Steps in risk management process. In: P.A. Luning, F. Devlieghere and R. Verhe (eds.) Safety in the agri-food chain. Wageningen Academic Publishers, Wageningen, the Netherlands, pp. 355-396.

MacFie, H. (2007). Consumer-led food product development. Woodhead Publishing Ltd., Cambridge, UK. 279 pp.

Maga, J.A. (1974). Influence of colour on taste thresholds. *Chemical Senses & Flavour*, 1, 115-119.

Man, C.M.D. and Jones, A.A. (2000). Shelf life evaluation of foods. 2nd edition. Springer, Berlin, Germany, pp. 52-71.

Maraz, A., Marin, F., Cava, R. and Rajkovic, A. (2006). Microbial analysis of food. In: P.A. Luning, F. Devlieghere and R. Verhe (eds.) Safety in the agri-food chain. Wageningen Academic Publishers, Wageningen, the Netherlands, pp. 471-524.

Marcelis, W.J., 1984. Maintenance management development (in Dutch). Kluwer, Deventer, the Netherlands.

Maroulis, Z.B. and Saravacos, G.D. (2003). Food process design. CRC Press, Boca Raton, FL, USA. 536 pp.

Matthews, R.K. (2006). Emerging issues in food safety. microbiology of fresh produce. ASM press, Washington DC, WA, USA. 239 pp.

McDermott, R.E., Mikulak, R.J. and Beauregard, M.R. (2008). The basics of FMEA. 2nd edition. Productivity Press Inc., Florence, KY, USA. 90 pp.

McDonough, W. and Braungart, M. (2002). Cradle to cradle: remaking the way we make things. North Point Press, San Francisco, CA, USA. 208 pp.

McKone, K.E, Schroeder, R.G., Cuab, K.O. (2001). The impact of total productive maintenance practices on manufacturing performance. *Journal of Operations Management*, 19, 39-58.

McMeekin, T.A., Baranyi, J., Bowman, J., Dalgaard, P., Kirk, M., Ross, T., Schmid, S. and Zwietering, M.H. (2006). Information systems in food safety management: review. *International Journal of Food Microbiology*, 112, 181-194.

Meiselman, H.L. and MacFie, H.J.H. (1996). Food choice, acceptance and consumption. Blackie Academic & Professional, London, UK. 397 pp.

Melnyk, S.A. and Denzler, D.R. (1996). Operations management, a value-driven approach. Mc-Graw-Hill, London, UK. 1000 pp.

Mgonja, J. (2007). Understanding traceability performance in fish supply chain. MSc thesis Wageningen University, Wageningen, the Netherlands. 101 pp.

Miele, M. (2001). Creating sustainability: the social construction of the market for organic products. PhD Dissertation, Wageningen University, Wageningen, the Netherlands.

Miller, S.S. and Mcnew, R.W. (2005). Effect of cultivar and site on fruit quality as demonstrated by the ne-183 regional project on apple cultivars. *Horticultural Technology*, 15 (4), 886-895.

Mingers, J. and Gill, A. (1997). Multimethodology: the theory and practice of combining management methodologies. John Wiley & Sons, Chicester, UK. 458 pp.

Mintzberg, H., Raisinghani, D. and Théorêt, A. (1976). The structure of 'unstructured' decision processes. *Administrative Science Quarterly*, 21, 246-274.

Montgomery, D.C. (2008). Introduction to statistical quality control. 6th edition. John & Sons Inc., New York, NY, USA. 734 pp.

Moskowitz, H.R., Beckley, J.H. and Resurreccion, A. (2006). sensory and consumer research in food product design and development. Blackwell Publishing, USA, 358 pp.

Much, P., Pichler, J. and Allerberger, F. (2007). Food borne infectious outbreaks, Austria 2005. *Wiener Klinische Wochenschrift*, 119 (5-6), 150-157.

Nestby, R., Lieten, F., Pivot, D., Raynal Lacroix, C. and Tagliavini, M. (2006). influence of mineral nutrients on strawberry fruit quality and their accumulation in plant organs: a review. *International Journal of Fruit Science*, 5 (1), 139-156.

Nettleton, J.A. (1995). Omega-3-fatty acids and health. Springer Verlag, Berlin, Germany. 359 pp.

Noori, H. and Radford, R. (1995). Production operations: total quality and responsiveness. McGraw Hill, New York, NY, USA. 676 pp.

Novak, S. and Eppinger, S. (2001). Sourcing by design: Product complexity and the supply chain. *Management Science*, 47, 1, 189-204.

Olsen, A. (1998). Regulatory action criteria for filth and other extraneous materials. *Regulatory Toxicology and Pharmacy*, 28, 181-189.

Otten, A. and Verdooren, L.R. (1996). Statistical process and quality control. Lecture book Agricultural University Wageningen. Wageningen, the Netherlands.

Panisello P.J. and Quantick P.C. (2001). Technical barriers to hazard analysis critical control point (HACCP). *Food Control*, 12, 165-173.

Parmar, B., Miller, P.F. and Burt, R. (1997). Stepwise approaches for estimating the intakes of chemicals in food. *Regulatory Toxicology and Pharmacology*, 26, 44-51.

Parry, M.L., Rosenzweig, C., Iglesias, A., Livermore, M. and Fischer, G. (2004). Effects of climate change on global food production under SERS emissions and socio-economic scenarios. *Global Environmental Change*, 14, 53-67.

Peri, C. (2006). The universe of food quality. *Food Quality and Preferences*, 17, 3-8.

Perrow, C. (1967). A framework for the comparative analysis of organizations. *American sociological review*, 32 (2), 194-208.

Piñeiro, M. and Diaz Rios, L.B. (2004). Improving the quality and safety of fresh fruits and vegetables: a practical approach. FAO, Rome, Italy. Available at: http://www.fao.org/docrep/007.

Poelman, A., Mojet, J., Lyon, D. and Sefa-Dedeh, S. (2008). The influence of information about organic production and fair trade on preferences for and perception of pineapple. *Food Quality and Preferences*, 19 (1), 114-121.

Pomeranz, Y. and Meloan, C.E. (2000). Food analysis: theory and practice. 3rd edition. Springer, Berlin, Germany. 778 pp.

Pons-Sanchez-Cascado, S., Vidal-Carou, M.C., Marine-Font, A. and Veciana-Nogues, M.T. (2005). Influence of the freshness grade of raw fish on the formation of volatile and biogenic amines during the manufacture and storage of vinegar-marinated anchovies. *Journal of Agricultural and Food Chemistry*, 53 (22), 8586-8592.

Popkin, B. (1998). Worldwide trends in obesity. *The Journal of Nutritional Biochemistry*, 9 (9), 487-488.

Porter, M.E. (1985). Competitive advantage. Creating and sustaining superior performance. The Free Press, New York, NY, USA. 592 pp.

Porter, M.E. (1980). Competitive strategy. Techniques for analyzing industries and competitors. The Free Press, New York, NY, USA. 397 pp.

Prajogo, D.I. (2005). The comparative analyse of TQM practices and quality, performance between manufacturing and service firms. *The International Journal of Service Industry Management*, 16, 217-228.

Prajogo, D.I. and Sohal, A.S. (2001). TQM and innovation: a literature review and research framework. *Technovation*, 21, 539-558.

Prasanna, V. Prabha, T.N. and Tharanathan, R.N. (2007). Fruit ripening phenomena - an overview. *Critical Reviews in Food Science and Nutrition*, 47 (1), 1-19.

Pugliese, C., Bozzi, R., Campodoni, G., Acciaioli, A., Franci, O. and Gandini, G. (2005). Performance of Cinta Senese pigs reared outdoors and indoors. 1. Meat and subcutaneous fat characteristics. *Meat Science*, 69 (3), 459-464.

\\\implementation in the vegetable industry. *Journal of Purchasing and Supply Management*, 12 (1), 39-50.

Ramdane, D. and Mohan, J.S. (2004). Production practices and quality assessment of food crops, Vol. 1: preharvest practice. Kluwer AcademicPublishers, Deventer, the Netherlands. 295 pp.

Ramirez, A.F., Corser. P.I., Leal, K.V., Cagnasso, M.A., Gonzalez, M.P. and Urdaneta, A.G. (2006). Effect of time and storage temperature on microbiological quality of hamburger meat. *Revista Científica-Facultad de Ciencias Veterinarias*, 16 (4), 428-437.

Rayas-Duarte, P., Majewska, K. and Doetkott, C. (1998). effect of extrusion process parameters on the quality of buckwheat flour mixes. *Cereal Chemistry*, 75 (3), 338-345.

Rennie, D.M. (1995). Health education models and food hygiene education. *The Journal of the Royal Society for the Promotion of Health*, 115, 75-79.

Renwick, A.G. (2004). Risk characterisation of chemicals in food. *Toxicology Letters*, 149, 163-176.

Richardson, R. (2004). Improving the thermal processing of foods. Woodhead Publishing Ltd., Cambridge, UK. 520 pp.

Riquelme, F. Pretel, M.T., Martinez, G., Serrano, M. Amoros, A. and Romojaro, F. (1999). Packaging of fruits and vegetables: recent results. In: Mathlouthi, M. (ed.) Food packaging and preservation. Chapman & Hall, London, UK, pp. 141-158.

Robertson, G.L. (1998). *Food packaging: principles and practices*. CRC Press, Boca Raton, FL, USA. 568 pp.

Robbins, S.P. (2000). Essentials of organizational behavior. Prentice Hall, Upper Saddle River, NJ, USA. 330 pp.

Robbins, S.P. and Coulter, M.K. (2007). Management. Pearson Prentice Hall, Upper Saddle River, NJ, USA. 672 pp.

Ropkins, K. and Beck, A.J. (2000). Evaluation of worldwide approaches to the use of HACCP to control food safety. *Trends in Food Science and Technology*, 11, 10-21.

Rosenthal, A.J. (1999). Food texture: measurement and perception. Aspen Publication, Gaitersburgh, MA, USA. 311 pp.

Ross, J.E. (1999). Total quality management. St. Lucie Press, Boca Raton, FL, USA. 568 pp.

Ross, T. and McMeekin, T.A. (2003). Modelling microbial growth within food safety risk assessments. *Risk Analysis*, 23, 179-197.

Roussel, P., Saad, K. and Erickson, T. (1991). Third generation R&D, managing the link to corporate strategy. Harvard Business Presss, Boston, MA, USA. 224 pp.

Rovira, J., Cencic, A. Santos, E. and Jacobsen, M. (2006). Biological hazards. In: P.A. Luning, F. Devlieghere and R. Verhe (eds.) Safety in the agri-food chain. Wageningen Academic Publishers, Wageningen, the Netherlands, pp. 67-144.

Sahin, N., Onderci, M., Balci, T.A., Cikim, G., Sahin, K. and Kucuk, O. (2007). The effect of soy isoflavones on egg quality and bone mineralisation during the late laying period of quail. *British Poultry Science*, 48 (3), 363-369.

Saint-Eve, A. Paçi Kora. E. and Martin, N. (2004). Impact of the olfactory quality and chemical complexity of the flavouring agent on the texture of low fat stirred yogurts assessed by three different sensory methodologies. *Food Quality and Preferences*, 15 (7-8), 655-668.

Salunkhe, D.K.; Bolin, H.R.; Reddy, N.R. (1991). Storage, processing, and nutritional quality of fruits and vegetables. Volume I: Fresh Fruits and vegetables. Salunkhe, D.K.; Bolin, H.R.; Reddy, N.R. (eds.). CRC Press, Inc. Florida, 195 pp.

Sans, P., Andrade, M.J., Ventanas, S. and Ruiz, J. (2004). Quality characteristics of fresh meat from pigs of the Gascon breed. *Food Science and Technology International*, 10 (1), 29-34.

Saravacos, G.D., Kostaropoulos, A.E. and Barbosa-Cánovas, G.V. (2003). Handbook of food processing equipment. Springer, Berlin, Germany. 720 pp.

Sarli, T.A., Carlino, D., Ferrante, S., Fioretto, F., Santoro, A.M. and Cortesi, M.L. (2004). Microbiological monitoring of slaughter hygiene in a cattle slaughterhouse. *Industrie Alimentari*, 43 (436), 527-532.

Schein, E.H. (1990). Organizational culture. *American Psychologist*, 45, 109-119.

Schermerhorn, J.R. (2007). Management. John Wiley, New York, NY, USA. 656 pp.

Schilling, E.G. and Neubauer, D.V. (2009). Acceptance sampling in quality control. 2nd edition. CRC, Boca Raton, FL, USA. 520 pp.

Schreiner, D.A. and Ruegg, P.L. (2003). Relationship between udder and leg hygiene scores and subclinical mastitis. *Journal of Dairy Science*, 86, 3460-3465.

Schroeder, R.G. (2000). Operations management: contemporary concepts and cases. Irwin, Boston, TX, USA. 528 pp.

Scipioni, A., Saccarola, G., Centazzo, A. and Arena, F. (2002). FMEA methodology design, implementation and integration with HACCP system in a food company. *Food Control*, 13 (8), 495-501.

Scott, V.N., Swanson, K.M.J., Freier, T.A., Pruett (jr), W.P., Sveum, W.H., Hall, P.A., Smoot, L.A. and Brown, D.G. (2005). Guidelines for conducting *Listeria monocytogenes* testing of foods. *Food Protection Trends*, 25 (11), 818-825.

Scott, V.N. (2005). How does industry validate elements of HACCP plans? *Food Control*, 16 (6), 497-503.

Scott, B.S., Wilcock, A.E. and Kanetkar, V. (2009). A survey of structured continuous improvement programs in the Canadian food sector. *Food Control*, 20, 209-217.

Senge, P.M. (1990). The fifth discipline. The art & practice of the learning organization. Harper, New York, NY, USA. 445 pp.

Shapiro, A. and Mercier, C. (1994). Safe food manufacturing. *The Science of Total Environment*, 143, 75-92.

Shaw, V. and Shaw C.T. (1998). Conflict between engineers and marketers. *Industrial Marketing Management*, 27, 279-291.

Shaw, V., Shaw, C.T. and Enke, M. (2003). Conflict between engineers and marketers: the experience of German engineers. *Industrial Marketing Management*, 32, 489-499.

Shepardson, D.P., Wee, B., Priddy, M. and Harbor, J. (2006). Students' mental models of the environment. *Journal of Research in Science Teaching*, 44 (2), 327-348.

Shewart, W.A. (1980). Economic control of quality of manufactured product. American Society for Quality, Milwaukee, USA. 501 pp.

Shirose, K. (1992). TPM for workshop leaders. Productivity Press, Portland, OR, USA. 176 pp.

Simchi-Levi, D., Kaminsky, P. and Simchi-Levi, E. (2000). Designing and managing the supply chain. McGraw-Hill, New York, NY, USA. 544 pp.

Simon, H.A. (1960). The new science of management decision. Harper & Row, New York, NY, USA. 175 pp.

Simopoulos, A.P. (2002) Omega-3 fatty acids in inflammation and autoimmune diseases. *Journal of the American College of Nutrition*, 21 (6), 495-505.

Singh, R.P. (2000). Scientific principles of shelf life evaluation. In: Man, C.M.D. and Jones, A.A. (eds). Shelf life evaluation of foods. 2nd edition. Aspen Publishers Inc., Gaiterburgh, MA, USA, pp. 3-22.

Sitkin, S.B. (1994). Distinguishing control from learning in total quality management: a contingency perspective. *Academy of Management Review*, 19 (3), 537-564.

Smith, D.J. and Edge, J. (1990). Essential quality procedures. In: D. Lock (ed.) Handbook of quality management. Gower, Aldershot, UK.

Smith, L., Gilhooly, K. and Walker, A. (2003). Factors influencing prescribing decisions in the treatment of depression: a social judgement theory approach. *Applied Cognitive Psychology*, 17, 51-63.

Soccaciu, C. (2007). Food colorants, CRC Press, New York, NY, USA. 648 pp.

Sofo, F. (1999). Human resource development: perspectives, roles and practice choices. Allen and Unwin publishers, London, UK. 365 pp.

Song, X.M., Thieme, R.J. and Xie, J. (1998). The impact of cross functional joint involvement across product development stages: an exploratory study. *Journal of Product Innovation Management*, 15, 289-303.

Sousa, R. (2003). Linking quality management to manufacturing strategy: an empirical investigation of customer focus practices. *Journal of Operations Management*, 21, 1-18.

Scientific Steering Committee (SSC) (2000a). First report on the harmonisation of risk assessment procedures. Part 1. The report of the Scientific Steering Committee's working group on harmonisation of risk assessment procedures in the Scientific Committees advising the European Commission in the area of human and environmental health. Available at: http://ec.europa.eu/food/fs/sc/ssc/out83_en.pdf.

Scientific Steering Committee (SSC) (2000b). First report on the harmonisation of risk assessment procedures. Part 2. Appendices. The report of the Scientific Steering Committee's working group on harmonisation of risk assessment procedures in the Scientific Committees advising the European Commission in the area of human and environmental health. Available at: http://ec.europa.eu/food/fs/sc/ssc/out84_en.pdf.

Staples, G. (1990). Quality uudits and reviews. In: Lock, D. (ed.) Handbook of quality management. Gower, Aldershot, UK.

Steenkamp, J-E.B.M. (1989). Product quality: an investigation into the concept and how it is perceived by consumers. Van Gorcum, Assen, the Netherlands. 288 pp.

Steenkamp, J-B.E.M. (1990). A conceptual model of the quality perception process. *Journal of Business Research*, 21, 309-333.

Stemerding, P. and Souleymane, N. (2003). FAO integrated pest management programm: experiences of participatory IPM in West Africa. In: Maredia, K.M. Dakouo, D. and Mota-Sanchez, D. (eds) Integrated pest management in the global arena. CABI Publishing, Wallingford, UK, pp. 397-406.

Stevenson, W.J. (2006). Operations management. McGraw-Hill, New York, NY, USA. 906 pp.

Stollman, U., Hohansson, F. and Leufven, A. (1994). Packaging and food quality. In: C.M.D. Man and A.A. Jones (eds.) Shelf life evaluation of foods. Blackie Academic and Professional, New York, NY, USA, pp. 52-71.

Swanson, A.R. and Falkman, K.S. (1997). Training delivery problems and solutions: identification of novice trainer problems and expert trainer solutions. *Human Resource Development Quarterly*, 8 (4), 305-314.

Swiegers, J.H. and Pretorius, I.S. (2007). Modulation of volatile sulfur compounds by wine yeast. *Applied Microbiology and Biotechnology*, 74 (5), 954-960.

Tarrant, V. and Grandin, T. (1993). Cattle transport. In: Grandin, T. (ed.). Livestock handling and transport. CAB International, Wallingford, UK, pp. 109-125.

Teale, M., Dispenza, V., Flynn, J. and Currie, D. (2003). Management decision-making. Towards an integrative approach. FT Prentice Hall, Essex, UK. 384 pp.

Tehrany, E.A. and Desobry, S. (2004). Partition coefficients in coefficients in food/packaging systems: a review. *Food additives and contaminants*, 21 (12), 1186-1202.

Tennant, D.R. (2002). Risk analysis of food additives. In: Watson, D.H. (ed.) Food chemical safety. Volume 2: additives. Woodhead Publishing Ltd., Cambridge, UK, pp. 61-78.

Tenner, A.R. and De Toro, J.J. (1992). Total quality management: three steps to continuous improvement. Addison Wesley, Reading, UK. 192 pp.

Terlouw, C. (2005). Stress reactions at slaughter and meat quality in pigs: genetic background and prior experience. A review of recent findings. *Livestock Production Science*, 94, 125-135.

Terziovski, M. and Sohal, A.S. (2000). The adoption of continuous improvement and innovation strategies in Australian manufacturing firms. *Technovation*, 20, 539-550.

Theuvsen, L., Spiller, A., Peupert, M. and Jahn, G. (eds.) (2007). Quality management in food chains. Wageningen Academic Publishers, Wageningen, the Netherlands. 509 pp.

Thomas, K.W. (1976). Conflict and conflict management. In: M.D. Dunnette, (ed.) Handbook of industrial and organizational psychology. John Wiley, New York, NY, USA, pp. 889-935.

Thompson, K. (2003). Fruit and vegetables. Harvesting, handling and storage. 2nd edition. Blackwell Publishing, New York, NY, USA. 480 pp.

Thompson, M., Sylvia, G. and Morrissey, M.T. (2005). Seafood traceability in the U.S. Current trends, system design, and potential applications. *Comprehensive Reviews in Food Science and Food Safety,* 4 (1), 1-7.

Tijskens, L.M.M., Konopacki, P., Simcic, M. and Hribar, J. (2000). Biological variance in agricultural products. Colour of apples as an example. In: Artés, F., Gil, M.I. and Conesa, M.A. (eds.) Proceedings IIR Conference: improving postharvest technologies of fruits, vegetables and ornamentals, 19-21 October 2000, Murcia, Spain, vol. 1, pp. 25-43.

Tijskens, L.M.M., Hertog, M. and Nicolai, B.M. (2001). Food process modelling. Woodhead Publishing Ltd., Cambridge, UK. 416 pp.

Torjusen, H., Lieblein, G., Wandel, M. and Francis, C.A. (2001). Food system orientation and quality perception among consumers and producers of organic food in Hedmark County, Norway. *Food Quality and Preferences,* 12 (3), 207-216.

Tricker, R. (2006). ISO 9001: 2000 the quality management process. 3rd edition. Van Haren Publishing B.V., Zaltbommel, the Netherlands.

Trienekens, J. and Van der Vorst, J. (2006). Traceability in food supply chains. In: P.A. Luning, F. Devlieghere and R. Verhe (eds.) Safety in the agri-food chain. Wageningen Academic Publishers, Wageningen, the Netherlands, pp. 439-470.

Trienekens J.H. and Zuurbier, P.J.P. (2008). Quality and safety standards in the food industry, developments and challenges. *International Journal for Production Economics,* 113 (1), 107-122.

Troeger, K. (1996). Transportation of slaughter animals. Treatment during storage and its consequences for product quality. *Fleischwirtschaft,* 76, 157-158.

Tsuchiya, S. (1992). Quality maintenance: zero defects through equipment management. Productivity Press, Cambridge, MA, USA. 223 pp.

Tucker, G.A. and Woods L.F.J. (1995). Enzymes in food processing. Blackie Academic and Professional, London, UK. 319 pp.

Tukker, A. and Jansen, B. (2006). Environment impacts of products - A detailed review of studies, *Journal of Industrial Ecology,* 10 (3), 159-182.

Urban, G.L. and Hauser, J.R. (1993). Design and marketing of new products. 2nd edition. Prentice Hall, Englewood Cliffs, NJ, USA. 670 pp.

US FDA (2008). *Salmonella* Saintpaul outbreak. Available at: http://www.fda.gov/oc/opacom/hottopics/tomatoes.html.

Van Boekel, M.A.J.S. (1998). Developments in technologies for food production. In: Jongen, W.M.F. and Meulenberg, M.T.G. (eds.) Innovation of food production systems. Wageningen Pers, Wageningen, the Netherlands, pp. 87-116.

Van Boekel, M.A.J.S. (2005). Technological innovation in the food industry: product design. In: Jongen, W.M.F. and Meulenberg, M.T.G. (eds.) Innovation in agri-food systems. Wageningen Academic Publishers, Wageningen, the Netherlands, pp. 149-174.

Van Boekel, M.A.J.S. (2008). Kinetic modelling of reactions in foods. CRC Press Ltd, Boca Raton, FL, USA. 767 pp.

Van den Brandt, P., Voorrips, L., Hertz-Picciotto, I., Shuker, D., Boeing, H., Speijers, G., Guittard, C., Kleiner, J., Knowles, M., Wolk, A. and Goldbohm, A. (2002). The contribution of epidemiology. *Food Chemistry Toxicology,* 40 (2-3), 387-427.

Van den Hurk, K., Van Dommelen, P., Van Buuren, S., Verkerk, P.H. and HiraSing, R.A. (2007). Prevalence of overweight and obesity in the Netherlands in 2003 compared to 1980. *Archives of Disease in Childhood,* 92, 992-995.

Van der Meulen, B. and Van der Velde, M. (2004). Food safety law in the European Union. An introduction. Wageningen Academic Publishers, Wageningen, the Netherlands. 270 pp.

Van der Meulen, B. and Van der Velde, M. (2007). Modern European food safety law. In: P.A. Luning, F. Devlieghere and R. Verhe (eds.) Safety in the agri-food chain. Wageningen Academic Publishers, Wageningen, the Netherlands, pp. 559-618.

Van der Meulen, B.M.J. and Van der Velde, M. (2008). European food law handbook. Wageningen Academic Publishers, Wageningen, the Netherlands. 632 pp.

Van der Spiegel, M., De Boer, W.J., Luning, P.A., Ziggers, G.W. and Jongen, W.M.F. (2003). Validation of the instrument IMAQE-Food to measure effectiveness of food quality management. *International Journal of Quality & Reliability Management,* 24 (4), 386-403.

Van der Vorst, J.G.A.J. and Beulens, A.J.M. (2002), Identifying sources of uncertainty to generate supply chain redesign strategies. *International Journal of Physical Distribution and Logistics Management,* 32 (6), 409-430.

Van der Vorst, J.G.A.J. (2003). Performance levels in food traceability and the impact on chain design: results of an international benchmark study. In: Bremmers, H.J., Omta, S.W.F., Trienekens, J.H., Wubben, E.F.M. (eds.) Dynamics in chains and networks: Proceedings of the sixth International Conference on chain and network management in agribusiness and the food industry, Ede, 27-28 May 2004. Wageningen Academic Publishers, Wageningen, the Netherlands.

Van der Vorst, J.G.A.J., Van Kooten, O., Marcelis, W.J., Luning, P.A., and Beulens, A.J.M. (2007). Quality controlled logistics in food supply chain networks: integrated decision-making on quality and logistics to meet advanced customer demands. In: Proceedings of the Euroma 2007 conference, Ankara, Turkey, 18-20 June 2007.

Van Kleef, E., Van Trijp, H. and Luning, P. (2004). Consumer research in the early stages of new product development: a critical review of methods and techniques. *Food Quality and Preferences,* 16, 181-201.

Van Trijp, J.C.M. and Steenkamp, J-B.E.M. (1998). Consumer-oriented new product development: principles and practice. In: W.M.F. Jongen and M.T.G. Meulenberg (eds.) Innovation of food production systems. Wageningen Pers, Wageningen, the Netherlands, pp. 37-66.

Van Trijp, J.C.M. and Steenkamp, J-B.E.M. (2005). Consumer-oriented new product development. In: W.M.F. Jongen and M.T.G. Meulenberg (eds.) Innovation in agri-food systems: product quality and consumer acceptance. Wageningen Academic Publishers, Wageningen, the Netherlands, pp. 87-124.

Vasta, V., Pennisi, P., Lanza, M., Barbagallo, D., Bella, M. and Priolo, A. (2007). Intramuscular fatty acid composition of lambs given a tanniniferous diet with or without polyethylene glycol supplementation. *Meat Science,* 76, 739-745.

Vela, A.R. and Fernandéz, J.M. (2003). Barriers for the development and implementation of HACCP plans: results from a Spanish regional survey. *Food Control,* 14, 333-337.

Verbeke, W., Scholderer, J. and Frewer, L. (2006). Consumer perception of safety in the agri-food chain. In: P.A. Luning, F. Devlieghere and R. Verhe (eds.) Safety in the agri-food chain. Wageningen Academic Publishers, Wageningen, the Netherlands, pp. 619-646.

Vestergaard, M., Oksbjerg, N. and Henckel, P. (2000a). Influence of feeding intensity, grazing and finishing feeding on muscle fibre characteristics and meat colour of semitendinosus, longissimus dorsi and supraspinatus muscles of young bulls. *Meat Science,* 54 (2), 177-185.

Vestergaard, M., Therkildsen, M., Henckel, P., Jensen, L.R., Andersen, H.R. and Sejrsen, K. (2000b). Influence of feeding intensity, grazing and finishing feeding on meat and eating quality of young bulls and the relationship between muscle fibre characteristics, fibre fragmentation and meat tenderness. *Meat Science,* 54(2), 187-195.

Viljoen, H.F., De Kock, H.L. and Webb, F.C. (2002).Consumer acceptability of dark, firm and dry (DFD) and normal pH beef steaks. *Meat Science*, 61 (2), 181-185.

Vlek, P.L.G. (1985). Micronutrients in tropical food crop production. Martinus Nijhoff/Dr W. Junk Publishers, Dordrecht, the Netherlands. 272 pp.

Voilley, A. (2006). Flavour in food. Woodhead Publishing Ltd., Camebridge, UK. 320 pp.

Wachira, A.M., Sinclair, L.A., Wilkinson, R.G., Enser, M., Wood, J.D. and Fisher, A.V. (2002). Effects of dietary fat source and breed on the carcass composition, n-3 polyunsaturated fatty acid and conjugated linoleic acid content of sheep meat and adipose tissue. *British Journal of Nutrition*, 88, 697-709.

Walter, D.T. (2002). Training on the job. A new team-driven approach that empowers employees. American Society for Training and Development, Alexandria, VA, USA. 202 pp.

Wandel, M. and Bugge, A. (1997). Environmental concern in consumer evaluation of food quality. *Food Quality and Preference*, 8, 19-26.

Warde, A. (1999). Convenience food: space and timing. *British Food Journal*, 101 (7), 518-527.

Warriss, P.D., Kestin, S.G., Brown, S.N. and Collins, L.J. (1984). The time required for recovery from mixing stress in young bulls and prevention of dark cutting beef. *Meat Science*, 10, 53-68.

Watson, D.H. (1998). Natural toxicants in food. A manual for experimental foods, dietetics and food scientists. Blackwell Publishing, London, UK. 349 pp.

Watson, D.H. (2004). Pesticide, veterinary and other residues in food. Woodhead publishing Ltd., Cambridge, UK. 686 pp.

Watson, D.H. (2005a). Food chemical safety: volume 1 contaminants. Woodhead publishing Ltd., Cambridge, UK. 320 pp.

Watson, D.H. (2005b). Food chemical safety: volume 2 food additives. Woodhead publishing Ltd., Cambridge, UK. 308 pp.

Weber, C.L. and Matthews, H.S. (2008). Food-miles and the relative climate impacts of food choices in the United States. *Environmental Science Technology*, 42 (10), 3508-3513.

Weihrich H. and Koontz, H. (1993). Management, a global perspective. McGraw Hill, New York, NY, USA. 640 pp.

Wheelwright, S. and Clark, K. (1992). Revolutionising product development. Quantum leaps in speed, efficiency, and quality. The Free Press, New York, NY, USA. 364 pp.

White, H.M.F. (2000). Buyer-supplier relationships in the UK fresh produce industry. *British Journal*, 102 (1), 6-17.

Whitehurst, R. and Law, B.A. (2002). Enzymes in food technology. Blackwell Publishing, London, UK. 255 pp.

Whiting, R.C. and Buchanan, R.L. (1994). Microbial modelling. *Food Technology*, 48 (6), 113-120.

Wilkinson, A., Redman, T., Snape, E. and Marchington, M. (1998). Managing with total quality management – theory and practice. MacMillan Business, Hong Kong, China. 224 pp.

Williams, S.C. (1993). Perceptions of quality in the Queensland seafood industry. *Asia Pacific Journal of Quality*, 2, 46-58.

Wong, D.W.S. (1995). Food enzymes: structure and mechanisms. Springer Verlag, Berlin, Germany. 390 pp.

Zuurbier, P.J.P., Trienekens, J.H. and Ziggers, G.W. (1996). Vertical cooperation: concepts to start partnerships in food and agribusiness. Kluwer, Deventer (in Dutch).

Internet sites

http://www.isixsigma.com Site: Six sigma visited 2007.

http://www.praxiom.com/iso-definition.htm Site: English terms of ISO standards, visited 2007.

http://www.asq.org/ Site: American Society for Quality Control (ASQC), visited 2007.

http://www.ifst.org/ Site: Institute of Food Science and Technology (IFST), visited 2007.

http://www.cdc.gov/flu/avian/outbreaks/asia.htm. Site: Avian influenza. Centre for disease control and prevention. Atlanta, visited February 2006.

http://www9.minlnv.nl/servlet Site: BSE cases. Ministerie van Landbouw, Natuur & Voedselkwaliteit (LNV), visited September 2006.

http://www.asq.org/ Site: American Society for Quality Control (ASQC), visited 2007.

http://www.ifst.org/ Site: Institute of Food Science and Technology (IFST), visited 2007.

http://www.praxiom.com/iso-definition.htm) Site: English terms of ISO standards, visited 2007.

http://wagyu.net/home.html Site: Japanese Wagyu cows in Europe, visited 2009.

http://www.fda.gov Site: U.S. Food and Drug Administration, visited 2009.

http://www.ehedg.org/guidelines/doc8.htm Site: Hygienic Equipment Design Criteria, visited 2009.

Index

A

acceptable quality level (AQL) 219, 221
acceptance sampling 217, 218
accreditation 323, 325
acidity pH 61, 63
administrative
- concepts 107, 322
- conditions 27, 95, 128
- infrastructure 95
- model 104, 105

adversary 209
aggregate project plan 184
agricultural products 19
agri-food production chain 19
AIDC *See:* automated identification and data collection/capture
ambiguity 22, 23
analytical model of food quality 34
analytical sensory research 168
animal
- breed 73
- feeding 74
- health 75
- housing 75
- production 73

appearance 52
AQL *See:* acceptable quality level
aroma 52
assess product characteristics 168
assigned quality 56
attitudes 127
auditing 322, 323
authority concept 107, 131, 322
automated identification and data collection/capture (AIDC) 319
availability and variability of raw materials 154
available water (aw) 61

B

bacteria 61
behavioural approaches 123
behavioural model 105
benchmarking 264
biochemical
- processes 67
- reactions 54, 66

biological
- hazards 44
- variance 156

Boston portfolio matrix 360
branch hygiene codes 291
BRC *See:* British Retail Consortium
British Retail Consortium (BRC) 329
bureaucratic concept 107, 322
bureaucratic control 240
business concept 322
business reengineering 263

C

capability index Cpk 227
cause-and-effect diagram 260
causes of variation 203
CCP *See:* critical control points
CCP decision tree 296
CE *See:* Conformité Européenne (CE)
centralisation 132
certification 322, 323, 324
chain quality of milk 331
challenge testing 171
change strategy 322
check sheet 259
chemical
- hazards 46, 47
- modifications 65
- reactions 54, 63

classical model 105
climacteric 68
Codex
- Alimentarius 84, 285, 290
- general principles of food hygiene 285, 291
- standards 85

cognitive process of individual decision-making 126
collaboration 185, 385
colouring agents 60
commitment 129
common causes 181, 203, 204
communication 101

competencies 129
complaint procedure 215
complexity food quality management 21
complexity of matrixes 155
compliance with procedures 138
compositional analyses 233, 234
concurrent engineering 185, 186, 188, 189
conflicting relationships 188
conflict management
– 102
conflict styles 103
Conformité Européenne (CE) 177
consequences for managing quality 20
consumer 31, 39, 56
– research 163
– risk 220
contaminants 47, 60
contingency approaches 124
continuous flow 211
continuous improvement 249, 377
contractual relationships 386
control and information systems 156
control charts 260
control circle 199
– feedback 201
– feed-forward 201
convenience 55
cooling 71
core competence philosophy 362
corrective action 201, 212
– plan 299
cost-benefit model 242
costs models 372
CQP *See:* critical quality points
cradle-to-cradle 154
credence attributes 161
critical aspects of design process 147
critical control points (CCP) 156, 293, 295
critical limits 297
critical points 279, 337
critical quality points (CQP) 393
crop varieties 77
Crosby
– Crosby's 14 steps for quality improvement 258
– Crosby's philosophy 113
– Crosby's quality management maturity grid 322, 367, 368
cross-functional teams 185, 188, 191, 270
cultivation practices 77
customer 30, 39
– driven 161
– focus 249, 376
– information 163
– oriented design 160, 162
customer-supplier relationships 140, 266
cusum chart 228, 229

D

dashboard metrics 261
data communication 320
decision-making process 94, 97, 99
decision support system 130, 315, 316
define, measure, analyse, improve, control (DMAIC) 263
define, measure, analyze, design, verify (DMADV) 264
Deming
– cycle 255
– philosophy 110
– wheel 255
design process 143
developing an HACCP plan 292
dielectric heating conditions 71
direct forcing strategy 275
directive 85
distribution control 213, 214
distributor integration (DI) 217
divisional structure 133
DMADV *See:* define, measure, analyze, design, verify
DMAIC *See:* define, measure, analyse, improve, control
documentation and record-keeping 283
dynamic biochemical processes (kinetics) 67
dynamic character of micro-organisms 62
dynamic character of the Maillard reaction 63
dynamic physical processes 67
dynamic physiological processes 68
dynamic systems 156

E
Early Warning & Response System (EWRS) 335
ECR *See:* efficient consumer response
efficient consumer response (ECR) 237
EFSA *See:* European Food Safety Authority
EFSIS *See:* European Food Safety Inspection Service
EFSN *See:* European Food Safety Network
EGEDG *See:* European Hygienic Equipment Design Group
employee commitment 123
empowerment 137, 378
environment 83
environmental conditions 72
enzyme 66
enzyme analyses 233, 234
equipment evaluation 212
escalator effect 192
European Food Law 87
European Food Safety Authority (EFSA) 85
European Food Safety Inspection Service (EFSIS) 330
European Food Safety Network (EFSN) 85
European General Food Law 85
European Hygienic Equipment Design Group (EHEDG) 176
European Quality Award 383, 384
EWRS *See:* Early Warning & Response System
experiential learning 254
exposure assessment 308
extended quality triangle 35, 36
external audit 324
external failure costs 371
extraction conditions 71
extrinsic attributes 42, 55
extrusion and high-shear conditions 71

F
factors in cultivation practices 77
factors in farming practices 73
factors influencing food quality management 21
factors in harvest practices 78
factors in post-harvest storage 79
factors in storage and subsequent preparation of raw materials 80
factors in the storage and distribution of final products 81
factors in the transformation of raw materials and packing of final products 80
factors in transport and slaughtering 76
failure mode and effect analysis (FMEA) 174, 224, 312
FDA *See:* Food and Drug Administration
Feigenbaum's philosophy 112
flavouring agents 60
flexible concept 107, 322
flow chart 259
food
- additives 58, 59
- allergy 46
- analyses for quality control 231
- behaviour 24, 26, 37
- crises 20
- dynamics 26, 27
- health concept 49
- infection 45, 46
- intolerances 46, 47
- intoxicants 47
- matrix 60
- poisoning 45, 46
- processes 57, 61, 62
- processing system 70
- production system 24
- sensitivities 46
- systems 22, 24, 57, 58

Food and Drug Administration (FDA) 88
foodborne
- illnesses 44
- microbial intoxications 45
- pathogens 44
- toxico-infections 45

food quality 21, 39, 40
- decisions 119
- decisions grid 120
- management 116
- management functions 118
- management functions model 27, 28
- management research 23, 24
- perception 40

– relationship 26
food safety
– concept 44
– education 268
– management system (FSMS) 290
– objectives 313
food shelf life concept 53
force-coercion strategies 275
foreign objects 49
formal communication 101
formalisation 132
freezing 71
functional structures 133

G
General Food Law (GFL) 86
generic strategies 364
GFL *See:* General Food Law
GFSI *See:* Global Food Safety Initiative
Global Food Safety Initiative (GFSI) 328
GlobalGAP 334
Global Standard for Food Safety (GSFS) 329
GMP *See:* Good Manufacturing Practice
GMP 5 288
GMP regulation 288, 289
GMP+ scheme 331
Good Manufacturing Practice (GMP) 287, 288
GSFS *See:* global standard for food safety

H
HACCP *See:* Hazard Analysis Critical Control Points
harvest time 78
Hazard Analysis and Critical Control Point (HACCP) 109, 224, 286, 290, 291, 294
– characterisation 310
– identification 308
health 49
heating and cooling conditions 70
hedonic sensory research 168
Hersey-Blanchard model 124
histogram 260
house of quality 165, 166
human
– behaviour 25, 26, 37
– dynamics 26, 27
– systems 22, 25
hundred-percentage inspection 217, 218
hygienic design
– concept 176
– requirements 177, 178

I
ICM *See:* Integrated chain management
IFS *See:* International Food Safety Standard
Imai 114, 257
incidental variation 203
informal agreements 386
informal communication 101
information 100
information systems 130, 313
innovation management 96
– innovation (mid term) management 95
inspection costs 371
instability quality attributes 151
in statistical control 203
integrated chain control 332
Integrated chain management (ICM) 394
integrated quality assurance system 334
interests 100
inter-functional communication barriers 190
intermittent flow 211
internal audit 324
internal failure costs 371
International Food Safety Standard (IFS) 330
International Standards Organisation *See:* ISO
intrinsic attributes 42
intrinsic-extrinsic quality attributes model 42
intrinsic properties 64
Ishikawa 114, 257
ISO 109
– 9000:2005 324
– 9001:2000 304
– 9001:2000 standard 302
– 9004:2000 306
– 22000 325, 326

– 22000 family of standards 327

J
JIT just in time delivery 157, 237
job-shop production 211
joint ventures 386
Juran 256
– Juran's philosophy 111

K
Kaizen 257

L
leadership 123, 377
lean manufacturing 263
learning process 128, 253
Lewin's model for implementing change 274
lifetime failure rates 213
limiting quality level (LQ) 220, 221
lipid oxidation 65
logistics 237
logistics supplier relationships 216
lot tolerance percent defective (LTPD) 220
LQ *See:* limiting quality level
LTPD *See:* lot tolerance percent defective

M
macro-nutrients 59
Maillard reaction 63
maintenance 212
Malcolm Baldrige Award 381, 382
management 93, 99
– concerns 95, 96
– responsibility 304
– system 25
managerial
– approach 25
– functions 29
– processes 95
manufacturing 186
marketing 186
marketing professionals 185
material and stock design 157
mathematical models 172
matrix of technologies by Perrow 370
matrix structure 133
mean time between failures (MTBF) 213
measuring procedures 200, 211
mechanical injury 78
mechanistic organisation designs 137
membrane conditions 71
mental models 127
methods and measurements for analyses 233
microbial
– analyses 233, 234
– processes 61
microbiological deterioration 54
micronutrients 59
Mintzberg process model 104, 106
mission 357
modelling areas 173
monitoring system 298
moulds 62
MTBF *See:* mean time between failures

N
network structure 134
new food products 150
non-climacteric 68
non-enzymatic browning 63
non-programmed decisions 98
non-thermal heating strategies 71
np-Chart 228
nutrients 57
nutritional
– additives 58
– value 57

O
OC-curves *See:* operating characteristic curve
odour 52
operating characteristic (OC) curve 219, 220, 221
operations management 97
– short term 95
organic control 240
organisation 93
– concepts 133
– of production 157
– structure 136
organisational

- change strategies 273
- control 240, 242
- maturity 367
- structure 130

out of control 203
over-the-wall approach 187
oxidation reactions 65
oxidation-reduction potential 63

P

packaging 69
packaging conditions 69
parasites 62
pareto analysis 224, 260
partnership 387
PDSA *See:* Plan Do Study Act cycle
perception 126
perishable character 156
personality traits 125
PERT *See:* programme evaluation and review technique
physical
- analyses 232
- evaluation 233
- hazards 48
- processes 54, 67

physicochemical properties food product 40, 41
physiological processes 68
physiological reaction 54
3PL *See:* Third party logistics
Plan Do Study Act cycle (PDSA) 255
plant 72
policies 355
policy orientations 373, 374
political
- behaviour 104
- manoeuvring 275
- model 104, 105

Porter's five forces model 361
power 101, 103
precautionary principle 86
predictive food microbiology 313, 314
prerequisite programmes (PRP) 286, 287, 293
preservatives 58
prevention costs 370
prion 62
process
- capability 224, 225, 227
- capability analysis 212
- complexity 251
- control 210
- design variables 154, 155
- equipment design 72
- improvement 249
- organisation 211
- parameters 70
- performance index 226
- potential index 224

processing equipment 70
producer's risk 220
product
- complexity 153, 251
- control 210
- control activity 207
- decisions 205
- design process model 162
- design variables 151, 152
- development 145
- development strategy 183
- preferences 163
- properties 57

production
- control 210
- planning 210
- professionals 185

product-packaging interactions 153
programmed decisions 98
programme evaluation and review technique (PERT) 191
project 190
- control 192
- management 190
- management system 191
- manager 191
- planning 191
- team 191, 271, 273

PRP *See:* Prerequisite programmes

Q

QCL *See:* quality controlled logistics
QMS *See:* quality management system
quality

– analyses 230, 232
– attributes 42, 252
– behaviour 94, 121, 122
– circles 270, 271
– concepts 31
– controlled logistics (QCL) 237, 238, 239
– costs 370, 371, 373
– cues 160
– definitions 30
– department 134, 136
– dimensions concept 32
– function deployment 162, 165
– guidance model 160
– gurus 110
– learning 379
– manual 321
– measurements 230, 232
– perception 39, 41, 55, 160
– perspectives concept 31
– philosophies 365, 366
– planning 115
– procedures 321
– requirements 35
– strategy 29
– trilogy 112, 115
quality affecting factors in food supply chain 74
quality assurance 29, 116, 119, 283
– activities 282
– decisions 283
– department 321
– function 279, 280
– process 280
– process and activities 281
quality based control 241
quality control 29, 115, 118
– activities 199
– decisions 202
– function 198
– process 198
quality design 29, 118
– activities 145
– decisions 149
– function 143, 144
– process 144, 145
quality engineering of food processes 157
quality improvement 29, 116, 118, 245, 248, 256, 267
– activities 247, 248
– decisions 250
– function 246
– process 247, 248, 249
quality management 21, 94, 97, 115
– history 108
– process model 303
– system (QMS) 284, 304
quality policy and strategy 117
– activities 357
– decisions 359
– function 356
– process 357
quality policy evaluation 381
quantitative risk assessment 306, 307

R
rational model 104, 105
rational persuasion 276
R&D 185
– professionals 185
record keeping and documentation 301
regulating 201
relationship marketing 216
repetitive flow 211
residues 47, 60, 61
resource-control activity 207
resource decisions 205
responsibility 131
retailer supplier partnerships 217
reversed chain viewpoint 82
risk
– analysis 307
– assessment 309
– characterisation 312
– communication 308
– management 308

S
SAFE *See:* Safe Food in Europe
Safe Food in Europe (SAFE) 85
Safe Quality Food (SQF) 327
safety 44
– risks 152
sample preparation 230

sampling 230
- design 219
scatter diagram 260
sector hygiene codes 301
self-control 240
self-managing teams 271, 272
sensory 50
- attributes 41, 50
- evaluation 232, 233
- experience 41
- modality 41
- panels 169
- quality 41
- systems 41
- techniques 168
sequential approach 187
sequential sampling 222
setting requirements on the quality system 282
shared power strategies 276
shelf life 53
- accelerated 170
- direct monitoring 170
- limiting food processes 53
- surveillance system 171
- tests 170
Shewart chart 228
simple structure 132
simultaneous engineering 188
six sigma programmes 263
SPC *See:* Statistical process control
special causes 203, 204
spot-check procedure 217, 218
SQF *See:* Safe Quality Food
statistical process control (SPC) 223, 224, 225
steps in decision-making process 98
Stewart cycle 255
strategic alternatives 363
strategic decision-making 356
strategic directions 364
strategic management 96
- process 358
- strategic (long term) management 95
strategic planning 377
strategic relationships 386, 387
strategy 355
strenghts weaknesses opportunities threats (SWOT) 358, 359
structural variation 203
supplier
- audits 208
- certification 208
- partnership 208
- relationships 216
supply chain
- concept 237
- management 139
supply control 205, 206
SWOT *See:* strenghts weaknesses opportunities threats

T

Taguchi 114
- method 180
- quality loss function 181, 182
taste 52
team structures 134
teamwork 269, 378
technological
- approach 25
- complexity 367
- conditions 27
- functions 28
technology assessment 183
techno-managerial approach 24, 25, 36
testing 200
texture 50
texturising agents 60
third party logistics (3PL) 216
total involvement 249
Total Productive Maintenance (TPM) 261
total quality control 379
Total Quality Management (TQM) 109, 137, 373, 375, 376, 394
total quality perspective 35
TPM *See:* Total productive maintenance
TQM *See:* Total Quality Management
traceability system 316, 317
- objective 318
- strategy 318
traceable resource unit (TRU) 319
tracing 316
tracking 316

traditional control 241
training 129, 267, 268
trait approach 123
transaction marketing 216
transformational leadership theory 124
TRU *See:* traceable resource unit

U
uncertainty 22, 23, 197

V
validating 282
value chain 362
value system 384, 385
variability 250
variable food composition 57
variation 202, 203
vendor analysis 208
verification 282
viruses 62
visual inspection/measurement *See:* visual measurement
visual measurement 236
voice of the customer 143

W
water activity (aw) 63
White paper on food safety 85, 316
work instructions 321
work units 272

Z
zero defects 114